Plant Tissue Culture: New Techniques and Application in Horticultural Species of Tropical Region

Duong Tan Nhut • Hoang Thanh Tung •
Edward Chee-Tak YEUNG
Editors

Plant Tissue Culture: New Techniques and Application in Horticultural Species of Tropical Region

Springer

Editors
Duong Tan Nhut
Molecular Biology and Plant Breeding
Tay Nguyen Institute for Scientific
Research, Vietnam Academy of Science
and Technology
Dalat City, Vietnam

Hoang Thanh Tung
Molecular Biology and Plant Breeding
Tay Nguyen Institute for Scientific Research,
Vietnam Academy of Science and Technology
Dalat City, Vietnam

Edward Chee-Tak YEUNG
Biological Sciences
University of Calgary
Calgary, AB, Canada

ISBN 978-981-16-6497-7 ISBN 978-981-16-6498-4 (eBook)
https://doi.org/10.1007/978-981-16-6498-4

This Springer imprint is published by the registered company Springer Nature Singapore Pte Ltd.
The registered company address is: 152 Beach Road, #21-01/04 Gateway East, Singapore 189721, Singapore

Preface

In the past decades, plant cell biotechnology combining genetic engineering and molecular biotechnology has emerged as a leading scientific field to create innovative and higher grade agricultural products that are increasingly demanded agricultural and horticultural systems in many countries around the world. Advances in plant cell biotechnology are making significant contributions in a number of areas, such as the production of proteins, antibodies, vaccines, and other substances, and also plant breeding. However, this would not have happened without the development of plant cell culture techniques, which provide appropriate tools to resolve related problems. No single published document has been known yet to provide an overview of recent studies conducted in the field of plant cell, tissue, and organ culture. Especially, reference material about new technologies and practical methods currently applied in research and production is unavailable.

The main purpose of this publication is to provide comprehensive information on up-to-date practical studies in the field of plant cell tissue culture with various experimental methods and to highlight some successful examples of their application. This book is not only aimed to present optimal techniques but to provide also a range of available ones that can be beneficial for further exploration in the research and development of plant tissue culture. Particularly, practical applications presented therein are hoped to be useful for developing countries.

The contributions of various researchers included in this book will present an overview of the plant cell technology field, including various developments and some achievements in the plant cell, tissue, and organ cultures. Their impacts on a number of practical aspects and research are noted, equally. In research, proper analysis, and interpretation of cytological, structural, and organizational characteristics of cells, tissues, and organs are essential for the success of any scientific experiment. In this book, we choose to detail the standard paraffin embedding (PE) the method as the primary method for histological studies of plant samples, especially in vitro cultures. Despite limitations in adapting this methodology, the PE techniques do have many advantages: they are easy to master, inexpensive, and their quality results are guaranteed. As a microscopic examination of plant tissue is basic to address biological questions and procedures for improving the quality of the specimens has been developed, we also cover the benefits of making good quality cut sections of samples with inexpensive tools for microscopic examination and most importantly, provide step-by-step procedures which are easy to follow. We describe most important techniques used in plant cell culture to improve and enhance the quality of cultured plants and their productivity as well as to overcome some abnormal phenomena encountered during culture. We include several techniques, such as wounding technology, to improve regeneration and propagation efficiency through a culture of elongated stem and somatic embryos in *Paphiopedilum* and many other plants. In this regard, we also provide basic technical information and demonstrate several positive effects of using various metal nanoparticles on the growth and development of economically important plants cultured in vitro. The emphasis is on the effectiveness of different light-emitting diode systems in saving energy and optimizing the required area in plant tissue culture. It is noted that not all techniques mentioned are completely suitable in every situation; therefore, we record a number of simple but highly effective culture systems that can easily be practically applied in many countries, especially in developing ones. Some of them are nylon film culture, hydroponic, microponic, and thin cell layer culture systems with the latest application in regeneration and micropropagation of selected plants species; we outline a few studies which are potential for application in plant growth and development under conditions of microgravity and in vitro flowering. Finally, the possible role of polyploidy is also covered in this book.

We would like to thank all of the authors for their outstanding contributions with solid research into various aspects of the field of plant cell culture. We are especially indebted to Professor C. C. Chinnappa and grateful to other authors for their collaboration in the preparation and edition of this book. We hope that the contents of this publication be widely shared with readers and facilitate their crosstalk. Researchers, plant breeders, biologists, gardeners, commercial companies, and students interested in micropropagation will benefit from the consultation of this scientific literature.

Dalat City, Vietnam — Duong Tan Nhut
Dalat City, Vietnam — Hoang Thanh Tung
Calgary, AB, Canada — Edward Chee-Tak YEUNG

Contents

Editors and Contributors

About the Editors

Duong Tan Nhut graduated from Kagawa University (Japan). He is the vice director and president of the Scientific Advisory Council of Tay Nguyen Institute for Scientific Research (VAST, Vietnam). He has more than 400 peer-reviewed articles in international and national publications and contributes approximately 20 book chapters. He has delivered numerous oral and poster presentations in numerous international meetings. Dr. Nhut was president of Biology–Agriculture–Life Sciences Council (NAFOSTED, Vietnam) in the years 2015–2017; is vice president of Vietnam Plant Physiology Association from 2003 to date; and is a member of editorial boards of certain journals in Vietnam.

Hoang Thanh Tung graduated from Hue University of Sciences (Vietnam). He is a researcher at Tay Nguyen Institute for Scientific Research (VAST, Vietnam). Dr. Tung trained in histomorphology and histochemistry at Calgary University (Canada) and in the field of bio-industrial science at Tsukuba University (Japan). He has more than 85 peer-reviewed articles in international and national publications, has written 2 book chapters, and has delivered numerous oral and poster presentations in numerous national and international meetings.

Edward Chee-Tak YEUNG graduated from Yale University (USA). He is a professor at Calgary University (Canada). He has more than 200 peer-reviewed international publications and has delivered numerous oral presentations at numerous international meetings. He is an editorial board member of several international journals in plant physiology, biotechnology, and histology. He was awarded the University of Calgary Teaching Award in 2002 and is recipient of the Visiting Sr Scientist Award by the Japanese Society for the Promotion of Science in 2007.

Contributors

Truong Thi Lan Anh University of Dalat, Dalat City, Vietnam

Huynh Gia Bao Tay Nguyen Institute for Scientific Research, VAST, Dalat City, Vietnam

Le The Bien Tay Nguyen Institute for Scientific Research, VAST, Dalat City, Vietnam

Ngo Quoc Buu Institute of Envrionmental Technology, VAST, Hanoi City, Vietnam

Nguyen Hoai Chau Institute of Envrionmental Technology, VAST, Hanoi City, Vietnam

Do Manh Cuong Tay Nguyen Institute for Scientific Research, VAST, Dalat City, Vietnam

Le Thi Diem Tay Nguyen Institute for Scientific Research, VAST, Dalat City, Vietnam

Nguyen Thi My Hanh Dalat Vocational Training College, Dalat City, Vietnam

Nguyen Thi Thanh Hien Ton Duc Thang University, Hochiminh City, Vietnam

Vu Thi Hien Tay Nguyen Institute for Scientific Research, VAST, Dalat City, Vietnam

Tran Hieu Tay Nguyen Institute for Scientific Research, VAST, Dalat City, Vietnam

Trinh Thi Huong Ho Chi Minh City University of Food Industry, Hochiminh City, Vietnam

Hoang Dac Khai Tay Nguyen Institute for Scientific Research, VAST, Dalat City, Vietnam

Le Van Thuc Dalat Nuclear Research Institute, Dalat City, Vietnam

Vu Quoc Luan Tay Nguyen Institute for Scientific Research, VAST, Dalat City, Vietnam

Nguyen Thi Nhu Mai Tay Nguyen Institute for Scientific Research, VAST, Dalat City, Vietnam

Mohamed M. Mira University of Manitoba, Winnipeg, MB, Canada

Vu Thi Mo Nhatrang Institute of Technology Research and Application, VAST, Nhatrang City, Vietnam

Nguyen Ba Nam University of Dalat, Dalat City, Vietnam

Ha Thi My Ngan Tay Nguyen Institute for Scientific Research, VAST, Dalat City, Vietnam

Phan Le Ha Nguyen Tay Nguyen Institute for Scientific Research, VAST, Dalat City, Vietnam

Duong Tan Nhut Tay Nguyen Institute for Scientific Research, VAST, Dalat City, Vietnam

Truong Hoai Phong Tay Nguyen Institute for Scientific Research, VAST, Dalat City, Vietnam

Truong Thi Bich Phuong Hue University of Sciences, Hue City, Vietnam

Claudio Stasolla University of Manitoba, Winnipeg, MB, Canada

Do Thi Thuy Tam Tay Nguyen Institute for Scientific Research, VAST, Dalat City, Vietnam

Michio Tanaka Kagawa University, Takamatsu, Kagawa, Japan

Nguyen Xuan Tuan Tay Nguyen Institute for Scientific Research, VAST, Dalat City, Vietnam

Tran Trong Tuan Institute of Tropical Biology, VAST, Hochiminh City, Vietnam

Hoang Thanh Tung Tay Nguyen Institute for Scientific Research, VAST, Dalat City, Vietnam

Le Thi Van Anh Graduate University of Science and Technology, VAST, Hanoi City, Vietnam

Nguyen Van Binh University of Dalat, Dalat City, Vietnam

K. Tran Thanh Van International Centre for Interdisciplinary Science and Education, Quynhon City, Vietnam

Bui Van Le University of Science, VNU—HCMC, Hochiminh City, Vietnam

Bui Van The Vinh Ho Chi Minh City University of Technology—HUTECH, Hochiminh City, Vietnam

Edward Chee-Tak YEUNG University of Calgary, Calgary, AB, Canada

General Information: Some Aspects of Plant Tissue Culture

1

Duong Tan Nhut

Abstract

An overview of the field of plant tissue culture is presented in this chapter, which includes a general history, successive developments, and key contributions by several leading scientists. The development of cell and tissue culture of plant organs has contributed to promoting research in several areas of plant biotechnology. It has made great progress and significant contributions to the advancement of modern agriculture, production of secondary substances used in pharmaceutical and biochemical industries, food security, and conservation of plant genetic resources. This chapter briefly presents the current status of plant tissue culture applications in Vietnam (as a developing country) and the world as well. Low-cost tissue culture technology will be one of the top priorities in the development of agriculture, horticulture, and forestry in many developing countries in order to produce affordable high-quality crop materials without affecting the quality of products created. Last but not least, some future developments in this field, including the great potential of plant micropropagation and the increasingly important role of new biological techniques, are covered in this chapter.

Keywords

Agriculture · Food security · Pharmaceutical · Micropropagation · Low-cost tissue culture · Trends in plant tissue culture · Vietnam

D. T. Nhut (✉)
Tay Nguyen Institute for Scientific Research, VAST, Hanoi, Vietnam

D. T. Nhut et al. (eds.), *Plant Tissue Culture: New Techniques and Application in Horticultural Species of Tropical Region*,
https://doi.org/10.1007/978-981-16-6498-4_1

1.1 Introduction

Plants play an essential role in addressing the three major challenges facing humanity in the twenty-first century: food, energy, and environment. Therefore, research in plant biotechnology becomes a foundation for possible solutions. Modern plant biotechnology has made certain achievements in many major areas, such as agriculture, pharmaceuticals, germplasm conservation, food security, and also in plant breeding. It is continuing to develop rapidly through many effective supporting methods such as genetic engineering, molecular biology, and nanotechnology in biological engineering. While effective biotechnology methods have been available, the plant cell culture technique continues to be an important tool for the production of many economically important plants. Because modern techniques for research often come with complexity and high investment costs, choices of low-cost technology to incorporate into several stages of research and producing process should be considered. In this chapter, we would like to present key information for the readers to have an overview of the developments in this field.

1.2 The Formation and Development of Plant Tissue Culture

1.2.1 History of Plant Tissue Culture

Plant tissue culture (PTC) research began with the discovery of the cell and the subsequent introduction of the cell theory. Schleiden and Schwann proposed that a cell is the basic structural unit of all living organisms and that each cell would be able to regenerate into a complete organism (Schleiden 1838; Müller-Wille 2010). In 1902, Gottlieb Haberlandt (a German physiologist) tried for the first time to culture leaf cells of *Lamium purpureum* and *Eichhornia crassipes* and epidermis of *Ornithogalum* species and *Pulmonaria mollissima*; they observed an increase of growth in cells, but no cells divided (Bhojwani and Razdan 1996). Despite being unsuccessful, he laid the foundations for tissue culture technology and he is considered the father of PTC. Subsequent landmark discoveries that took place in plant biotechnology are summarized in Fig. 1.1.

In the mid-1960s, the influence of these techniques had increased in many areas such as basic biology, agriculture, horticulture, and forestry. These applications can be divided into five major directions, including cell behavior, plant modification and improvement, pathogen-free plants and germplasm storage, clonal propagation, and product formation (Thorpe 2007). PTC has contributed greatly to the field of science over the past 100 years, especially in the second half of the twentieth century after the discovery of the process of somatic embryogenesis and the application of innovative techniques such as protein engineering and molecular biology in tissue culture (Gulzar et al. 2020). The further development of plant cell tissue culture technology has made great progress, bringing important contributions to research and production. Some specific studies will be discussed in the following chapters of this book.

1904 **Cultured embryos from several cruciferous species**
1922 **Cultured root and stem tips**
1926 **Discovered Indole acetic acid – first plant growth hormone**
1934 **First successful report of continuously growing cultures of tomato root tips**
1939 **Set up interminable proliferation of callus cultures**
1941 **Added coconut water to divide plant cells**
1946 **Raised plant by shoot tip culture**
1952 **Developed some virus-free plants using shoot meristem culture**
1954 **Breaked callus tissues into single cells**
1957 **Auxin: cytokinin ratio of plant organ formation**
1959 **First evidence of somatic embryogenesis with carrot cell**
1960 **Filtered cell suspension and isolated single cells by plating**
1960 **Developed test tube fertilization technique**
1962 **Developed Murashige and Skoog medium**
1964 **Produced first haploid plants from pollen grains**
1966 **Regenerated carrot plants from single cells of tomato**
1971 **Regenerated first plants from isolated mesophyll protoplasts**
1972 **Produced first interspecific hybrid by protoplast fusion**
1974 **Introduced biotransformation in plant tissue culture**
1977 **Integrated Ti plasmid DNA from *Agrobacterium tumefaciens* into plants**
1978 **Developed intergenic hybrid between potato and tomato**
1983 **Cytoplasmic hybridization between generations in Radish and Grapes**
1984 **Transfer of *Agrobacterium* mediated gene to create transgenic plants**
1987 **Plant biolistic gene transfer method**
1994 **First commercialization of transgenic crops (delayed-ripening tomato)**
2000 **First plant genome (*A. thaliana*) decoded**
2002 **Omics technologies adopted**
2005 **Rice genome sequenced**
2012 **Integrated breeding platform**
2013 **CRISPR first applied to plants**

Fig. 1.1 Summarized milestones that have taken place in plant biotechnology (Pantchev et al. 2017; Sussex 2008; Vasil 2008; Thorpe 2007)

The scientific development of plant cell tissue culture is associated with some famous scientists in specific areas. During each stage of our research in the field, it was our fortune to meet, exchange, and learn with many leading scientists (Figs. 1.2, 1.3, 1.4, 1.5, 1.6, 1.7, 1.8, 1.9, 1.10, and 1.11); this provided a solid foundation for us to develop confidence and the right direction to conduct further research in this field.

1.2.2 The Role of Plant Tissue Culture Technology

Tissue culture has become a powerful tool used in a number of areas: plant cell biology, metabolic pathways, elucidate cellular processes, genetic improvement, genetic engineering, generating cell lines against biotic and abiotic stress, studying complex cellular genomes, producing plants for synthesis of secondary metabolites,

Fig. 1.2 Professor Duong Tan Nhut (left) with Professor Oluf Lind Gamborg (right) (*1924–2007*) at World Congress on In Vitro Biology, 2000 Meeting of the Society for In Vitro Biology; San Diego, California, USA *(Professor Oluf Lind Gamborg devoted himself to establishing PTC as an agricultural tool essential and viable in many developing parts of the world, especially in Asia)*

Fig. 1.3 Professor Duong Tan Nhut (right) with Professor Trevor Alleyne Thorpe (*1936–2020*) (The University of Calgary, Canada) (left) at Transplant production in the twenty-first century in Japan (2000) *(Professor Trevor Alleyne Thorpe had many impressive studies on aspects of organized growth and development of plants, mainly emphasizing on organogenesis. He pioneered tissue culture in woody and coniferous micropropagation)*

Fig. 1.4 Professor Duong Tan Nhut (middle) with Professor Geert-Jan de Klerk (second from right) at first International Symposium on Acclimatization and Establishment of Micropropagated Plants, Sani-Halkidiki, Greece 2001 *(Professor Geert-Jan de Klerk is an experienced senior scientist with a long history in research at Wageningen University and Research, the Netherlands. He has worked in many fields such as Life Science, Genetics, Micro-propagation, Molecular Biology, and Plant Physiology)*

etc. In general, the achievements in this field have contributed significantly to advancing various fields of plant biology and promoting the development of other fields.

1.2.2.1 Plant Tissue Culture in Agriculture

PTC has contributed significantly to the advancement of agricultural science and has become an irreplaceable tool in modern agriculture. It has strong impact on the fields in many ways.

Rapid Propagation with Tissue Culture Technology

Plant tissue culture techniques provide the ability to establish cultures from various plant parts such as leaves, stems, roots, and meristem tissue, and this technique can be applied continuously year-round without seasonal constraints guaranteeing a reliable and predictable level of production. They allow mass production of clones at a much greater rate than does the collection of shoots from a growing plant. Tissue culture allows producing and propagating plants through micropropagation (Lone et al. 2020) and thus provides the industry with a huge volume of plantlets that no other current methods can afford. In fact, in vitro propagation is a very developed and commercialized field all over the world. A large number of laboratories produce millions of plants annually, mainly through vegetative propagating plants, for flowers, ornamentals, fruit trees, and rootstocks (Yancheva and Kondakova 2016). For example, tissue culture has revolutionized the cultivation of bananas and has

Fig. 1.5 Professor Duong Tan Nhut (right) with Professor Paul E. Read (University of Nebraska-Lincoln, USA) (left) at first International Symposium on Acclimatization and Establishment of Micropropagated Plants, Sani-Halkidiki, Greece 2001 *(His main specialty is tissue culture for horticulture crop improvement and vineyard management issues in support of Nebraska's developing grape and wine industry. He assists the horticultural industry in adopting practices that facilitate the rapid introduction of new and improved genotypes of horticultural crops)*

replaced the use of conventional vegetative shoots in many intensive banana-growing regions. It is estimated that up to 50 million tissue culture plants are produced annually (FAO 2001).

Genetics and Plant Breeding

PTC techniques are an effective tool for genetic improvement and plant breeding programs. Any particular variety can be produced in bulk and the development time of new varieties is reduced by 50% (Bhatia and Sharma 2015a). Genetic alterations detected in callus and cell cultures may be due to genetic or epigenetic changes and exhibit an ability to restore somatic variants or mutations with some agronomic traits. The intervention of biotechnology approaches to in vitro regeneration, mass micropropagation techniques, and genetic engineering studies in plant species has been encouraged. In vitro culture of mature or immature zygotic embryos has been used to restore fertility in plants obtained from crossbreeding between plants that do not produce fertile seeds. Genetic engineering could produce a number of improved cultivars with high yield potential and resistance to pests and diseases.

Homozygous plants are an important material for breeding programs because they are used as parental lines to produce hybrid seeds. The creation of homozygous or heterozygous lines can take up to ten times of self-fertilization compared with

Fig. 1.6 Professor Duong Tan Nhut (left) with Professor Atsushi Komamine (*1929–2011*) (right) at Tay Nguyen Institute for Scientific Research, Vietnam *(Professor Atsushi Komamine is one of the pioneers in plant tissue culture in Japan. His goal is to determine the potential of plant cells and he has made a significant contribution to plant tissue culture research by establishing many unique culture systems that control cell division and differentiation)*

traditional breeding techniques. However, by using microspores or other cultures, the time to produce isomorphic lines can be significantly reduced, since haploid plants can be regenerated in just one culture cycle and can be diploidized by colchicine treatment to produce diploid plants with fixed homozygous sets of chromosomes. In vitro microspores or other cultures can also be used to correct the traits of hybrid plants produced by conventional techniques. Embryo saving and

Fig. 1.7 Professor Duong Tan Nhut (right) with Professor Toyoki Kozai (Chiba University, Japan) (left) at Potato cutting farmer in Vietnam *(Professor Toyoki Kozai is a leading expert in research to control* in vitro *environments under artificial light and sugar-free medium micropropagation. In particular, Professor Toyoki Kozai is known as "Father of the Japanese Plant Factory")*

Fig. 1.8 Professor Duong Tan Nhut (left) with Professor Michio Tanaka (Kagawa University, Japan) (right) *(Professor Michio Tanaka is a leading expert in the field of LED light applications in plant tissue culture and orchid micropropagation)*

Fig. 1.9 Professor Duong Tan Nhut (right) with Professor K. Tran Thanh Van (Institute of Plant Biotechnology, Paris-Sud University, Orsay Cedex, France) (left) *(Professor K. Tran Thanh Van was the world's first person who introduced the concept of "Thin Cell Layer," which has created a revolutionary development in plant biotechnology. One of her other shocking works was a study to discover the laws of flowering in plants. She was the author of hundreds of scientific articles, with three most notably in Nature)*

culture allow for the recovery of hybrid plants from partially sexually compatible species. After cross-pollination between two different species, a hybrid embryo development may occur, but the endosperm does not necessarily accompany the entire seed development. At certain stages, the hybrid embryo is destroyed, and during that time the embryo can be saved and cultured for further development. In tissue culture laboratories, doubled haploid populations are produced by regenerating plants by the induction of chromosome doubling from pollen grains, which greatly shortens the line fixation stage (Mishra and Rao 2016). This method has been used in rice breeding for decades and has led to the release of many rice varieties (Pauk et al. 2009).

Genetically Modified (GM) Crops

After the commercialization of the first transgenic crops, the agricultural revolution has begun. Biotechnological innovations have been rapidly integrated into agricultural technology to meet global demands. GM technology is based on technical aspects of PTC and molecular biology to create transgenic plants in agriculture. Transgenic agricultural crops continue to be grown on large scale around the world. Tissue culture allows for the production and propagation of genetically homogeneous disease-free plants (Taşkın et al. 2013; Pe et al. 2020).

Fig. 1.10 Professor Duong Tan Nhut (left) with Professor Meira Ziv (The Robert H. Smith Faculty of Agriculture, Food and Environment, the Hebrew University of Jerusalem, Rehovot, Israel) (right) at first International Symposium on Acclimatization and Establishment of Micropropagated Plants, Sani-Halkidiki, Greece 2001 (*Professor Meira Ziv is famous in the field of plant tissue culture*)

The regeneration of transformed plants is essential for the success of genetic engineering. Among the most commonly used PTC techniques is the transformation of the protoplast and calli. The genetic variation caused by tissue culture can be used as a source of mutation to obtain new stable genotypes. Until CRISPR was introduced; editing genomes, especially in plants, was laborious and inefficient. This technology is currently revolutionizing genome editing because it is easy, fast, inexpensive, and powerful. Genome editing technologies such as CRISPR/Cas can help make agriculture more efficient and environmentally friendly. Researchers advocate for the responsible use and support of these new technologies. The CRISPR/Cas9 system has been used to modify various plant species, such as Arabidopsis, Tobacco, Rice, Wheat, Corn, Sorghum, Chinese white poplar (*Populus tomentosa*), Sweet orange, Soybean, Tomato, Potato, Apple, Cucumber, Broccoli, Rapeseed, Common liverwort, and Camelina (Loyola-Vargas and Ochoa-Alejo 2018b).

1.2.2.2 Tissue Culture in Pharmaceuticals

The substances used in pharmaceuticals today are often metabolites of plant origin, with new products discovered continuously (Table 1.1). However, this kind of plant source is often limited in quantity and unstable. An alternative method for the production of important plant compounds is using in vitro cultured plants, as it ensures independence from geographic conditions by eliminating the reliance on wild plants. The use of in vitro tissue culture remains a viable strategy for the production of natural products with complex structures and high value, especially

Fig. 1.11 Professor Duong Tan Nhut (right) with Professor Kee Yoeup Paek (Research Center for the Development of Advanced Horticultural Technology, Chungbuk National University, Cheongju, Korea) (left) (*Professor Kee Yoeup Paek is a leading researcher in the field of bioreactor applications in plant tissue culture*)

if the plant material source is overexploited, slowly growing, or in low productivity (Espinosa-Leal et al. 2018).

Current in vitro culture techniques are indispensable for the rapid multiplication of plant genotypes, plant genome conversion, and commercially valuable plant-derived metabolites (Debnarh et al. 2006; Altpeter et al. 2016; Cardoso et al. 2019). In vitro tissue culture is an important tool that can be used for the rapid production of key metabolites. Once a commercially important compound has been identified and isolated, measures can be taken for large-scale production. Currently, many systems, such as the culture of silk suspensions and roots, allow the large-scale

Table 1.1 List of some plants used in pharmaceutical production

Plant	Pharmaceutical/chemical	Use
Adonis vernalis	Adoniside	Heart medication
Aesculus hippocastanum	Aescin	Antiinflammatory
Agrimonia eupatoria	Agrimophol	Anthelmintic
Ammi visaga	Kheltin	Bronchodilator
Anabasis sphylla	Anabesine	For skeletal muscle
Anamirta cocculus	Picrotoxin	Analeptic
Ananas comosus	Bromelain	Antiinflammatory, proteolytic
Andrographis paniculata	Neoandrographolide	For dysentery
Anisodus tanguticus	Anisodamine	Anticholinergic
Ardisia japonica	Bergenin	Antitussive
Artemisia maritima	Santonin	Ascaricide
Atropa belladonna	Shikonin	For stomach ulcers
Brassica nigra	Allyl isothiocyanate	Rubefacient
Camellia sinensis	Caffeine	CNS stimulant
Camptotheca acuminata	Irinotecan	Antitumor agent
Camptotheca acuminata	Camptothecin	Anticancerous
Carica papaya	Chymopapain	Proteolytic, mucolytic
Carica papaya	Papain	Mucolytic
Cassia spp.	Danthron	Laxative
Cinchona ledgeriana	Quinidine	Antiarrhythmic
Cinchona ledgeriana	Quinine	Antimalarial, antipyretic
Cissampelos pareira	Cissampeline	Skeletal muscle relaxant
Citrus spp.	Hesperidin	For capillary fragility
Colchicum autumnale	Colchiceine amide	Antitumor agent
Colchicum autumnale	Colchicine	Antigout
Coptis japonica	Palmatine	Antipyretic, detoxicant
Crotalaria sessiliflora	Monocrotaline	Topical antitumor agent
Cynara scolymus	Cynarin	Choleretic
Datura spp.	Scopolamine	Sedative
Duboisia spp.	Rosmarinic acid	Antioxidant
Echinacea purpurea	Echinacea polysaccharides	For immune system
Ephedra sinica	Pseudoephedrine	Sympathomimetic
Ephedra sinica	Ephedrine	Antihistamine
Erythroxylum coca	Cocaine	Anesthetic
Fraxinus rhynchophylla	Aesculetin	Antidysentery
Glycyrrhiza glabra	Glycyrrhizin	For Addison's disease
Gossypium spp.	Gossypol	Male contraceptive
Hemsleya amabilis	Hemsleyadin	For bacillary dysentery
Hydrangea macrophylla	Phyllodulcin	Sweetener
Hydrastis canadensis	Hydrastine	Hemostatic
Larrea divaricata	Nordihydroguaiaretic acid	Antioxidant
Lithospermum erythrorhizon	Scopolamine	Anesthesia
Lobelia inflata	a-Lobeline	Respiratory stimulant

(continued)

Table 1.1 (continued)

Plant	Pharmaceutical/chemical	Use
Lonchocarpus nicou	Rotenone	Insecticide
Lycoris squamigera	Galanthamine	Cholinesterase inhibitor
Mucuna spp.	L-Dopa	Anti-parkinsonism
Nicotiana glauca	Coumarins	Flavoring, antioxidant
Nicotiana tabacum	Nicotine	Insecticide
Octea glaziovii	Glasiovine	Antidepressant
Panax ginseng	Ginseng saponins	Analeptic, anticancerous
Papaver somniferum	Codeine	Analgesic, antitussive
Papaver somniferum	Papavarine	Smooth muscle relaxant
Pilocarpus jaborandi	Pilocarpine	Parasympathomimetic
Piper methysticum	Kawain	Tranquilizer
Podophyllum peltatum	Etoposide	Antitumor agent
Podophyllum peltatum	Podophyllotoxin	Antitumor, anticancer agent
Podophyllum spp.	Podophyllotoxin	For genital warts
Rauvolfia canescens	Deserpidine	Tranquilizer
Rauwolfia serpentina	Ajmalicine	For circulatory disorders
Sanguinaria canadensis	Sanguinarine	Dental plaque inhibitor
Simarouba glauca	Glaucarubin	Amoebic killer
Sophora pachycarpa	Pachycarpine	Oxytocic
Stephania sinica	Rotundine	Analagesic, sedative
Strophanthus gratus	Ouabain	Cardiotonic
Tabebuia spp.	Lapachol	Anticancer
Taxus spp.	Paclitaxel	Anticancerous
Thalictrum minus	Berberine	For diabetes
Veratrum album	Protoveratrines	Antihypertensives

production of plant compounds (Xu et al. 2012; Chandran et al. 2020). However, as mentioned earlier, the high cost associated with this technology makes it uncompetitive when compared to less expensive but environmentally unsustainable processes such as wild plant collection or synthesis of chemicals.

The advent of new molecular tools reveals new possibilities for the production of key metabolites using plant systems. Most of these use targeted genome engineering, in particular the aforementioned genome editing by CRISPR/Cas9. Gene editing can be used to create new alleles, replace starter sequences or create new pathways, all of which can lead to the creation of expressive plant-based systems that produce useful bioactive molecules (Nogueira et al. 2018). Once the plant has been engineered (either by gene editing using CRISPR/Cas9 or by traditional methods) and grown in vitro, genetic material can be stored using primary seed banks and undergoes activities (Sack et al. 2015). The plants can then be used for the large-scale production of desired metabolites, including pharmaceuticals. In vitro culture of some medicinal plants is able to increase the production of secondary metabolites of plant cells, such as the use of precursors and stimulants. Adjustment of environmental factors, including high/low temperatures, drought, ultraviolet rays, alkalinity, salinity, exposure to heavy metals and others, is currently emphasized. These conditions, potentially harmful to the crop, often increase the yield or even induce de novo synthesis of secondary metabolites in the plant in vitro culture. Therefore, it depends on the object to propose the strategies for the expression of plant biologically active compounds (Table 1.1).

Plant cell and tissue cultures hold promises for the controlled production of a multitude of useful secondary metabolites. Plant cell culture combines the value of a whole plant system with a microbiological or animal cell culture system to produce therapeutic secondary metabolites. In search of alternatives to the production of medicinal compounds from plants, biotechnology approaches, in particular PTC, have been found to have potential as an adjunct to traditional agriculture in the industrial production of biologically active plant metabolites.

Similarly, using plant culture systems can offer significant advantages for the production of pharmaceuticals, including cost reduction, rapid production, burden, and scalability. All of these advantages are factory specific and depend on production efficiency versus the advantages provided by alternative sources. In the next decade, tissue culture will reach its highest potential with the use of new technologies such as gene editing and manipulation of the effects of environmental factors. Plant transformation also allows more use of plants to produce genetically modified compounds, such as vaccines and many pharmaceuticals.

Commercial production of plant cells or tissue biomass requires extensive process expansion to ensure a continuous supply. Currently, applying different bioreactor systems to culture in vitro of undifferentiated or differentiated plants can solve this problem (Lehmann et al. 2014; Ramawat 2020). The first commercial application of large-scale plant cell culture was carried out in 200-L and 750-L paddle-stirred reactors for shikonin production by the culture of *Lithospermum erythrorhizon* cells. Cell cultures of *Catharanthus roseus*, *Dioscorea deltoidea*, *Digitalis lanata*, *Panax notoginseng*, *Taxus wallichiana,* and *Podophyllum hexandrum* were carried out in different bioreactors to produce secondary plant products. Several medically important alkaloids, anticancer drugs, recombinant proteins, and food additives are produced in different cultures of plant cells and tissues. Advances in cell culture for the production of medicinal compounds have led to the production of a wide variety of pharmaceuticals such as alkaloids, terpenoids, steroids, saponins, phenols, flavonoids, and amino acids. Currently, 20 different types of recombinant proteins have been produced in plant cell culture, including antibodies, enzymes, edible vaccines, growth factors, and cytokines. Advances in scaling approaches and immobilization techniques contribute to a significant increase in the number of applications of plant cell culture for the production of high value-added compounds. Due to the growth and desire to develop large-scale production over the past several decades, tissue culture techniques have been employed to improve plant growth, biological activities, and the production of secondary metabolites.

1.2.2.3 Germplasm Conservation

One of the most pressing problems in biology is preserving the genetic diversity of living organisms. Previous estimates suggest that about 20% of the world's plants are threatened with extinction (Johannes et al. 2019). The available evidence indicates that overall, the diversity of crops grown in farmers' fields has decreased (FAO 2010). Genetic conservation is important not only for economically important crops but also for rare and endangered plants. Their disappearance can seriously affect other species, thereby causing an ecological imbalance. Modern biotechnology tools have been developed as the fastest and most effective strategies for conserving plant genetic resources.

Currently, in vitro culture is widely used to solve the problems of conserving and restoring the genetic capital of plants. Besides, this method can provide material in larger quantities for plant breeding programs in specific locations. In vitro plant collections can be considered as a form of effective ex situ conservation of local biodiversity. PTC offers excellent alternatives to the survival of plants propagated as thousands of seedlings can be conserved in small spaces under controlled conditions. The growth of PTC-derived material may be maintained dormant (cryopreservation) (Henrik and Johannes 2021). Cryopreservation is one of the effective methods of germ preservation and is used for a wide variety of explants of both temperate and tropical plants such as cell suspensions, callus, somatic embryos, and zygotes (Gonzalez-Arnao et al. 2008). Cryopreservation has also been used to eradicate viruses from affected plants by cryotherapy (Wang et al. 2018; Reed 2011).

More than seven million samples of seeds, tissues, and other plant propagation materials from food crops, along with their wild relatives, are protected in approximately 1750 gene banks (FAO 2014). Most attention has focused upon 30 crops that currently feed the world, and within those 9 species provide 67% of global crop production by weight (FAO 2020). Plant genetic resources are collected and conserved for plant breeding and improvement programs. Plant shoots of important crops such as corn and wheat are maintained under low temperatures at the Centro Internacional de Mejoramiento de Maíz y Trigo (Center for Wheat Improvement and International). Rice germs are concentrated at the International Rice Research Institute. Vegetative germs of propagated plants such as potatoes and sweet potatoes are preserved as tubers or under tissue culture conditions at Centro Internacional de la Papa (International Potato Center; IPC) (Loyola-Vargas and Ochoa-Alejo 2018a).

1.2.2.4 Tissue Culture and Food Security

Plant tissue culture plays an important role in improving crops and food security in many countries. In India, the annual net imports of cereals amounted to 4.1 million tons in 1950–1951, but it became an exporter of cereals after 1995–1996. In the last 50 years, there has been an increase in the per capita availability of cereals to the extent of 9%. This improvement was believed to be associated with the application of biotechnology techniques, especially micropropagation and genetic transformation of different crop varieties including Rice, Ragi, Sorghum (Ragavendran and Natarajan 2017).

An estimated 820 million people around the world are still hungry today, underscoring the enormous challenge in achieving the United Nations' Sustainable Development Goal of Zero Hunger by 2030. The lack of regular access to nutritious and adequate food that these people experience puts them at greater risk of malnutrition and poor health (FAO 2019). The supply of food to the world will become a global problem. The world population is growing at an alarming rate and is estimated to be 9.7 billion by 2050 (FAO 2021). As result, residential space and agricultural land will decrease significantly. Besides, global climate change will be an issue to consider (Zhao et al. 2017; Van Oort and Zwart 2018; Ray et al. 2019; Leisner 2020). With these in mind, we must ensure a sustainable world to secure the next generation. With the success of the green revolution and the advancement of biotechnology, especially in plant cell tissue culture, the agricultural output has increased dramatically over the years. Plant tissue culture offers significant opportunities for in vitro propagation, improved crop quality, and production of crops of desirable agronomic quantity and quality (Gubser et al. 2021).

The trend of producing crops for both food and medicine has also attracted a lot of attention (Sands et al. 2009). Plant cell and tissue cultures offer an alternative to the controlled production of these products. A variety of food ingredients are produced using cultural media. Progress has been made in improving technology to the extent that large-scale production is possible. Plants are bred for improving the nutritional value of staple crop plants. Major advances in plant transformation have come from the development and improvement of technology combined with plant tissue culture and regeneration from transformed cells or tissues (Espinosa-Leal et al. 2018). However, in addition to process-efficiency challenges, the complex regulatory landscape is another limiting factor to the commercialization of these products. As countries have their own legal frameworks, definitions and approval processes can vary, which can lead to increased time and costs to bring products to the market (Gubser et al. 2021).

1.3 Current Status of Plant Cell Culture Applications and Future Trends

1.3.1 Application of Plant Tissue Culture in the World

Large-scale commercial plant breeding based on PTC is pioneered in the United States. In recent years, plant breeding through tissue culture has emerged as one of the leading agricultural technologies globally. From 1986 to 1993, the global production of tissue cultured crops increased by 50%. In 1997, there were 800 million plants produced. During the period 1990–1994, the micropropagation industry declined in Europe, mainly due to production shift to developing countries. Since 1995, output has increased by 14% in Asian countries, mainly due to the entry of the Chinese market. There is an increase in plantlets in South America and Central America due to increased imports from Cuba. Recently, many companies from Israel, the United States, and the UK have moved their production requirements to

Costa Rica and India. Tissue culture plants are reaching a wide variety of growers and farmers in developing countries and demonstrating their superiority in fast production and disease-free and uniform crops. Plants can be multiplied in a controlled environment, anywhere, and year-round, regardless of seasons and weather. Currently, more than 700 seed companies around the world have applied plant tissue, organ, and cell culture technology annually producing hundreds of millions of plantlets free of diseases, including those providing medicine, foods, flowers, ornaments, and reforestation seedlings). They have brought about high economic benefits and contributed to food security in the context of global climate change. The global cell culture market is predicted to have a year-on-year growth of 11.8% during the forecast period of 2020 to 2026, reaching a valuation of $42.9 billion (Market Study Report 2020).

The current PTC industry is estimated to be worth about 150 billion USD yearly (accounting for 50–60% of the whole agriculture), of which about 10% is assigned for the demand for tissue culture products (i.e., USD 15 billion) with a 15% growth rate. Most industries based on PTC are completely limiting the use of time-consuming labor or manual processes by switching to more robust and automated technologies. The advent of such powerful and automated technologies has led to the rapid replication of seedlings at a large-scale production level (Bhatia and Sharma 2015b).

1.3.1.1 The Need for Low-Cost Production Methods in the Field of Tissue Culture

Micropropagation technology used to be more expensive than conventional plant breeding methods and it requires some skills and the unit cost per facility became unaffordable in some cases. The main reasons were production cost and technology. Now, these problems have been solved by inventing reliable and cost-effective tissue culture methods of producing plants without compromising on quality. This requires constant monitoring of the input costs of chemicals, vehicles, energy, labor, and capital. In industrialized countries, labor is a major contributor to the high production costs of tissue culture crops. In less developed countries such as Africa, Asia, and Latin America, where labor is relatively cheaper, consumable materials such as culture media, culture tanks, and energy are the main causes of increased production costs. For example, average preparation costs (chemicals, energy, and labor) can account for 30–35% of the production of micropropagated plants. However, automated production processes based on presterilized membrane capsules, bioreactors, mechanized sample transfer, and container sealing are not commercially viable proposals in many developing countries. Therefore, low-cost alternatives are adapted to reduce the production costs of tissue culture plants.

The low-cost options will reduce production costs without compromising the quality of the micropropagation and crops. The reduction in technology costs is achieved by improving process efficiency and better utilization of resources. Low-cost tissue culture technology will be a top priority in agriculture, horticulture, forestry, and flower growing in many developing countries and produce affordable high-quality crop materials without compromising the quality of products created.

1.3.2 Actual Situation of Plant Tissue Culture Application in a Developing Country—Vietnam

In recent years, the application of plant cell, tissue, and organ cultures in Vietnam has become more and more popular. Many new varieties have been produced using methods such as gene transfer, fusion, and single-cell culture for many purposes such as creating antistress plants, accumulating bioactive substances, adapting to different growing conditions. Especially, micropropagation has been applied for many crops.

In Vietnam, there are currently over 100 laboratories using these techniques and have produced nearly 30 million clonal plantlets. Da Lat is the country's leading city in the field, which has produced about 70% of the plantlets. It is home to a private laboratory with the largest scale production in the world. The company, namely Dalat Flower Forest Biotechnology Corporation, was the first place in Southeast Asia to produce seedlings by tissue culture method (in vitro technology) and now becomes an important seedling "bank" in Vietnam. It exports a lot of new flower varieties to the world market. Possessing a laboratory scale of 5000 m^2 equipped with more than 140 transplanting boxes, it employs more than 350 active workers. Every year, this company provides about 100 million tissue culture-derived seedlings for both domestic needs and export to many countries around the world. Flowers that are favored in the international market such as Coin, Statice, Lily, Chrysanthemum, Hydrangeas, and some ornamental plants are focused on their production. The company began exporting to Belgium in 2006, initially with one million plantlets, and in 2008 the number increased to four million. In 2010, it exported seven million plantlets to New Zealand, the Netherlands, Belgium, and China.

The potential of PTC in increasing agricultural outputs and creating rural jobs has been recognized by both investors and policymakers in developing countries. However, in many developing countries, the cost of building facilities and the unit production cost of micropropagation are high, and often the return on investment does not commensurate with the potential economic advantages of micropropagation technology. These problems can be solved by standardizing more precise agronomic practices or by reducing production costs or both. Currently, to reduce production costs, many simple but effective methods combined with plant cell tissue culture have also been carried out in some facilities in Vietnam, for example: using the nonautoclave or sugar-free environment to reduce production costs, alternative LEDs to reduce consumed energy, microponic system for easy transportation, other techniques to improve propagation efficiency, etc. In particular, the replacement of glass culture flasks with plastic bags is becoming increasingly common (Fig. 1.12).

Fig. 1.12 Plantlet production using plastic bags instead of glass bottles at the Dalat Flower Forest Biotechnology Corporation, Dalat, Vietnam

1.3.3 The Trend of Using Tissue Culture in the Future

PTC is one of the most promising fields of application today and has great potential for development in the future. In particular, it plays a key role in sustainable development and is competitive in many fields, especially agriculture and forestry.

1.3.3.1 Huge Growth Potential of Micropropagation

Plant cell, tissue, and organ cultures have been used for a variety of purposes, and among these micropropagation is the most extensive and successful commercialized application. This achievement is sure to continue in the future. Plantlets derived from PTC are superior to those vegetatively propagated from conventional materials in terms of uniformity, yield, speed, and quality. The superior performance of tissue-cultured seedlings and the rising global food demand have fueled the need for rapid micropropagation systems of cultivated materials, including tissue culture.

1.3.3.2 High Production Costs as an Obstacle to Tissue Culture Adoption

Tissue culture applications and their importance are on the rise and are likely to develop further in the future. High production costs are an obstacle to the adoption of tissue culture, especially to resource-constrained farmers. Therefore, the reduction of production costs is an issue that needs to be considered. Challenges for producers involve the development of innovative in vitro techniques that cut production costs per plant.

Necessity of Low-Cost Plant Growth Systems

Compared with conventional propagation, the high production cost per unit appears to be the main challenge for the expansion of tissue culture applications. Therefore, the application of production cost reduction measures is very necessary. Depending on the production situation, it can be achieved by improving process efficiency and incorporating simple yet efficient production methods. Some of them like the nylon film culture system, hydroponic system, microponic system, TCL culture system will be discussed in the following chapters.

Labor as the Main Factor That Should Be Considered

To reduce the cost of tissue culture labor, many micropropagation companies tend to move operations to places with abundant low-cost labor force (mainly Southeast Asia and Latin America). However, it should be noted that the cost of labor in these regions is also increasing. Consequently, automated mass propagation systems for in vitro seedling production and the development of automated/robotized implant production methods are also being developed (Ducos et al. 2005).

1.3.3.3 Development of New Technologies as Essential

New technologies, such as genomic editing in combination with tissue culture, have opened up a new and vast pathway for the second green revolution and will undoubtedly allow the creation of new plant varieties with useful agronomic traits, such as plants that are resistant to pests and herbicides and thus allow for increased yields with less use of harmful pesticides and herbicides, or crops with resistance to abiotic factors, for example, resistance to drought, salinity or cold as results of the global warming. Besides, through PTC, plant breeding and crop improvement using embryo rescue, double haploids, or somatic hybrids will become easier. This will contribute to make improved hybrids with increased yield.

Gene technology will inevitably be associated with the development of many fields: from micropropagation of ornamental and forest plants, production of pharmaceutical compounds, vaccines derived from plants and plant breeding to improvement of the nutritional value of major crops, development of high-flux transcription and gene sequencing techniques, application of protein separation and sequencing, and extraction, segregation and identification of metabolites. Practice shows that biotechnology is often combined with plant breeding for optimal crop development. In this context, genetic markers mapped near genes responsible for important agronomic traits are used to select desired plants. Thus, a new wide door has been opened up with new possibilities for plant research through the approach of contemporary molecular biology techniques.

Multiomics-based systems are evidence for understanding the pathways or molecules that play a role in certain plant functions (Langridge and Fleury 2011; Federico et al. 2021). The application of omics to PTC certainly offers an incredible advantage in the investigation of plant morphogenesis processes in vitro, which may allow for improved efficiency of regenerative processes for species difficult to regenerate (Neelakandan and Wang 2012; Loyola-Vargas and Ochoa-Alejo 2018a). Besides, the combination of plant tissue culture and omics in breeding plants resistant to adverse abiotic conditions has also been considered. Tolerance to abiotic stress in plants through conventional breeding programs has had limited success, mainly due to the polygenic nature of the abiotic stress response in plants (Peleg et al. 2011). Therefore, to consider the polygenic nature of abiotic stress tolerance, detailed studies of plant transcription and protein are needed to fully analyze the stress response pathway. Omics technology plays an important and potential role in discovering profiles of specific links to abiotic stress tolerance in plants (Rensink and Buell 2005).

References

Altpeter F, Springer NM, Bartley LE, Blechl AE, Brutnell TP, Citovsky V, Conrad L, Gelvin SB, Jackson D, Kausch AP, Lemaux PG, Medford JI, Orozco-Cardenas M, Tricoli D, VanEck J, Voytas DF, Walbot V, Wang K, Zhang ZJ, Stewart CN (2016) Advancing crop transformation in the era of genome editing. Plant Cell 28:1510–1520

Bhatia S, Sharma K (2015a) Micropropagation. In: Modern applications of plant biotechnology in pharmaceutical sciences, pp 361–368

Bhatia S, Sharma K (2015b) Plant tissue culture-based industries. In: Modern applications of plant biotechnology in pharmaceutical sciences, pp 405–417

Bhojwani SS, Razdan MK (1996) Plant tissue culture, theory and practices, a revised edition. Studies in plant science, 755p

Cardoso JC, Oliveira MEBS, Cardoso FCI (2019) Advances and challenges on the *in vitro* production of secondary metabolites from medicinal plants. Hortic Bras 37(2):124–132

Chandran H, Meena M, Barupal T, Sharma K (2020) Plant tissue culture as a perpetual source for production of industrially important bioactive compounds. Biotechnol Rep 26:e00450

Debnarh M, Malik CP, Baisen PS (2006) Micropropagation: a tool for the production of high quality plant based medicines. Curr Pharm Biotechnol 7:33–49

Ducos JP, Florin B, Dupuis JM, Pétiard V (2005) A new approach for automation: sorting and sowing dehydrated somatic embryos of Daucus and Coffea using seed technologies. In: Hvoslef-Eide AK, Preil W (eds) Liquid culture systems for *in vitro* plant propagation. Springer, Dordrecht, pp 415–423

Espinosa-Leal CA, Puente-Garza CA, García-Lara S (2018) *In vitro* plant tissue culture: means for production of biological active compounds. Planta 248:1–18

FAO (2001) Intergovernmental group on bananas and on tropical fruits, Second Session. FAO, San José

FAO (2010) Global forest resources assessment 2010. Forestry research paper 163. FAO, Rome

FAO (2014) Genebank standards for plant genetic resources for food and agriculture. FAO, Rome
FAO (2019) The state of food security and nutrition in the world 2019. Safeguarding against economic slowdowns and downturns. FAO, Rome
FAO (2020) Global status of genetic resources for food and agriculture: challenges and research needs. FAO, Rome
FAO (2021) FAOSTAT online database. http://faostat.fao.org/
Federico S, Saleh A, Alisdair RF (2021) Integrating multi-omics data for crop improvement. J Plant Physiol 257:153352
Gonzalez-Arnao MT, Panta A, Roca WM, Escobar RH, Engelmann F (2008) Development and large scale application of cryopreservation techniques for shoot and somatic embryo cultures of tropical crops. Plant Cell Tissue Organ Cult 92:1–13
Gubser G, Vollenweider S, Eibl D, Eibl R (2021) Food ingredients and food made with plant cell and tissue cultures: state-of-the artand future trends. Eng Life Sci 21:87–98
Gulzar B, Mujib A, Malik MQ, Mamgain J, Syeed R, Zafar N (2020) Plant tissue culture: agriculture and industrial applications. In: Transgenic technology based value addition in plant biotechnology, 342p
Henrik N, Johannes FB (2021) Cryopreservation of plant cell cultures—diverse practices and protocols. New Biotechnol 62:86–95
Johannes JLR, Hui C, Castillo ML, Iriondo JM, Jan-Hendrik K, Khapugin AA, Médail F, Rejmánek M, Theron G, Yannelli FA, Hirsch H (2019) Recent anthropogenic plant extinctions differ in biodiversity hotspots and coldspots. Curr Biol 29(17):2912–2918
Langridge P, Fleury D (2011) Making the most of "omics" for crop breeding. Trends Biotechnol 9: 33–40
Lehmann N, Dittler I, Lämsä M, Ritala A, Rischer H (2014) Disposable bioreactors for cultivation of plant cell cultures. In: Paek KY, Murthy HN, Zhong JJ (eds) Production of biomass and bioactive compounds using bioreactor technology. Springer, Dordrecht, pp 17–46
Leisner CP (2020) Review: climate change impacts on food security—focus on perennial cropping systems and nutritional value. Plant Sci 293:110412
Lone SM, Hussain K, Malik A, Magray M, Hussain SM, Rashid M, Farwah S (2020) Plant propagation through tissue culture—a biotechnological intervention. Int J Curr Microbiol App Sci 9(7):2176–2190
Loyola-Vargas VM, Ochoa-Alejo N (2018a) An introduction to plant tissue culture: advances and perspectives. In: Plant cell culture protocols. Methods in molecular biology, vol. 1815, pp 3–13
Loyola-Vargas VM, Ochoa-Alejo N (2018b) Plant cell culture protocols. Plant tissue culture: a battle horse in the genome edtting using CRISPR/Cas9 1815, pp 131–148
Mishra R, Rao GJN (2016) In-vitro androgenesis in rice: advantages, constraints and future prospects. Rice Sci 23:57–68
Müller-Wille S (2010) Cell theory, specificity, and reproduction, 1837–1870. Stud Hist Phil Biol Biomed Sci 41(3):225–231
Neelakandan AK, Wang K (2012) Recent progress in the understanding of tissue culture-induced genome level changes in plants and potential applications. Plant Cell Rep 31:597–620
Nogueira M, Enfissi EM, Almeida J, Fraser PD (2018) Creating plant molecular factories for industrial and nutritional isoprenoid production. Curr Opin Biotechnol 40:80–87
Pantchev I, Rakleova G, Pavlov A, Atanassov A (2017) History of plant biotechnology development. In: Pavlov A, Bley T (eds) Bioprocessing of plant *in vitro* systems. Springer, Cham, pp 1–35
Pauk J, Jancsó M, Simon-Kiss I (2009) Rice doubled haploids and breeding. In: Advances in haploid production in higher plants. Springer, Dordrecht, pp 189–197
Pe PPW, Naing AHS, May T, Kang H, Park KI, Kim CK (2020) Establishment of meristem culture for virus-free and genetically stable production of the endangered plant Hosta capitata. Sci Hortic 272:109591
Peleg Z, Apse MP, Blumwald E (2011) Engineering salinity and water stress tolerance in crop plants: getting closer to the field. Adv Bot Res 57:406–432

Ragavendran C, Natarajan D (2017) Role of plant tissue culture for improving the food security in India: a review update. In: Dhanarajan A (ed) Sustainable agriculture towards food security. Springer, Singapore, pp 231–262

Ramawat KG (2020) An introduction to the process of cell, tissue, and organ differentiation, and production of secondary metabolites. In: Ramawat KG, Ekiert HM, Goyal S (eds) Plant cell and tissue differentiation and secondary metabolites: fundamentals and applications. Springer, Cham, pp 1–22

Ray DK, West PC, Clark M, Gerber JS, Prishchepovn AV, Chatterjee S, Jung YH (2019) Climate change has likely already affected global food production. PLoS One 14(5):e0217148

Reed BM (2011) Choosing and applying cryopreservation protocols to new plant species or tissues. Acta Hortic 908:363–372

Rensink WA, Buell CR (2005) Microarray expression profiling resources for plant genomics. Trends Plant Sci 10:603–609

Sack M, Rademacher T, Spiegel H, Boes A, Hellwig S, Drossard J, Stoger E, Fischer R (2015) From gene to harvest: insights into upstream process development for the GMP production of a monoclonal antibody in transgenic tobacco plants. Plant Biotechnol J 13:1094–1105

Sands DC, Morris CE, Dratz EA, Pilgeram AL (2009) Elevating optimal human nutrition to a central goal of plant breeding and production of plant-based foods. Plant Sci 117(5):377–389

Schleiden MS (1838) Beiträge zur Phytogenesis. In: Müller J (ed) Archiv für Anatomie. Physiologie und wissenschaftliche Medicin, Berlin, pp 137–176

Sussex IM (2008) The scientific roots of modern plant biotechnology. Plant Cell 20(5):1189–1198

Taşkın H, Baktemur G, Kurul M, Büyükalaca S (2013) Use of tissue culture techniques for producing virus-free plant in garlic and their identification through real-time PCR. Sci World J 2013:1–5

Thorpe TA (2007) History of plant tissue culture. Mol Biotechnol 37:169–180

Van Oort PA, Zwart SJ (2018) Impacts of climate change on rice production in Africa and causes of simulated yield changes. Glob Change Biol Bioenergy 24:1029–1045

Vasil IK (2008) A history of plant biotechnology: from the cell theory of Schleiden and Schwann to biotech crops. Plant Cell Rep 27(9):1423–1440

Wang M, Chen L, Zhang Z, Blystad D, Wang Q (2018) Cryotherapy: a novel method for virus eradication in economically important plant species. Methods Mol Biol 1815:257–268

Xu JF, Dolan MC, Medrano G, Cramer CL, Weathers PJ (2012) Green factory: plants as bioproduction platforms for recombinant proteins. Biotechnol Adv 30:1171–1184

Yancheva S, Kondakova V (2016) Plant tissue culture technology: present and future development. In: Pavlov A, Bley T (eds) Bioprocessing of plant *in vitro* systems, Reference series in phytochemistry. Springer, Cham, pp 1–26

Zhao C, Liu B, Piao S, Wang X, Lobell DB, Huang Y, Huang M (2017) Temperature increase reduces global yields of major crops in four independent estimates. Proc Natl Acad Sci U S A 114(35):9326–9331

The Use of the Paraffin Embedding Method in the Study of Cultured Explants I: Background Information

2

Edward C. Yeung, Hoang Thanh Tung, Claudio Stasolla, and Duong Tan Nhut

Abstract

A proper understanding of the cytological features and structural organization of cells, tissues, and organs is essential to the success of any experimental study. Careful analyses and interpretation of structural features will provide additional insight, generating hypotheses for further investigations. At present, many histological procedures are available in the literature. Moreover, in this chapter, we elected to detail the standard paraffin embedding (PE) method as a fundamental method to study the histology of botanical specimens, especially the in vitro cultured explants. Although there are limitations to this procedure, the PE technique is easy to master and with practice, quality results are guaranteed. This method also provides essential theoretical and practical training for other embedding protocols. When compared to other embedding methods, the PE procedure is not as costly in equipment and supplies. In this chapter, we first discuss PE's theoretical and technical aspects before detailing the protocol in a companion chapter.

Keywords

Anatomy · Histology · Paraffin embedding · Fixation · Staining · In vitro explants

E. C. Yeung (✉)
Department of Biological Sciences, University of Calgary, Calgary, AB, Canada
e-mail: yeung@ucalgary.ca

H. T. Tung · D. T. Nhut
Tay Nguyen Institute for Scientific Research, VAST, Dalat City, Vietnam

C. Stasolla
Department of Plant Science, University of Manitoba, Winnipeg, MB, Canada

D. T. Nhut et al. (eds.), *Plant Tissue Culture: New Techniques and Application in Horticultural Species of Tropical Region*,
https://doi.org/10.1007/978-981-16-6498-4_2

2.1 Introduction

Morphological and anatomical investigations provide background information on the experimental system of interest. "Observational" research, that is, a descriptive account of changes, is an essential first step in any research program, often preceding further physiological and molecular investigations. Since we are dealing with living objects, we must understand their developmental changes during an experiment. For in vitro studies, histological investigations can provide detailed insights on cell and tissue characteristics that might be altered by specific treatments. Most importantly, a histological study's real benefit is that the observation can provide an initial framework of information that can be expanded through the formulation of hypotheses for further experimental testing (Yeung 1999, 2015).

This chapter focuses on the paraffin embedding (PE) method. Although other embedding methods are available in the literature, the classical PE protocol has proven reliable and is still being used today (Jensen 1962; Berlyn and Miskche 1976; Stasolla and Yeung 2015; Tung et al. 2019). The PE method is also an essential component of modern-day cell and molecular biology techniques, such as in situ hybridizations (Javelle et al. 2011) and laser microdissection procedures. When compared to other embedding techniques, the operating expenses for PE are significantly lower. For beginners, the paraffin protocol provides essential foundation training for other embedding techniques. The method is easy to master, and quality results can be obtained in a short time.

The purpose of this chapter is to provide an introductory overview of tissue processing and PE procedure, with particular emphasis on in vitro explants, and to encourage their applications in the study of in vitro developmental processes. Once students and researchers are familiar with the basic techniques, other variants of the PE method from the literature can be tested and adapted for one's own needs. This chapter focuses on background information and technical notes. The companion chapter deals with protocols and hands-on procedures. These two chapters are an updated version of our earlier publications, that is, Yeung and Sexena (2005), Stasolla and Yeung (2015), and Tung et al. (2019).

2.2 Paraffin Embedding and In Vitro Culture Cells

Before embarking on a detailed discussion on PE, it is essential to remark that this method has limitations, and it might not be easily applicable to all cultured cells and tissues. For example, the PE protocol is difficult to apply to the study of cell suspensions, small cell clusters, and loose embryogenic calli. These culture cells can be free or loosely packed and tend to be vacuolated with thin walls. In order to process these cells for PE, cells need to be concentrated by pre-embedded in a medium such as agar or agarose before fixation. A procedure in processing small cell clusters for the PE method is presented in the next chapter. The other drawback of PE is that since cells are subjected to a high temperature during wax infiltration, the thin-walled, vacuolated cells tend to shrink or collapse, giving poor quality preparations.

To remedy the situation, culture cells and small cell clusters should be examined first using interference contrast microscopy to capture essential structural features. Histological and histochemical staining can reveal cellular features and components (Yeung 2012, 2015). If further details are required, instead of using the PE method, the plastic embedding procedure is recommended (Yeung 1999; Yeung and Sexena 2005; Yeung and Chan 2015a, b).

2.3 The Use of Paraffin Wax in Histology

The use of paraffin wax has a long history [for details, see Bracegirdle 1978 and Sanderson et al. 1988]. Paraffin was discovered and named by Karl Von Reichenbach in 1830, and it was produced by cracking mineral oil. Edwin Kleb was credited to be the first user of wax in the embedding of tissue in 1866; he submerged the tissue with molten wax, which subsequently hardened, allowing improved sectioning of specimens. Later, methods were developed, having paraffin infiltrated into fixed, dehydrated tissues to provide better support for sectioning. Since then, paraffin wax has become the most popular embedding medium used in histology, especially in laboratory medicine. Relatively inexpensive when compared to other embedding media such as glycol methacrylate and epoxy resins, paraffin wax is easy to handle and safe to use. Furthermore, with some practice, sections are readily obtained.

Over the years, the quality of paraffin wax for histology has improved. Paraffin wax is a mixture of long-chain hydrocarbons with a range of molecular weights and melting temperatures. While a wax with high melting points tends to be hard and brittle, wax with a low melting point tends to be too soft to section. A melting point of 54–58 °C is preferred. More delicate tissues can be embedded in wax with a slightly lower melting point, and more rigid tissues prefer a wax with a higher melting point.

In order to generate consistency when handling and improving sectioning, different manufacturers have modified paraffin wax by mixing different types of waxes, adding plasticizers, and penetration-enhancing compounds such as dimethyl sulfoxide. This results in waxes having different melting points, hardness, and sectioning quality, which enable investigators to choose the appropriate mix for their experiments. For example, the commonly used paraffin with the trade name Paraplast® has four types of waxes used explicitly in tissue embedding with melting points range from 52 to 60 °C. Paraplast Plus® appears to be the most popular. This wax contains purified paraffin with plastic polymers and approximately 0.8% dimethyl sulfoxide to enhance infiltration and sectioning. This wax is recommended for tissues that are difficult to process. Paraplast Plus® has a melting point of 56–57 °C. In warm countries such as those in South East Asia, wax with a higher melting point is preferred as thinner sections are easier to obtain with harder wax blocks. Moreover, one can also briefly store the paraffin specimen blocks in a refrigerator before sectioning. Low temperature hardens the blocks and enables the use of wax with lower melting points. For optimal results, “there is a local need to investigate the most suitable mixture of waxes and resins that will give the best results" (Anderson and Bancroft 2002).

One of the most desirable paraffin wax features is that blocks are easy to section, forming ribbons readily compared to other embedding media such as polyester wax. Ribbons of sections allow for three-dimensional reconstruction of the specimen if desired. The major drawback of paraffin wax is the relatively high melting point (about 55–60 °C), which might not be compatible for those cells and tissue sensitive to elevated temperatures, which induce shrinkage in the size of the specimen.

2.4 The Paraffin Embedding Method: General Information and Technical Comments

The following information and comments, specific to the PE protocol only, provide some theoretical background information, technical details, and precautionary notes on the methodology. For additional information on other related techniques, readers are invited to consult other sources, e.g., Berlyn and Miksche (1976); O'Brien and McCully (1981); Sanderson (1994); Ruzin (1999); Kiernan (2008) and Yeung et al. (2015).

2.4.1 Fixation and Fixatives

Fixation is the first step in any embedding procedure. To prepare specimens for sectioning, tissues need to be killed immediately as soon as they are removed from their natural environment. The aims of fixation are: to (1) prevent cellular autolysis, (2) preserve cellular structures, and (3) harden cells and tissues, minimizing distortion during subsequent processing steps. Ideally, a fixative penetrates the tissue and acts quickly by preventing the loss of molecules while preserving cellular components. However, these desirable effects are difficult to achieve, given the absence of a "perfect" fixative able to fix all macromolecules and cellular components. Soluble materials, such as sugars and ions, are difficult to retained and often lost. Any material that is not fixed will be extracted during fixation and subsequent steps of preparation. Hence, it is imperative to understand the chemistry of fixation and interpret the results accordingly. For example, lipids are not fixed using common fixatives, and therefore, the absence of lipids in a preparation does not reflect their actual presence in cells and tissue under natural conditions. Specific preparations and protocols must be employed when studying lipids (Gahan 1984). Therefore, one must select specific fixative formulations based on the experimental system and the embedding method used. Plant cells react to different fixatives and fixing agents differently during fixation (O'Brien et al. 1973). A proper understanding of fixation is key to our evaluation and interpretation of results. Although our knowledge of fixation is incomplete, we need to strive for the best results possible through comparative studies and testing various recipes. Fixative recipes are not formulae. Recipes can be changed or modified to suit one's own needs. The willingness to try and experimenting with different recipes is the key to a good quality histological study.

For best results, the first rule of fixation is to collect and fix the sample simultaneously (Berlyn and Miksche 1976). Cells and tissues should be fixed immediately after being dissected, as cellular autolysis begins as soon as they are removed from the culture medium or excised from the plant body (Mersey and McCully 1978). Rapid fixation of the specimens will minimize artifacts. It is important to note that all subsequent processing steps cannot revert the mistakes made during fixation.

Formalin-acetic acid-alcohol (FAA) is the most common and popular botanical fixative used in conjunction with PE. FAA works well for most botanical specimens, and it can serve both as a fixative and a storage fluid as it does not over harden the tissues. It is a coagulant fixative that coagulates proteins and can destroy or distort cytoplasmic organelles. Moreover, in general, coagulant fixatives produce a "sponge-like proteinaceous reticulum that is easily permeated, after dehydration and clearing, by the large hydrophobic molecules of melted paraffin wax" (Kiernan 2008). FAA penetrates tissue quickly, and the swelling effect of acetic acid counterbalances the shrinkage effect of ethanol. The cross-linking property of formaldehyde stabilizes the tissue. After fixation, the extensive washing of tissues is unnecessary as the fixing agents are readily removed during the dehydration process. The FAA's main drawback is that plant cells tend to be plasmolyzed with the combination of fixing agents in this fixative. Plasmolysis can be minimized by reducing ethanol concentration in the formulation or using a buffered formalin-based fixative. Depending on the types of cells and tissues, a very low concentration (0.2%) of glutaraldehyde can be added to the fixative to harden the tissue further before processing.

Besides FAA, other fixative formulations for PE are also available. The Nawaschin (CRAF—chromic acid, acetic acid, and formalin) formulations were popular among plant anatomists in the early to mid-1900s; however, they are less frequently used nowadays. This fixative produces a high-quality cytological fixation. Moreover, a longer fixation time is needed as chromium ions penetrate the tissue slower than FAA. The fixative is prepared from stock solutions and mixed just before use. They cannot be premixed and stored as chromium ions react with formaldehyde to form chromic acid and the color of fixative changes to green (Berlyn and Miksche 1976). After fixation, the tissues also need to be washed carefully so that residual chromium ions do not interfere with the staining protocols.

Microscopists must evaluate the merit of each fixative carefully before use and select an appropriate formulation depending on the experiment's objective, e.g., gross morphology vs. detailed cytological details. For example, FAA is more suitable for fixing large tissue pieces for gross histological changes as it rapidly penetrates the specimen.

To appreciate the function of a fixative, one needs to understand the chemical properties and actions of all its components. FAA comprises three fixing agents, that is, formalin, acetic acid, and ethanol, and each has different actions on cellular components. The chemical properties of the fixing agents are briefly summarized below. For a more detailed discussion of various fixing agents and additives used in biological fixation, readers are referred to published literature, e.g., Kiernan (2008); Huang and Yeung (2015).

Formaldehyde has long been used as a fixing agent in different fixatives. It has a small molecular weight, which results in rapid penetration and good stabilization of cells and tissues. It is a noncoagulant fixing agent, and its reaction toward proteins and other macromolecules is complex. The aldehyde group reacts primarily with the basic amino acid lysine to form methylene bridges. The initial binding to protein is relatively fast, but the formation of methylene bridges occurs slowly. Hence, a longer fixation time is needed. But once reacted with proteins, they appear to be stable. As formaldehyde is the most commonly used fixing agent, a proper understanding of its reactions toward proteins and other macromolecules is essential to our understanding of biological fixation.

Commercial formalin solution is a saturated solution of formaldehyde (about 37% by weight) in water, and 10% of methanol is added as a stabilizer to prevent self-polymerization forming paraformaldehyde on prolonged standing. The addition of methanol, especially in old stock solutions, can form formic acid. As methanol and formic acid can have adverse effects on tissues' fixation, it is always recommended to use fresh solutions when preparing fixatives. To avoid using methanol-stabilized formalin for fixation, one can prepare a new formaldehyde solution from paraformaldehyde powder and use it immediately.

Ethanol is a coagulant fixing agent that denatures proteins, making them insoluble in water. Nucleic acids are indirectly stabilized by ethanol as they are closely associated with proteins. It is important to note that lipids are not preserved by ethanol and thus can be extracted during fixation and subsequent processing.

Acetic acid is a coagulant fixing agent; it does not fix proteins but stabilizes nucleic acids through unknown mechanisms. It is commonly used in conjunction with ethanol as a cytological fixative and aids in preserving nucleic acids. It is not a lipid solvent and does not react with lipids. When it is used alone, it can cause swelling of cells. Hence, acetic acid serves to counteract the shrinkage effect of other fixing agents, such as ethanol. Acetic acid penetrates fast into tissues and is used as a fixing agent for preparations to be visualized by light microscopy.

Many other chemicals have proven to be useful as fixing agents (Huang and Yeung 2015). Chromic acid is the key fixing agent for the CRAF fixatives. It is a strong oxidizing agent and reacts with tissue components, likely through oxidation. It functions as a coagulant fixing agent at a low pH, that is, results in precipitation of macromolecules, especially proteins. Tissues are well preserved morphologically, and the hardening of tissue is not excessive with minimal shrinkage. Moreover, the penetration of chromic acid into tissues is slow. Thus, small tissue pieces are processed when the CRAF fixative is used. Since chromium ions are used in conjunction with reducing agents, that is, formaldehyde and alcohol, the fixative is prepared at the time of fixation, and washing is necessary to remove residual chromium ions from the tissues before further processing.

Glutaraldehyde and osmium tetroxide are both noncoagulant fixing agents and are occasionally used in conjunction with the PE method. Glutaraldehyde is commonly used as a fixing agent for ultrastructural studies. This compound has two aldehyde groups separated by three methylene bridges. Glutaraldehyde is a much more efficient cross-linker for proteins than formaldehyde, and its reactions with macromolecules are believed to be irreversible. Since the penetration rate into tissue is slow (0.34 mm/h), large tissue pieces cannot be adequately fixed. Nowadays, besides being used as a fixative component for electron microscopy, glutaraldehyde is commonly used in conjunction with paraformaldehyde for high-resolution light microscopy (Yeung and Chan 2015a, b). Osmium tetroxide is an excellent fixing agent, especially for fixing lipid components of cells. It is mainly used in conjunction with glutaraldehyde for electron microscopy. For PE, osmium tetroxide is used in the study of animal tissue, such as in the myelin sheath study, and is rarely used in the study of plant histology. Similar to glutaraldehyde, osmium tetroxide penetrates tissues slowly. Although glutaraldehyde and osmium tetroxide are excellent fixing agents, they are rarely used in conjunction with the PE method. They render the tissues very dense and prevent the infiltration of the large paraffin wax molecules, resulting in tissue blocks difficult to section. Moreover, depending on the tissues' cellular content, a lower percentage of these fixing agents can be used with formaldehyde. The conditions are best determined empirically.

Regardless of the fixative used, it is important that the tissue is fixed as soon as it is excised from a specimen or collected from culture vessels. If immediate fixation of samples is not possible, the tissue should be stored on ice and fixed immediately upon arrival in the laboratory. Immediate fixation of the sample is a prerequisite for good quality histological preparations.

2.4.2 Excision of Tissue

For larger specimens, identify and only excise the tissue of interest. It is advisable to cut large tissues into smaller pieces or thin slices less than 5 mm in thickness. This ensures rapid penetration of fixative and the subsequent wax infiltration. If a certain type of section, such as a cross or longitudinal section, is required, excise the tissue with the desired orientation. During tissue excision, sharp knives and/or new double-edged razor blades should be used. The tissue must be cut cleanly and not being squashed, as often happens when a dull knife is used.

The preferable way to fix plant tissues is to submerge and excise the samples directly in the fixative. This method will ensure the immediate death of the tissue while minimizing artifacts. However, this excision method should only be done in a fume hood as fixative fume is toxic. This is especially true if osmium tetroxide is used. Osmium tetroxide sublimates from liquid to vapor form. One can be blinded by osmium tetroxide fume! For routine preparation of samples in conjunction with the PE method, the tissues can be excised quickly underwater or a buffer solution and transferred at once to vials containing a fixative. During the tissues' transfer, handle the tissues as gently as possible; instead of using forceps, a brush or toothpick can be used to minimize damage to the tissues.

2.4.3 Removal of Air from Botanical Specimens

Plant tissues are characterized by the presence of intercellular spaces, which can be large, such as the case of leaf mesophyll tissue, or small, such as cultured explants. Regardless of their size, air must be removed from the apoplast to favor the fixative penetration.

Air extraction is best done with a dedicated vacuum system such as a vacuum oven or a desiccator connected to a vacuum pump with sufficient suction to remove air from the tissues. A vacuum oven is preferred as it can withstand a high vacuum, and the temperature can be regulated. Temperature regulation is a useful feature when removing air/solvent from samples in molten wax at 60 °C during wax infiltration. From a safety standpoint, it is essential to use a vacuum chamber that can withstand the vacuuming force delivered by the vacuum pump. Otherwise, injury of laboratory personnel is possible if the chamber happened to shatter during vacuuming. Furthermore, the entire system should be housed in a fume hood to disperse exhaust fume from the vacuum pump. Although not ideal, if a vacuum system is not available, one can generate a mild vacuum using a laboratory water aspirator.

In general, it is best to fix the samples for several hours before performing the first vacuuming step. This allows the initial hardening of the tissues. Vacuum the samples gently until no more air bubbles escape from the cells, and the sample sinks to the bottom of the vial. Occasionally, the samples remain afloat after vacuuming, especially when a fixative without ethanol is used. This is likely caused by the presence of a hydrophobic cuticular surface. A few microliters of a wetting agent, that is, Tween 20, can be added to the fixative to reduce the surface tension. Additionally, one can repeat the vacuuming procedure at the 50% step of ethanol dehydration. This will aid in the complete removal of air from samples.

It is crucial to control the vacuuming rate to avoid rapid air extraction from the tissue. Rapid extraction can result in the collapse of tissues. It is advisable to replace fixative with a new solution, as some fixing agents such as ethanol evaporate under vacuum, reducing the fixative effectiveness. Depending on the tissue's size, the tissue is fixed for a total of 24–48 h in FAA before dehydration. Fixation is best to carry out at 4 °C. This is to minimize autolytic processes at the time of fixation, especially for large tissue pieces.

2.4.4 Dehydration

Dehydration of samples is necessary as cellular water is not miscible with wax. Although different solvents such as acetone and methyl cellosolve can be used as dehydrating agents, ethanol in conjunction with tert-butyl alcohol (TBA) is the most common dehydration combination used in PE of botanical specimens. TBA gradually replaces ethanol during dehydration. If FAA is used as the fixative, dehydration begins at the 50% step since FAA contains 50% ethanol. If an aqueous fixative such as CRAF is used, the fixative must be removed, followed by careful washing before commencing the dehydration at 30% ethanol. Dehydration should be gradual, especially for vacuolated plant cells. Rapid dehydration can cause rapid diffusion of water resulting in plasmolysis and collapsing of cells. Each dehydration step's duration ranges from 30 min to overnight, depending on the fixed tissues' size and cytological features. Highly vacuolated tissues will require a longer dehydration time. A rotary mixer can be used to facilitate fixation and dehydration processes. If necessary, the rotary mixer can be housed inside a refrigerator at 4 °C during the entire course of fixation and the first three steps of TBA dehydration. This can minimize the extraction of cell constituents.

One can take advantage of the dehydration steps to prestain the tissues. As tissues are being dehydrated, the pigments such as chlorophyll will be extracted, and the specimens will appear translucent. This makes them difficult to locate within a solid wax block. By prestaining at the 100% step of dehydration, the specimens can be easily seen during embedding and subsequent sectioning. The prestaining solution is composed of 0.05% safranin O or 0.1% eosine in the 100% TBA. Other stains can be used as long as they do not interfere with the subsequent staining process.

Anhydrous ethanol is expensive. In the preparation of an alcohol series, 95% ethanol can be used instead. Anhydrous ethanol (100%) is hygroscopic. It must be appropriately stored with an added molecular sieve to ensure it remains "dry".

2.4.5 Transitional Fluids and Wax Infiltration

Wax is not miscible with ethanol; hence, a transitional fluid is required for wax infiltration. One needs to choose a transitional fluid that is compatible with the dehydrating agent and wax simultaneously. The transitional fluid must be miscible and, better yet, a solvent of wax so that it can infiltrate into the specimens gradually. The criteria for choosing a transitional agent are speed in replacing the dehydrating agent, ease of removal by molten paraffin wax, minimal tissue damage, flammability, toxicity, and cost (Anderson and Bancroft 2002). For animal tissues, xylene, toluene, chloroform, xylene substitutes have been used. Xylene is still commonly used in processing pathological medical specimens. It is an excellent transitional fluid as it is miscible with absolute alcohol and a solvent for paraffin wax. It has a high refractive index and renders the specimen "clear." Hence, transitional fluids are also known as clearing fluids. However, xylene is toxic, and the liquid is dense, requiring a few more exchanges of pure wax to remove it entirely from the tissue. Furthermore, xylene can result in further extraction of lipidic substances from cells and may weaken cellular structures. Because of the toxicity of xylene, xylene substitutes, such as Histoclear®, can be used. This is a by-product of the citrus juice industry and appears to be a safer alternative. However, these compounds have a fixed shelf life and will deteriorate over time. Hence, for practical purposes, xylene is a more cost-effective product to use. Lower-cost technical grade xylene (mix isomers) is acceptable for histological uses.

For botanical specimens, TBA is the most commonly used transitional fluid. Johansen (1940) introduced TBA as a transitional fluid for botanical specimens since it is less hygroscopic and miscible with wax. As wax does not dissolve in TBA, wax's infiltration is not as "smooth" as xylene. Moreover, TBA is a less toxic solvent and is easily exchanged and removed from the tissues. A small or thinner specimen will ensure a more efficient exchange of TBA with wax. TBA is a solid at room temperature; hence, it needs to be kept at 28 °C to maintain it in a liquid state.

In countries with high humidity, traces of water in absolute ethanol and some transitional fluids can have a negative impact on tissue embedding. Moreover, traces of water can be removed using Drierite© and molecular sieves. When moisture is present in the solvent, the copper sulfate indicator in Drierite© will turn from blue to pink, indicating water in the solvent. When molecular sieves are used, they should be changed regularly, ensuring the solvent remains "dry." Always remove ethanol carefully from a bottle containing the drying agent to avoid contamination. One can insert several pieces of filter paper as a barrier, separating the ethanol from the drying agent. Ethanol needs to be "dried" when used in conjunction with xylene, as the latter will react with traces of water, giving rise to an oily white product that will ruin the specimen.

To ensure complete removal of air and transitional fluid from the tissues, we routinely perform a vacuuming step when the samples are in the molten wax at 60 °C. This step ensures the extraction of the transitional fluid and aid in the infiltration of wax into the tissues. Proper wax infiltration gives a firm consistency to the specimen, preventing damage to the tissue during sectioning.

2.4.6 Embedding

Embedding is a process in which the well-infiltrated specimen is placed into an embedding mold in the desired orientation. During embedding, the tissue can be placed singly or in groups. Disposable embedding molds of different sizes and shapes, such as the Peel-Away-Molds®, can be purchased commercially. One can also prepare paper boats or use small aluminum dishes and plastic weighing dishes for embedding as long as the solidified wax blocks can be removed easily. Paper boats are easy to prepare, inexpensive, and different sizes can be made to suit one's need (Jensen 1962).

The molten wax begins to solidify once removed from the oven. During embedding, one needs to work rapidly to arrange all the specimens in a desirable orientation before wax solidification. To allow for more time for specimen embedding, one can place the embedding mold or paper boat on a slide warmer at the wax's melting point. It is advisable to practice embedding ahead of time. Embed a few specimens first; with improving skills, more specimens can be embedded at one time.

During embedding, use a hot probe such as a blunt needle (which can also be prewarmed in the oven) to arrange and orient the samples in molten wax. Once the tissues are correctly placed, the entire embedding mold should be floated on ice water for the rapid solidification of wax. This is to avoid air bubbles trapped near or around the tissues from expanding during the embedding process. Another reason is to prevent the formation of large crystals that form if the mold is not cooled quickly.

Paraffin wax of a small crystalline structure fits the embedded tissue closely, providing adequate support for sectioning. In contrast, wax of larger crystalline structures adheres poorly to tissue, giving less support. The faster paraffin is cooled, the smaller will be the size of crystals formed. It is essential to ensure that all the wax blocks with embedded specimens are appropriately labeled.

2.4.7 Mounting of Wax Blocks

Depending on the types of embedding molds used, the wax block can be removed from the mold and mounted directly onto a suitable mounting support. If a paper boat method is used with several tissue pieces inside, the tissues need to be separated by cutting using a hot knife/spatula. Warming (by flaming with a Bunsen burner) the knife/spatula minimizes the forces required to cut the wax block, preventing uncontrolled "cracking" of the block.

Traditionally, simple wood cubes can be prepared to suit a particular microtome for mounting purposes. The newly made wood cubes need to be "cured" by submerging them in molten wax for a few days, creating a better bonding between the cube and the specimen. We prefer using plastic mounting rings [see figures in Stasolla and Yeung 2015; Tung et al. 2019]. The prepared blocks are stored at room temperature.

2.4.8 Microtomy

A microtome is necessary for the sectioning of tissue blocks. There are different types of microtomes designed for various purposes. All microtomes for paraffin sectioning have a rotary mechanism. This rotary microtome has a fine advance system controlled by the ratchet and pawl mechanism. The key external features are: (1) a specimen holder with orientation adjustments/screws, (2) a knife holder with adjustable knife angle, (3) specimen thickness selection knob, and (4) a hand wheel with a handle and a locking device. Before embarking on sectioning, it is imperative that one has a good understanding and working knowledge of the microtome to be used. Read the manufacturer's instructions for proper maintenance and care of a microtome. If an older type of rotary microtome, such as the American Optical Spencer microtome is used, one must regularly lubricate the recommended parts with oil for a smooth operation. This is essential to ensure a trouble-free operation.

The price of the rotary microtomes varies wildly, depending on the features included in each model. For wax sectioning, a conventional rotary microtome is adequate for routine work. Microtomes with retraction during the return stroke are currently available. The retraction function prevents the wax block from touching the back of a knife during a return stroke, enhancing ribbon formation. Unfortunately, the cost of a retraction-type microtome is twofold that of a conventional one. Since microtomes are solidly built, purchasing used microtomes for paraffin sectioning can be a cost-saving alternative. Furthermore, microtomes manufactured in Asia are less expensive than the brands from Europe and North America.

2.4.8.1 Section Thickness Selection

The thickness of sections can be selected and adjusted using the thickness adjustment knob. The general guideline is that the smaller the cell size, the thinner should be the section thickness. This is to avoid sections having overlapping cell layers, which will reduce the clarity of cytological details. For example, meristematic cells tend to be small with dense cytoplasm. Sections between 5 and 7 μm are preferred. Moreover, sections with a thickness between 10 and 15 μm can provide suitable structural information for tissues with large vacuolated cells, reducing the need to prepare many slides. A thickness between 7 and 10 μm is generally recommended for paraffin-embedded specimens.

2.4.8.2 Microtome Knives

The knife is the soul of a microtome. Given a reliable microtome, the key in obtaining good sections is the knife. In the past century, steel knives were sharpened with a special knife sharpener and reused. With proper care and regular resharpening, a microtome knife can last a long time. Based on the shape of the knife edge, the C-profile wedge-shaped steel knife is the most common. It is a general-purpose knife for sectioning a majority of tissues. For hard tissue, a D-profile plane-shaped tungsten carbide knife is preferred. A solid steel knife can be secured tightly to the knife holder avoiding vibration during sectioning. Moreover, knife sharpening needs to be done correctly, and it is time-consuming. As a result, disposable blades are becoming common. In recent years, the overall quality of disposable blades has improved. For a given brand, the blades' characteristics, that is, high or low-profile, thickness, and coating material, are well specified. The narrower and thinner low-profile blades are designed for softer tissues, and the wider and thicker high-profile blades are for more rigid tissues. For a majority of plant materials, the latter type is preferred. The coated disposable blades reduce friction during sectioning and increase the longevity of the cutting edge. The main disadvantage of using disposable blades is that a dedicated, matching blade-holder is needed. The blade needs to be tightly secured, and the cutting edge cannot be overly extended beyond the edge of the knife holder by more than 2 mm. Furthermore, both the blade holder and blades are relatively expensive. It is important to note that the disposable blades' edge is sharp but thin and easily damaged. The users need to handle the blades with care. To maximize the knife edge's sharpness, use an older blade or a previously used part of the blade for rough trimming before using a new portion of the same blade for sectioning.

2.4.8.3 Understanding the Knife Angle

All microtomes have an engraved clearance angle setting on the side of the knife holder. This allows the user to test and set up the optimum knife angle for the blade selected. Each blade has two facets that form the cutting edge, and the facet angle is unique to a blade type. Hence, for each blade selected, the clearance angle must be determined empirically. The correct setting of the knife angle is key to successful sectioning. If one notices a build-up of wax at the back of the blade and the block face appears crushed, this is due to improper adjustment of the clearance angle. A clearance angle has to be large enough so that the block face only touches the knife edge as it sections. The angle should not be too large that the knife "scrapes" the block's surface instead of making a sharp, clean cut. The clearance angle for paraffin sectioning is usually about 5–7°. One needs to find the best angle for the type of wax, tissue, and blade used.

2.4.8.4 Block Trimming and Orientation

Just before sectioning, examine and trim the wax block containing the sample carefully. The upper and lower edges must be parallel to one another. This ensures a straight ribbon formation during sectioning. To economize the space on a slide for research purposes, it is desirable to place as many sections on a slide as possible. This can be achieved only if the ribbon is straight. Furthermore, it is useful to trim away excess wax and unnecessary tissue before sectioning to maximize the number of sections placed on a slide.

After trimming, insert and secure the specimen block into the specimen holder. Adjust the orientation of the specimen block using the orientation screws. Once the orientation is fixed, tighten all screws ensuring the block is secured before sectioning.

2.4.8.5 Common Problems in Sectioning and Remedies

Beginners can encounter different difficulties during sectioning. Some of the common issues and remedies are described below. Additional discussion can be found in Berlyn and Miksche (1976); Sanderson (1994); Ruzin (1999).

The technique of sectioning is detailed in the next chapter. In general, once the specimen block is properly trimmed with parallel edges, together with a sharp blade, serial sections are readily obtained. From time to time, a ribbon of sections fails to form. If one notices that sectioning is not consistent, with skipped or missing sections, or the sections are of uneven thickness, check the specimen block and tighten all screws and clamps, ensuring there are no loose parts. Any loose fitting will generate vibrations compromising the quality of the ribbon and sections.

The next common problem is that sections are compressed with apparent wrinkles. Severely wrinkled sections are difficult to be flattened during the flattening step (see next section). This is often due to an improper cutting speed and/or a dull blade. Sectioning too fast can cause compression of sections as the specimen block passes through a blade. One needs to reduce the speed of sectioning to minimize wrinkle formation. A blade will become dull upon prolonged sectioning, and this is another cause of the compression. As a remedy, a new part of the same blade or a new blade should be used. When possible, avoid changing blades while sectioning the same block, as this requires new adjustments that might result in the loss of the initial sections.

Lines on sections are due to scratches caused by a damaged knife. The knife can be damaged due to ergastic substances such as oxalate crystals within the tissue. If this is the case, move the blade to a new spot and continue sectioning. Scratches can also appear due to dirt particles adhering to the knife edge. In this case, one can try to clean the knife using a Kimwipe® with some xylene; wiping upward toward the edge will prevent damages to the knife blade. Again, if scratches persist, move the blade to a new spot, or a new blade should be used. Care should be taken when moving or exchanging a new blade as the edge can easily be damaged.

After section for a while, one can notice that the ribbon of sections begins to "stick" to the microtome's surface. This is due to the generation of static electricity during sectioning, especially in a dry environment. An increase in the humidity around the microtome using a vaporizer or placing the wax block and the knife assembly in the fridge for a short time before sectioning allows moisture to condense on the block face. This can minimize the generation of static electricity.

2.4.9 Flattening of Sections on Slides

Glass slides need to be selected and labeled before section flattening. Nowadays, microscope slides' cost and quality differ depending on the material use and the manufacturing process. Slides with different dimensions and with and without coatings are available from commercial suppliers. For general histological studies, and if the prepared slides are not subjected to extensive washing and high-temperature treatments, good quality precleaned slides can be used as such, without the need of prewashing and coating with a binding agent. Moreover, for techniques such as in situ hybridization, coated slides are needed, as the sections are subjected to high temperature and chemical treatments. In this case, slides should be cleaned, followed by coating with an adhesive, such as gelatin–chrome alum or silane (for protocol, see next chapter).

Before placing the sections on slides, slides should be labeled with appropriate information. One can etch minimal information using a diamond pencil near one end of the slide. If the slides have a frosted end, label the slides using a pencil as the markings will not be removed during staining. If necessary, relabel the slides after the slides are made permanent.

Once the slides are prepared, proceed to mount the ribbon of sections on the slides. There are two sides to a ribbon, a matt upper surface, and a smooth, shiny lower surface. Always place the shiny lower surface onto the slide; this enables a more secure attachment once sections are dried.

For any given ribbon, sections are always compressed to different degrees and need to be flattened or stretched to remove wrinkles. If wrinkles are not removed, the cells will be out of focus when viewed with a microscope. To eliminate wrinkles, heat is used to expand the ribbon, removing the wrinkles. Place a short ribbon of sections (maximum 3/4 length of a slide) with the shiny side down on a warm water bath or float sections directly on slides with water on a slide warmer. The former procedure requires a reliable thermostat-controlled water bath. This procedure is intended for processing single sections or short ribbons of sections. The water needs to be very clean to avoid contaminants adhering to the slide. The slide warmer procedure is intended for processing serial sections and the placement of multiple ribbons on a slide. We prefer the latter method as the sections are readily arranged and manipulated directly on a slide.

2.4.10 Staining

After processing, most cells have no color. Staining is necessary to increase the contrast and identify different cell types within the section. Besides studying structural details, known chemical reactions can provide additional histochemical information about a specimen. For more information on stains and the theory of staining, consult the monographs by Baker (1966), Gahan (1984), Culling et al. (1985), Ruzin (1999), Bancroft and Gamble (2002), and Horobin and Kiernan (2002). For a detailed and critical discussion on histochemical staining, see Horobin (1982). The following provides a brief introduction to this topic.

Stains can be classified as natural and synthetic dyes. Natural dyes are extracted from plant and animal sources. The active components need to be oxidized before they can be used as a stain and require a mordant, usually a heavy metal serving as a "go-between." Natural dyes are basic stains and are excellent nuclear and chromosome stains. The best-known natural dyes are hematoxylin and carmine. Hematoxylin is still a preferred nuclear stain for medical histopathological specimens. The majority of dyes we use today are synthetic stains, coal tar products, and petroleum extracts. Once prepared, modification is unnecessary before staining, and a mordant is not required. There are numerous synthetic dyes, which are classified according to chromophores. Chromophores are groupings that give color to a stain. The synthetic dyes can also be broadly classified according to their reactivity to tissues such as acidic, basic, and amphoteric stains. The commonly used botanical stains, that is, safranin O, crystal violet, and basic fuchsin, are basic dyes, while fast green FCF, orange G, and aniline blue are acidic dyes.

The stains interact with cells and tissues through the following mechanisms: (1) adsorption via ionic interactions, (2) differential solubility, and (3) covalent bonding. A majority of histological stains, such as safranin and fast green, binds to cell components via ionic interactions. A majority of lipid stains at the light microscope level works on the principle of differential solubility. More lipidic dye molecules can dissolve in a "like"-environment, e.g., lipid, highlighting the structure. Specific histochemical reactions generate irreversible reactions through covalent bonds, which provide specific information on cell components. The commonly used Schiff's reagent is generated using pararosaniline, the main component of basic fuchsin. Leuco-fuchsin is formed when reduced with sulfur dioxide under an acidic environment. Under this condition, leuco-fuchsin interacts with aldehyde groups giving a permanent red reaction product. Through specific oxidation of cellular components, insoluble carbohydrates can be identified using Schiff's reagent after the periodic acid treatment. DNA can be visualized after hydrolysis using 1 M hydrochloric acid solution at 60 °C.

At present, there are numerous staining recipes available in the literature, e.g., Jensen (1962), Clark (1981), and Gahan (1984). Before using any staining schedule, the investigator must understand the chosen stain's chemical properties and the staining mechanism. It is important to note that many factors, such as fixation and staining conditions, can alter the outcome.

Based on our experiences, four straining protocols are recommended, that is, (1) safranin mixture (safranin O—crystal violet —basic fuchsin) and fast green FCF as a general histological stain, (2) amido black 10B as a protein stain, (3) the periodic acid-Schiff's (PAS) reaction in identifying total insoluble carbohydrates, and (4) the Fuelgen reaction for the visualization of DNA. The staining protocols are detailed in the companion chapter.

The safranin O-fast green procedure is commonly used for paraffin-embedded tissues (Jensen 1962; Berlyn and Miksche 1976). Safranin O is a basic dye and, when used alone, will color the specimen red. Moreover, it will bind strongly to the cell's acidic components such as chromosomes, nucleoli, lignified walls, and phenolic compounds. Fast green FCF is an acidic stain that can replace safranin from those components that bind only weekly to it. Therefore, a carefully timed differentiation counterstaining with fast green will give a red and green preparation, highlighting different cellular features. With practice, the result is reproducible and pleasing. The traditional safranin-fast green recipe requires a long staining time, several hours to overnight. To speed up the process, especially when many slides are needed to be processed, a mixture of red, basic dyes was developed and used successfully (Yeung and Peterson 1972). This mixture enables the completion of safranin staining within 30 minutes. The excess stain is removed in water and ethanol washes before counterstaining with fast green. Differentiation using fast green is quick, requiring several seconds to a minute. Different types of tissues react to the stains differently. Therefore, the timing of staining is best determined by trial and error. The traditional fast green staining solution comprises absolute alcohol, xylene, and clove oil (Jensen 1962). However, clove oil is not readily available and expensive; an alcoholic solution of fast green is adequate for general staining purposes [see formulation by Berlyn and Miksche 1976].

Using this method, nucleoli, chromosomes, cuticles, lignified cell walls, and phenolic substances stain red, while the remaining structures such as the primary cell wall stain green (Jensen 1962; Berlyn and Miksche 1976).

Histochemistry combines histology and biochemical techniques in the identification and distribution of chemical components within cells and tissues. Histochemical analyses provide valuable information confirming biochemical measurements. Amido black 10B, also known as naphthol blue black, is a general protein stain. It is an anionic dye that binds strongly to proteins in an acidic environment. This is a reliable general protein stain for cells and tissues and is useful as a counterstain for the PAS staining procedure. The PAS reaction is a specific stain for insoluble carbohydrates such as cell walls and starch granules. The oxidation reaction of periodic acid on glucose units from carbohydrate polymers results in the cleavage of the carbon-to-carbon bonds to form dialdehyde. The aldehyde groups generated can react positively to the Schiff reagent, producing a red color that is specific and permanent. Similarly, for the Fuelgen reaction, the purine-deoxyribose glycoside bonds of DNA can be hydrolyzed using 1 M HCl at 60 °C generating aldehyde groups. The aldehyde groups are then reacted with the Schiff's reagent giving the characteristic pinkish-red color. Again, the reaction is specific to demonstrate the presence of DNA within cells. For the PAS and Fuelgen reactions, the staining outcomes are reliable in identifying insoluble carbohydrates and DNA, respectively. Moreover, one needs to optimize the timing of oxidation/hydrolysis for the best results. If necessary, control staining reactions, that is, enzymatic digestions, need to be conducted to ascertain the nature of the positively stained products.

When staining a few slides, a Coplin jar sequence is preferred. Each Coplin jar can hold five slides and 50 mL of liquid. The small volume can save the solvents required for the staining reaction. At present, a variety of staining containers of different sizes is available from commercial suppliers. For staining a large number of slides, a staining system such as that from Tissue-Tek® (www.sakuraus.com) is preferred. The Manual Slide Staining Set from Tissue-Tek® has 12 solution wells used for different staining protocols. Together with a 24-slide holder, many slides can be processed in a relatively short time.

For a majority of staining protocols, stains are prepared in water or low concentrations of ethanol. Hence, wax needs to be removed before section hydration and staining. The wax from sections is first removed by dissolving in xylene, and sections are gradually hydrated using a down series of ethanol solutions bringing the slides to water or a buffer before staining. After staining, the sections are dehydrated using an up-series of ethanol solutions, clear in xylene, and mount. Clearing in xylene and applying a mounting medium are necessary steps in preparing permanent preparations. Xylene has a similar refractive index as glass; this enables viewing the stained specimen through a microscope. Furthermore, as most mounting media are not compatible with water, xylene is needed as a transition solvent.

2.4.11 The Final "Cover-Up"

Once the sections are stained, a mounting medium (mountant) is applied, covered with a coverslip, making the preparation permanent. A mountant functions to fill out the spaces between tissue and coverslip, and it also provides a permanent seal for the preparation. Nowadays, resinous mountants are commonly used in conjunction with stained paraffin-embedded sections. The mountants' essential quality is a refractive index (RI) similar to glass (RI = 1.5). This enables the seeing of stained components without interference (Sanderson 1994). The synthetic mountants such as DPX in xylene (RI = 1.53), Euparal (RI = 1.48), Entellan (RI = 1.49), and Eukitt (RI = 1.49) are readily available from commercial sources. These media are xylene based and used explicitly for permanent preparations. For temporary mount and specific staining protocols, mountants with different additives and properties are also available such as the Citifluor antifadent mounting solutions designed to reduce photobleaching of dyes' fluorescence (see Electron Microscopy Sciences, for additional information).

Once a mountant is added to the slide, a coverslip is applied. One needs to pay attention to the thickness of the coverslip. Research light microscope objectives are designed to be used in conjunction with a specific coverslip thickness. The recommended thickness, 0.17 mm, is engraved on the objectives. This coverslip thickness is factored into the objective when the lens is designed. This value corresponds to 1.5 thickness of coverslip, which is readily available from commercial suppliers.

After staining, the slides are stored in xylene. Due to the xylene fume's toxic nature, the process of coverslipping should be done in a fume hood, whenever possible, or in a well-ventilated area. Allow the mountant to solidify by placing them horizontally on a slide warmer at 35–40 °C for several days before storing them in a slide box. One needs to ensure that all prepared slides are correctly labeled and organized before viewing and storage. Although applying a coverslip appears to be a simple process, cares and practice are needed for beginners. One needs to avoid trapping air bubbles during coverslipping and make sure that only one coverslip is used and placed on the correct side of a slide.

2.5 Safety in a Histology Laboratory

There are many safety concerns with the wax embedding method (Sanderson 1994; Dapson 2002). First and foremost, many chemical reagents and solvents used are hazardous, toxic with carcinogenic properties. Furthermore, reagents such as ethanol are flammable, and they must be handled with care and stored correctly. We must understand the reagents' chemical properties by reading the Material Safety Data Sheets before use. Avoid unnecessary contact by wearing gloves and preparing fixatives and xylene solutions inside a fume hood if possible.

Be sure to dispose of used fixatives, chemicals, xylene waste, and xylene–wax mixture in proper waste containers according to the protocols at one's institution. It is important to note that some fixatives contain heavy metal ions such as mercuric and chromium compounds. Extra care is needed when handling these fixatives, and proper disposal is required. We must protect our environment, avoiding unnecessary damage to our ecosystems.

It is advisable to keep fixatives, fixing agents, and fixed samples in the same refrigerators and not store them with tissue culture chemicals, media, and biochemicals. Some fixing agents are volatile compounds and can destroy expensive tissue culture chemicals and organics if kept together.

As a safety note for microtomy, be sure to lock the microtome's wheel handle before manipulating the specimen block and changing the microtome blade. This is to avoid accidental injury. The microtome blades are very sharp, and their edges can be damaged easily. One needs to handle microtome blades with care. The blade should be removed from the holder when not in use.

Be sure to clean the work area used for wax sectioning. A vacuum cleaner is best to remove small pieces of wax or wax ribbons. Waste wax can result in a slippery floor and can be dangerous to walk on.

2.6 General Perspective on Paraffin Embedding and Related Techniques

The paraffin embedding method is a well-established technique for the study of the plant body. Irrespective of which embedding method is used, we still do not fully understand the specific chemical process of fixation, dehydration, and embedding media on the tissues. Hence, we need to be aware of the limitations concerning the method's specificity and potential artifacts and interpret the results accordingly. Careful comparative studies using different methods are crucial to our understanding of cells and tissues' structural organization. Only by comparing to living cells, we can then ensure "that the image we see with the microscope is a good representation of what existed in life" (Baker 1966). Furthermore, we need to think in combined terms of morphology, physiology, and biochemistry (Jensen 1962) to benefit from a histological and histochemical study and gain insight into a biological process.

References

Anderson G, Bancroft J (2002) Tissue processing and microtomy. In: Bancroft JD, Gamble M (eds) Theory and practice of histological techniques. Churchill Livingstone, London, pp 85–107

Baker JR (1966) Cytological technique, 6th edn. Chapman and Hall, London, pp 67–112

Bancroft JD, Gamble M (2002) Theory and practice of histological techniques. Churchill Livingstone, London, pp 63–84

Berlyn GP, Miksche JP (1976) Botanical microtechnique and cytochemistry. Iowa State University Press, Ames, pp 1–18

Bracegirdle B (1978) A history of microtechnique. Cornell University Press, Ithaca, pp 2–37

Clark G (1981) Staining procedures, 4th edn. Williams and Wilkins, Baltimore, pp 199–200

Culling CFA, Allison RT, Barr WT (1985) Cellular pathology techniques, 4th edn. Butterworths, London, pp 512–610

Dapson RW (2002) Safety in the laboratory. In: Bancroft JD, Gamble M (eds) Theory and practice of histological techniques. Churchill Livingstone, London, pp 11–32

Gahan PB (1984) Plant histochemistry and cytochemistry, an introduction. Academic, London, pp 86–112

Horobin RW (1982) Histochemistry. Gustav Fischer, Stuttgart, pp 323–360

Horobin RW, Kiernan JA (2002) Conn's biological stains, 10th edn. Bios Scientific Publishers, Oxford, pp 110–214

Huang BQ, Yeung EC (2015) Chemical and physical fixation of cells and tissues: an overview. In: Yeung EC, Stasolla C, Sumner MJ, Huang BQ (eds) Plant microtechniques and protocols. Springer, New York, pp 67–82

Javelle M, Marco CF, Timmermans M (2011) *In situ* hybridization for the precise localization of transcripts in plants. J Vis Exp (57):e3328

Jensen WA (1962) Botanical histochemistry. Freeman, San Francisco, pp 223–306

Johansen DA (1940) Plant microtechnique. McGraw-Hill Book, New York, pp 511–518

Kiernan JA (2008) Histological and histochemical methods, theory and practice, 4th edn. Scion Publishing, Oxford, pp 318–405

Mersey B, McCully ME (1978) Monitoring of the course of fixation of plant cells. J Microsc 114: 49–76
O'Brien TP, McCully ME (1981) The study of plant structure: principles and selected methods. Termarcarphi, Melbourne, pp 107–118
O'Brien TP, Kuo J, McCully ME, Zee SY (1973) Coagulant and non-coagulant fixation of plant cells. Aust J Biol Sci 26:1231–1250
Ruzin SE (1999) Plant microtechnique and microscopy. Oxford University Press, New York, pp 189–245
Sanderson JB (1994) Biological microtechnique. BIOS Scientific Publishers Ltd, Oxford, pp 222–228
Sanderson C, Emmanuel J, Campbell P (1988) A historical review of paraffin and its development as an embedding medium. J Histotechnol 11:61–63
Stasolla C, Yeung EC (2015) Paraffin and polyester waxes. In: Yeung EC, Stasolla C, Sumner MJ, Huang BQ (eds) Plant microtechniques and protocols. Springer, New York, pp 45–66
Tung HT, Yeung EC, Cuong LK, Nhut DT (2019) The paraffin embedding technique in the study of plant histology. J Biotechnol 17:197–212
Yeung EC (1999) The use of histology in the study of plant tissue culture systems—some practical comments. *In Vitro* Cell Dev Biol Plant 35:137–143
Yeung EC (2012) The study of *in vitro* development in plants: general approaches and photography. In: Loyola-Vargas VM, Ochoa-Alejo N (eds) Plant cell culture protocols. Humana Press, New York, pp 95–108
Yeung EC (2015) A guide to the study of plant structure with emphasis on living specimens. In: Yeung EC, Stasolla C, Sumner MJ, Huang BQ (eds) Plant microtechniques and protocols. Springer, New York, pp 3–21
Yeung EC, Chan KWC (2015a) Glycol methacrylate: the art of embedding and serial sectioning. Botany 93:1–8
Yeung EC, Chan KWC (2015b) The glycol methacrylate embedding resins—Technovit 7100 and 8100. In: Yeung EC, Stasolla C, Sumner MJ, Huang BQ (eds) Plant microtechniques and protocols. Springer, New York, pp 67–82
Yeung EC, Peterson RL (1972) Studies on the rosette plant *Hieracium floribundum*. I. Observations related to flowering and axillary bud development. Can J Bot 50:73–78
Yeung EC, Sexena PK (2005) Histological techniques. In: Jain SM, Gupta PK (eds) Protocols for somatic embryogenesis in woody plants. Springer, Dordrecht, pp 517–537
Yeung EC, Stasolla C, Sumner MJ, Huang BQ (eds) (2015) Plant microtechniques and protocols. Springer, New York, pp 67–82

The Paraffin Embedding Method II: Protocols

3

Edward C. Yeung, Hoang Thanh Tung, Claudio Stasolla, and Duong Tan Nhut

Abstract

This chapter details the protocols and related procedures of tissue processing of the paraffin embedding method. Four staining protocols are detailed, allowing the visualization of histological features and histochemical components of cells and tissues. With a proper theoretical understanding of the methods used, continual practices and comparative studies will ensure good-quality histological studies and a better understanding of cellular processes.

Keywords

In vitro cultures · Histological and histochemical staining · Paraffin embedding

3.1 Introduction

The previous chapter provides basic information and discusses some technical issues of the paraffin embedding (PE) method. This chapter focuses on hands-on procedures of the PE method starting with fixation. Since this book emphasizes techniques that apply to in vitro cultured cells and tissues, comments are made in handling in vitro explants when appropriate. Because of space limitations, we only

E. C. Yeung (✉)
Department of Biological Sciences, University of Calgary, Calgary, AB, Canada
e-mail: yeung@ucalgary.ca

H. T. Tung · D. T. Nhut
Tay Nguyen Institute for Scientific Research, VAST, Dalat City, Vietnam

C. Stasolla
Department of Plant Science, University of Manitoba, Winnipeg, MB, Canada

D. T. Nhut et al. (eds.), *Plant Tissue Culture: New Techniques and Application in Horticultural Species of Tropical Region*,
https://doi.org/10.1007/978-981-16-6498-4_3

document those methods that are successfully employed in our laboratories. Readers are urged to consult other publications for additional references and protocols. With a proper theoretical understanding of the methods used, continual practices and comparative studies will ensure good-quality histological studies.

3.2 Know Your System

Before embarking on a histological study, scrutinize the specimen of interest. Be sure to know your experimental system. With clear objectives in mind, design a course of action and a fixation scheme for a histological study.

For studying live cultures or explants of an in vitro system, the experimental system needs to be optimized first. Only when consistent responses are obtained one can then design a histological study to investigate cellular changes associated with the developmental process if warranted.

Initial examination of suspension cultures, cells, and cell clusters can be carried out either using an inverted microscope or a stereomicroscope. Morphological changes can be documented using photographic equipment in conjunction with a quality stereomicroscope or an inverted microscope (Yeung 2012, 2015). When examining bulky tissues, such as large callus pieces, simple dissection or freehand sections and staining can be carried out for a closer look at the cells (Yeung 2012, 2015). Chapter 4 provides a useful technique that enables the sectioning of tissues using a compresstome for a quick structural evaluation.

3.3 Planning for a Histological Study

A quality histological study takes a lot of time to complete. Careful planning and preparation are needed. Histological studies usually involve the collection and fixation of a large number of specimens. For studying changes in explants, tissues need to be collected over the entire course of their development to obtain meaningful results. If it is haphazardly done, the results can generate more questions than answers.

3.3.1 Studying Explants from in In Vitro Systems

Histological studies are useful in providing information concerning structural changes of a developmental event such as organogenesis. Structural changes can generate additional hypotheses for further testing. Such an approach can be found in the study of shoot induction in *Arabidopsis* embryo explants treated with abscisic acid (Paulraj et al. 2014, 2015). In planning for histological analysis, one has to determine the number of explants/embryos/callus pieces needed. The total number depends on the number of sampling points. Changes usually begin to occur immediately after the placement of explants on an induction medium. In our laboratories, we typically sample the explants during the first few days of culture and then gradually lengthen the sampling time until such time as the appearance of somatic embryos, shoots, etc. A minimum of 10 explants or at least 50 somatic embryos should be processed for each sampling point. As a result, for each experiment, more than a hundred explants or somatic embryos are needed. A large number of explants ensure reproducible results. Furthermore, quality median sections are always required. Moreover, it is not easy to obtain perfect median sections to illustrate the results. Hence, many explants need to be processed to ensure one can obtain "perfect sections" to justify the inferences. Control treatments, for example, explants cultured in maintenance medium without growth substance, should also be fixed and processed simultaneously. The entire procedure should be repeated at least once to ensure the reproducibility of the observation.

3.3.2 General Studies on Plant Structures

For the study of plant histology in general, the tissues are easier to handle than tissue culture cells. Moreover, we still need to handle specimens carefully during the entire course of fixation and PE embedding. Certain plant materials are delicate such as thin roots and floral parts. Be gentle in the removal of plant tissues and organs from their natural habitat. Any unwanted adherent materials such as soil particles on root surfaces should be removed carefully using a brush. Bulky tissues need to be cut to the appropriate size before fixation. All tissue pieces should be thin in two dimensions, that is, no more than 5 mm in thickness. The desired planes of sections need to be determined before fixation. For example, suppose a median section through a shoot apical meristem is desired, the shoot tip is first excised, and an off-center longitudinal section is carefully removed from one edge of the shoot tip. This step can be repeated on the opposite side of the shoot tip if it is large. The cut surfaces allow speedy fixation and subsequent wax infiltration. During embedding, the cut surface of the shoot tip is laid down on the embedding mold. Serial longitudinal sections can be obtained readily from wax blocks, and the chance of getting a "perfect" section is high. Additional information on tissue collection can be found in monographs, for example, Sanderson (1994) and Ruzin (1999).

3.4 Materials

For the PE procedure, the equipment, supplies, and chemicals can be obtained readily from different sources. The suppliers listed in this section serve as sample sources and examples only.

3.4.1 Equipment

A rotary microtome, a vacuum pump with a compatible vacuum oven or a vacuum desiccator, a rotary mixer (PELCO® R2 Rotator #1050 with rotator head PELCO® #1051 for 24 mL vials and PELCO® #1054 for 4 mL vials), a dedicated refrigerator for histology use only, a laboratory oven with good thermostat control for wax infiltration, slide warmers, a compound microscope, and a stereomicroscope were used.

3.4.2 Laboratory Supplies

Disposable microtome blades and blade holders (see suppliers such as Electron Microscopy Sciences), glass vials (20 mL), pipettes, razor blades, forceps, needle and scalpel, brushes, tissue probes, Peel-A-Way® Disposable Histology Molds (different sizes), embedding ring, wood blocks for mounting wax tissue blocks, metal tray, Bunsen burner, slide boxes for storage, plain glass microscope slides, slides with an adherent coating, for example, Superfrost® Plus Microscope Slides (Fisherbrand® 12-555-15), and number 1.5 thickness coverslips of various sizes were used.

3.4.3 Chemical Reagents

Formalin (Sigma-Aldrich #F15587), paraformaldehyde (Sigma #30525-89-4), absolute (100%) and 95% ethanol (EtOH), glacial acetic acid, tert-butyl alcohol (TBA) (Sigma-Aldrich #471712), xylene, gelatin, agar, periodic acid, chromium potassium sulfate, aminopropyltriethoxysilane, molecular sieve for drying (Delta Adsorbents, USA) and drying agent with a moisture indicator such as Drerite® were used.

3.4.4 Certified Biological Stains

Amido black 10B, basic fuchsin, crystal violet, eosin, fast green FCF, safranine O, and pararosaniline (Sigma, USA) were used.

3.4.5 Embedding Wax

Paraplast® plus tissue embedding medium (see suppliers such as Electron Microscopy Sciences for further information and selection) was used.

3.4.6 Staining Supplies and Mounting Media

Staining jars such as Coplin jars that house a different number of slides are available from various companies. The Tissue-Tek® manual slide staining set (www.sakuraus.com) can be a useful apparatus for staining many slides at one time. Each slide holder can hold 24 slides, and the plastic staining dishes are designed to use with solvents such as xylene. Mounting medium such as Cytoseal®, Acrytol, and DPX can be obtained from different commercial sources.

3.4.7 Solution Preparation

1. Formalin-acetic acid-alcohol (FAA): Prepare 400 mL of FAA fixative in a beaker by adding 200 mL of 95% EtOH to 140 mL of water. Bring the beaker into the fume hood and add 40 mL of 37% formaldehyde solution and 20 mL of glacial acetic acid. Mix, transfer the contents to a glass bottle with a secure cap. Store the fixative in the fridge.
2. Paraformaldehyde solution (16% stock solution): Prepare a 100 mL stock solution of paraformaldehyde (16%) by adding 16 gm of paraformaldehyde into a beaker containing 60 mL hot distilled water (60–70 °C) in which a few drops of 1 N KOH have been added. Stirred the solution continuously with heat to dissolve the powder. After about 5 min, the solution should be clear with a few undissolved particles. Adjust the final volume of the solution to 100 mL. Filter the solution to remove a few undissolved particles. Store the solution in a tightly capped glass bottle.
3. FAA prepared using a 16% stock paraformaldehyde solution: Prepare 400 mL of FAA fixative in a beaker by combining 200 mL of 95% EtOH, 100 mL of 16% paraformaldehyde stock solution, 20 mL glacial acetic acid, and 80 mL of water. Mix well and transfer the contents to a glass bottle with a secure cap. Store the fixative in the fridge.
4. A 4% paraformaldehyde fixative in 0.05 M phosphate buffer, pH 6.8: For 100 mL of fixative, mix 25 mL of paraformaldehyde (16%), 50 mL 0.1 phosphate buffer (pH 6.8), and 25 mL of water. In preparing this fixative, the phosphate buffer should be able to maintain the fixative pH at about 6.8. However, if too much hydroxide was used in preparing the paraformaldehyde solution, the final fixative pH may need to be adjusted. For adjusting the pH, use 1 N sulfuric acid solution instead of hydrochloric acid, resulting in the production of a carcinogenic product (Goodbody and Lloyd 1994). Since the paraformaldehyde in the stock solution can repolymerize upon storage, a freshly prepared paraformaldehyde solution should be used in preparing a new batch of a fixative solution. The fixative should be used at once or soon after. Always prepare the paraformaldehyde solution and fixative solution in a fume hood to avoid inhaling the toxic aldehyde fumes.
5. TBA dehydration series:
   [D1] 50%: 50 mL H_2O + 40 mL EtOH + 10 mL TBA
   [D2] 70%: 30 mL H_2O + 50 mL EtOH + 20 mL TBA

[D3] 85%: 15 mL H_2O + 50 mL EtOH + 35 mL TBA
[D4] 95%: 45 mL EtOH + 55 mL TBA
[D5] 100%: 25 mL absolute EtOH + 75 mL TBA (0.05% safranin or 0.1% Eosin which can be added to prestain the tissue)
[D6] Absolute TBA: 100% TBA
[D7] Absolute TBA: 100% TBA

6. EtOH series for dehydration: 30, 50, 70, 85, 95, and 100% EtOH. Use 95% EtOH to prepare the lower percentage of the EtOH series. Ensure that the 100% EtOH is dry; if not, a molecular sieve should be added.
7. Down series of EtOH for staining: Xylene (X2), xylene: 100% EtOH (1:1), 100% EtOH, 95% EtOH, 70% EtOH, 50% EtOH, 30% EtOH, water.
8. Up series of EtOH after staining: same as the Down series but in reverse order.

3.4.8 Staining Solutions

1. Safranin mixture staining solutions: safranin O, basic fuchsin, and crystal violet: 0.5%, 0.2%, 0.2%, respectively, in 50% EtOH.
2. Fast green staining solution: 0.5% fast green FCF in 95% EtOH.
3. Amido black 10B: Dissolving 1 gm of amido black 10B (CI 20470) into 100 mL of 7% acetic acid. Filter the solution before use. A 0.5% acetic acid for detaining.
4. Schiff's reagent: Prepare Schiff's reagent by dissolving 1 gm basic fuchsin (CI 42510) in 200 mL boiling distilled water. Cool to 50 °C and add 30 mL 1 N HCl and then 3 g potassium metabisulfite. This combination of HCl and metabisulfite generates sulfur dioxide. This serves to reduce the pararosaniline to a colorless form. Therefore, the container must be tightly capped once the metabisulfite is added. Gently shake for 2 min. Leave in the dark for 24 h and then add 1 g decolorizing activated charcoal. Shake for 5 min and filter quickly through filter paper. Preferably, a Buchner funnel is used, and a mild vacuum can aid filtering. This is to minimize the escape of sulfur dioxide. The solution should be clear and colorless. Store it at 4 °C when not in use. This solution can be stored for months in a well-capped dark bottle. One can also purchase Schiff's reagent directly from a commercial source. Since sulfur dioxide is generated during the Schiff's reagent preparation, perform all the steps in the fume hood. The Schiff's reagent can be reused as long as the solution remains clear; moreover, when it turns pink after prolong used, discard the solution.
5. Bisulfite solution wash after staining: Prepare stocks of 10% potassium metabisulfite $K_2S_2O_5$ and 1 N HCl. Just before use, mix 5 mL of each stock solution with 90 mL distilled water. Always add stock solutions to water. Use and discard.
6. Periodic acid: Prepare a 0.5% periodic acid in distilled water just before use. Prepare the amount needed, use, and discard.
7. 1 N HCl for DNA hydrolysis: Add 85 mL of concentrated HCl to 1 L of distilled water.

3.4.9 Slide Cleaning and Subbing Solutions and Procedures

1. Slide cleaning solution: 1% HCl in 70% EtOH.
2. Chrome alum slide subbing solution (Pappas 1971): Dissolve 0.5% gelatin in distilled water at 35 °C and add 0.05% chromium potassium sulfate. In order to avoid microbial growth, use freshly prepared solution, and discard after 48 h. First clean slides by soaking in acidic 70% EtOH overnight, then wash in deionized water followed by distilled water. Dip individual slides into the subbing solution, dry vertically in a dust-free area.
3. Silane subbing solution (Ruzin 1999): Prepare a 1% silane in 100% EtOH and a 10% acetic acid solution. Mix silane and acetic acid in a ratio of 1:3 (v:v). Dip clean, dry slides into the subbing solution, and air dry. Rinse the slides again in 70% EtOH and allow them to dry. The slides are ready to use.

3.5 Methods

3.5.1 Handling of Plant Tissues Prior to Fixation

Identify and select the tissue to be fixed. For small explants, a few millimeters in size, such as somatic embryos and embryogenic calli, they can be fixed whole. As indicated in the previous chapter, for the ease of processing because of their size, these explants can first be embedded in agar just before fixation. For large explants >1 cm in length, cut the tissue into smaller pieces with the desired orientation. It is preferable to be 5 mm or less in thickness in one dimension and transfer quickly into the fixative. Use a stereomicroscope to aid in the excision and the selection of tissue. Be sure to handle the tissues carefully, avoiding unnecessary physical damage. Always trim the tissues with a sharp double edge razor blade.

Pre-Embedding Small Cell Clusters Before Fixation

1. Transfer cell suspension or cell clusters from the liquid medium into a 10 mL size beaker. Allow cells to settle.
2. Prepare a 1–1.5% agar solution. Allow it to cool to room temperature.
3. Remove excess liquid medium from the beaker and add an approximately equal volume of agar solution to the beaker. Swell the beaker gently to mix the content.
4. Transfer the beaker at once to a bath of ice water and allow it to gel quickly.
5. Once gelled, gently remove the agar block from the beaker and cut it into small pieces, for example, 5–7 mm^3, and place them into fixative. This method allows easy processing of small cell clusters or tiny somatic embryos.

3.5.2 Fixation

Prior to fixation, prepare, and ensure all necessary solutions, supplies, and tools for the fixation processes are available (see Tung et al. 2019 for illustrations). The fixation duration depends on the fixative used and the specimen's size, ranging from a few hours to several days. To minimize the extraction of macromolecules, the fixation and subsequent dehydration steps can be performed at 4 °C before subjecting them to a higher temperature during wax infiltration. To ensure an effective fixation, a minimum ratio of 1:10 specimen to fixative volume should be maintained.

General Fixation Protocol

1. Before fixing the material, aliquot about 15 mL of FAA fixative into glass vials. Insert a small piece of paper, written using a pencil with appropriate information of the specimen into the vial. The solvent cannot remove the pencil markings. Transfer the label along with the specimens during the entire course of fixation, dehydration, and embedding. One can also mark the vials with a marking pen, but the markings need to be protected using transparent tape. Cap the vials and store them on ice.
2. Using a double edge razor blade and forceps, cut the tissue (1–5 mm thick) and place it immediately in the fixative. For woody material, thin (1 mm thick) sections are preferred. Seeds are difficult to process. Scarify or remove the seed coat completely before fixation, allowing better penetration of the fixative and paraffin wax. For dry seeds, imbibe seeds for several hours before fixation.
3. Place the loosely capped vials in a small tray or box filled with ice. Transfer the tray/box to a vacuum chamber and gently vacuum (about 24″ Hg) the samples for 15–20 min. Gradual evacuation of air will cause air bubbles to form and escape from tissues. Most tissues will sink to the bottom of the vial. If this does not happen, re-vacuum for another 15 min.
4. Release the vacuum, replace the fixative with fresh solution, tighten the caps, and transfer the vials to a rotary mixer with gentle rotation overnight at 4 °C. The solution's volume must be in excess to cover the samples (10:1; liquid: tissue). Incubate on the rotary mixer overnight at 4 °C.

3.5.3 Dehydration

Before dehydration, remove the fixative and rinse with an appropriate washing solution. For FAA, a 50% EtOH rinse is sufficient. When a metal ion-based formula such as CRAF, rinse the tissues with several water changes and then in 30% EtOH before further dehydration. If a buffer is used as a fixative component, rinsed the sample with the same buffer several times before dehydration. Improper removal of fixative may interfere with subsequent staining procedures.

General Dehydration Protocol

1. For FAA fixed samples, gently remove the fixative from the vial with a pipette. Rinse several times with 50% EtOH, and add 15 mL of solution D1. Since dehydration is a gradual process, there is no need to remove the dehydrating solvent completely before adding the next solution. For large specimens, simply

decant most of the dehydrating solvent before adding the next solution. This prevents unnecessary contact and avoids damage to the specimen.
2. Repeat the procedure with the remaining solutions D2-7; the time of dehydration per step ranges from 30 min to overnight for each solution. The length of the dehydration steps can be shortened to a few hours for small (a few cubic millimeters) samples. The incubation time should not be shortened for larger and/or woody samples.

3.5.4 Infiltration of Paraffin Wax

Nowadays, commercial paraffin is clean and ready to use. The paraffin wax comes in a pellet form and will take some time to melt before using. It is essential to plan ahead, ensuring a sufficient quantity of wax is available before embedding. A dedicated oven for paraffin embedding should be used and the temperature fixed at 2–3 °C above the melting point of the selected wax. The wax should not be overheated, as it can denature the additives and alter its cutting properties. It is best to melt sufficient wax for each batch of embedding. Molten wax turns yellow upon prolong storage in an oven. Furthermore, all wax containers and sample vials should be placed on trays inside an oven to collect spills as waxes are a fire hazard from a safety standpoint. Wax is flammable. Used wax can be recycled one time. Simply melt waste wax pieces in a funnel with filter paper. The filtered melted wax can be reused for embedding purposes. This will reduce the cost of operation.

Infiltration Protocol

1. Fill a beaker (500 mL) with pellets of Paraplast® Plus tissue embedding medium. Melt the wax in an oven, set at a temperature of 56–60 °C, or 2 °C above the melting point of the wax selected. As it might take a long time to melt such a large volume of wax, it is recommended to melt the wax 2 days before use. Handle the hot glass beaker with care, and do not slip from one's hand.
2. Remove the vials from the rotary mixer and discard excess TBA from the vials but ensure that sufficient TBA is left to cover the tissue. Fill the vials with molten wax. The volume of wax should be more than the TBA and tissue. At this time, the wax will solidify.
3. Cap the vials tightly and return them to an oven. Incubate in the oven for 12 h to overnight. The tissues will sink to the bottom of the vials.
4. Remove the vials from the wax oven, quickly decant the TBA-wax mixture into a waste container in a fume hood. Refill the vials with fresh molten wax and return them to the oven without a cap.
5. Leave the uncapped vial in the oven overnight to allow the residual TBA to evaporate.
6. Repeat the changes with melted wax (step 5) twice.
7. To ensure no residual TBA and air are present within the samples, a vacuum step can be performed with the vacuum oven temperature set at about 4 °C above the wax melting temperature. If there is no visible sign of air bubbles, terminate the

vacuuming process and return the vials to the oven. The specimens are ready for embedding.

3.5.5 Embedding

The primary purpose of embedding is to arrange the specimen in the desired orientation to obtain appropriate sections. Before embedding, set up an embedding station with the necessary equipment and supplies. The embedding molds can be placed on a slide warmer set at 2 °C above the melting point of the wax. This can slow wax solidification during embedding. Use a hot needle or probe to orientate the specimens. A Bunsen burner or alcohol lamp can be used as a heat source. Be careful when an open flame is used in a histology laboratory as many flammable solvents are present. Select appropriate embedding molds. One needs to ensure that there is sufficient wax for embedding. A cold/ice water bath for rapid cooling of wax blocks is required.

Embedding Protocol

1. Set up an embedding station, that is, a slide warmer at 60–62 °C, tools such as a needle or probe, and an alcohol lamp or Bunsen burner. Prewarm paper boats, the Peel-A-Way® Disposable Histology Molds, or other forms of embedding molds by placing them on the slide warmer.
2. Remove a vial from the oven. Gently swirl the vial and quickly pour its content, that is, the molten wax with samples, into the mold. This step must be performed quickly as the wax will solidify rapidly. One can increase the oven temperature by several degrees at the time of embedding to prevent substantial lowering of temperature due to the opening and closing of the oven door. If dealing with many vials, process a few vials at one time, ensuring the oven temperature remains constant.

   Heat a metal probe using the alcohol lamp or Bunsen burner. While the probe is hot, use it to push the specimen gently to the bottom of the mold. Separate the samples 1 cm apart and away from the edge of the mold. Perform this procedure quickly as the wax begins to solidify soon.

   If many samples are present in a single vial, paper boats are preferred, as different size boats are easily prepared. For small tissue samples, the Peel-A-Way® Disposable Molds are more suitable. This is particularly useful for embedding cells generated from suspension cultures. Different types of embedding molds are illustrated in Tung et al. (2019).
3. Once the samples are arranged at the bottom of the mold in the desired position/orientation, place the mold directly into a tray of ice water for about 30 min, ensuring the blocks are quickly solidified. Once fully hardened, the molds/paper boats can be stored in a plastic bag at room temperature. Be sure to label the molds.

3.5.6 Block Mounting Protocol

1. A properly trimmed wax block with a specimen needs to be mounted onto a microtome chuck to be sectioned. We use the plastic embedding ring as our embedding chuck by filling their cavity with melted wax (this needs to be done at least a day in advance). Adhesive tape can be used to seal one end of the rings while the wax is being poured. Once solidified, the wax in the embedding ring will provide flat support for the wax block (see Stasolla and Yeung 2015, for illustration).
2. Using an alcohol burner or Bunsen burner, heat a spatula to melt and reduce the wax block's height and flatten its surface opposite to the samples by melting away excess wax. This procedure is preferably done in a fume hood to avoid inhaling smoke from melting wax and carried out on a sheet of aluminum foil for easy cleanup. If samples are embedded in a large paper boat, the wax block will need to be cut into the smaller piece using a hot spatula.
3. Remove the plastic rings with wax. Heat the spatula and place one side on top of the wax support of the embedding ring. Quickly place the sample-containing block of wax on the other side. As soon as the wax on both sides starts melting, slide the spatula away, and allow the two surfaces to melt together by gently pressing the top of the wax block. Release the pressure after a few seconds. Reheat the spatula and melt the edges together one more time, ensuring the block is secured to the plastic microtome chuck.
4. Store the blocks in the fridge to ensure all the wax solidifies. Depending on the procedure, the blocks can be stored in the refrigerator or at room temperature for several months.

3.5.7 Sectioning

Be familiar with the operation of the microtome. For beginners, one can practice sectioning with paraffin blocks without specimens.

1. Using a single-edged razor blade, trim the block to the appropriate size and ensure that the top and bottom edges parallel one another and the trims are clean. This enables the formation of a "straight" ribbon. It is useful to trim one corner of the block to register the correct orientation.
2. Be sure the microtome handle is locked; insert the plastic ring in the holder of a rotary microtome. Depending on the size of the block face, the cutting angle should be set at 5–10° while the section thickness can be set between 5 and 10 μM. Generally speaking, the smaller the block face, the smaller the clearance angle. Always place the wax block with the long axis against the knife edge.
3. Once the block is correctly mounted into the specimen holder. Place a knife holder with a blade into the slot and nudge gently forward until the blade edge almost touches the specimen block face. Secure and lock the knife-holder.

4. Unlock the handle and begin to section. If the microtome has a “macro-advance” feature, begin to section with the macroadvance on. The block will advance 20–50 μm at each turn of the handle. Turn the wheel handle slowly, at a slow and constant speed. Once sections begin to appear, turn off the macroadvance and begin to section at the desired thickness. Adjust the cutting speed to avoid obvious compressions or wrinkles appearing on the ribbon of sections.
5. As one begins to section, it is often difficult to generate a ribbon. The sections tend to curl up and may fall off from the knife edge. When this happens, increase the initial cutting speed; make a few quick turns of the microtome handle, and use a wet brush to hold down the sections. This will avoid the initial curling of sections and generate a ribbon. Try not to break the ribbon during sectioning, as it may be difficult to reform a ribbon easily. Adjust the cutting speed to ensure that the sections are smooth with minimal compressions. It is also a useful technique to grab the entire cutting wheel to better control the cutting speed instead of the wheel handle. Reduce the block face by trimming away unnecessary wax and tissues. A block with a small block face is easier to obtain serial sections.
6. With practice, a ribbon of sections is easily obtained. Use a wet brush to help to lift the ribbon. The ribbon will stick to the wet brush; gently lift it from the knife without detaching it from the knife edge. The brush supports the weight of the ribbon without stretching, avoiding the breaking of the ribbon. Continue to section.
7. Once the desired length is reached, using another wet brush, gently lift the ribbon at the knife edge and place it on black-colored cardboard with the matt side facing up. Using a sharp single-edge razor blade, divide the ribbon into segments of about two-thirds the length of a slide.

3.5.8 Slide Selection and Subbing Procedures

Select an appropriate type of slide for the study. Depending on the procedure, different types of slides are available commercially. For slides using general staining protocols such as without the need for a high-temperature treatment, good quality, precleaned slides can be used as-is. One can prepare one’s own subbed slides. First, clean the slides using the slide cleaning solution. Soak the slides overnight in 1% HCl in 70% EtOH. Wash slides in slow-running deionized water for at least 5 min, rinse twice in distilled water, and dry vertically in a dust-free area. Coat the slides using the chrome alum subbing solution or the silane subbing solution as detailed under Subheading 4.9.

3.5.9 Section Flattening

1. Label the slide and place them on a slide warmer. With a pipette, gently add distilled water onto each slide.

2. Set the slide warmer at 45–50 °C. Using a wet brush, gently apply each ribbon segment, shiny side down, to a slide with water on the slide warmer. Depending on the ribbon's size, if there is space for a second row, add an additional ribbon to the water. When applying the ribbon to the slide, start by placing one end of the ribbon to water; with a slow lateral movement, lower the whole ribbon segment. The ribbon will expand on the warm water removing the compression marks generated during sectioning. Be sure that the ribbon is free-floating and not touching the slide directly. This will allow the ribbon "free to stretch," removing wrinkles when present. If necessary, facilitate its expansion by adding additional water at the edge of the segment.
3. Place the remaining ribbon segments on the consecutive slides. Each segment will expand in water.
4. Once wrinkles disappear from the sections, drain off the water carefully by touching a paper towel with the help of a brush. Do not allow the water to evaporate on its own. After draining off excess water, allow the slide to dry at about 35 °C for 24–48 h in a dust-free area. Do not dry the slides at high temperatures. This can result in bubble formation due to the vaporization of residual trapped moisture underneath the sections, which can ruin the sections. Once thoroughly dried, store the slides in a slide box or proceed to stain.

3.6 Staining

Paraffin-embedded specimens usually undergo the following steps in the course of staining. Prepared slides are passed through a down series of EtOH to water or buffer after removing wax using xylene. Slides are stained using a selected staining scheme. After staining, sections are dehydrated using an up-series of EtOH, clear in xylene, and mount.

Staining is best carried out using the Tissue-Tek® manual slide staining set (www.sakuraus.com) with a 24-slide holder (see Tung et al. 2019, for illustrations). Each solvent/staining container holds 200 mL of liquid and can process about 100 slides before replacing them with new solvents. This enables a large number of slides to be stained at one time. One added advantage is that the slides within the same slide carrier are subjected to precisely the same processing and staining regime. This is important for comparative staining studies. Always agitate the slides from time to time, ensuring evenness in processing and staining.

The following details four staining schedules commonly use in general histological and histochemical studies of botanical specimens.

3.6.1 The Safranin Mixture and Fast Green Staining Procedure for General Histology

1. Remove the paraffin from sections by placing the slide in two xylene changes for 10 min and then in a 1:1 mixture of xylene and absolute alcohol for another 5 min. Keep solutions in Coplin stain jars or the Tissue-Tek® slide staining container.
2. Partially hydrate sections by passing through an EtOH series of decreasing concentrations: absolute, 95%, 70%, and 50% (about 5 min each). Agitate the slides occasionally.
3. Stain in the safranin mixture for about 20 min (minimal time).
4. Rinse in distilled water to remove the excess stain on slides. The slide background should be transparent.
5. Dehydrate quickly (a few seconds each) from 50 to 95% alcohol as the alcohol solution will extract the stain from sections. In order to obtain reproducible results, once the dehydrating solutions are intensely colored, they should be changed.
6. Counterstain and differentiate in fast green solution—3–4 dips (a few seconds in total). The staining time is usually very short, or else the entire section simply stains green. The time for this step is determined by trial and error. If the sections appear too green, they can be restained with the safranin staining solution by going back through the staining schedule.
7. Pass through absolute alcohol quickly, removing the excess stain from slide surface and into EtOH and xylene (1:1) for about 1 min., and into 2 changes of xylene—5 min each and store in xylene.
8. Mount in Cytoseal® mounting medium (VWR International) or other similar mounting media.

Results: Nuclei, phenolic substances, cuticle, and lignified elements stain red to purplish-red. Cytoplasm and nonlignified cell walls stain green (Fig. 3.1a).

3.6.2 Amido Black 10B for Total Proteins

This stain has a strong affinity for proteins. Moreover, most plant cells' protein content tends to be low, resulting in weak staining. For general tissue types such as vacuolated parenchyma cells, a longer staining time coupled with a short washing, is preferred. Moreover, for meristematic cells, a short staining time is sufficient. If necessary, the excess stain can be removed using 0.5% acetic acid.

1. Remove paraffin and hydrate sections through a graded alcohol series to water.
2. Stain in the amido black 10B staining solution for one to several minutes. The staining time is usually short for cells with dense cytoplasm.
3. Remove excess stain in 0.5% acetic acid.
4. Rinse in distilled water.

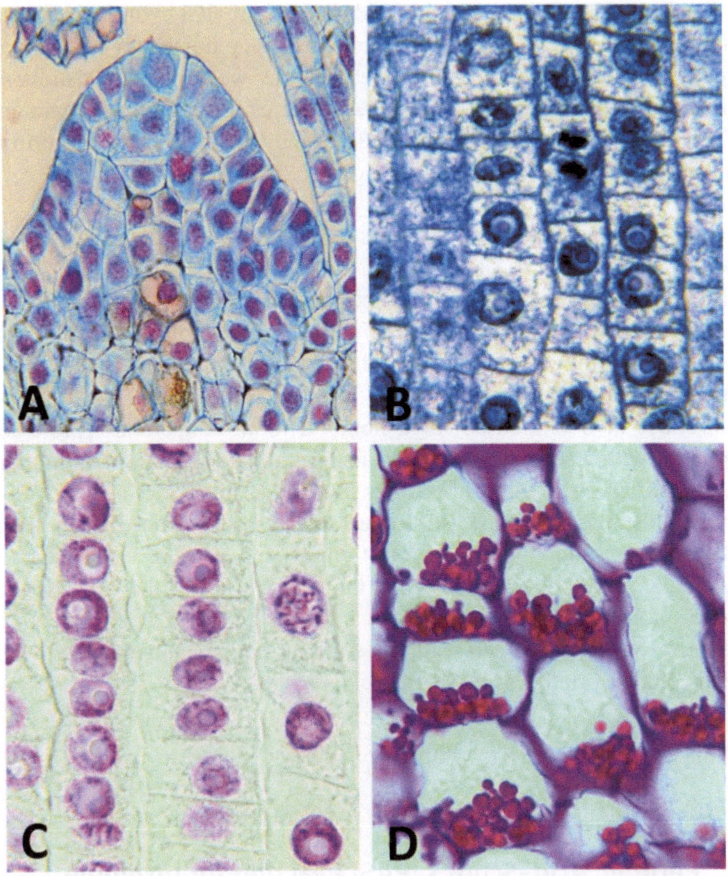

Fig. 3.1 Illustrates the staining results, as indicated in the text. (**a**) Safranin mixture and fast green stain of a radiata pine shoot apex. (**b**) Amido black 10B staining of corn root tip cells. (**c**) Fuelgen staining of corn root tip cells. (**d**) PAS staining of corn root cap cells showing large starch grains of statoliths

5. Place slides briefly in 95% EtOH.
6. Dehydrate through absolute EtOH, EtOH-xylene, clear in xylene and mount.

Results: Proteins stain deep blue (Fig. 3.1b).

3.6.3 Fuelgen Stain for DNA

The Fuelgen staining protocol is a reliable method for the visualization of DNA. Quantitation of DNA content can be performed using microspectrophotographic procedures (Berlyn and Miksche 1976) after Fuelgen staining. It is important to note that many variables can affect this staining procedure's outcome. The type of fixing agent, section thickness, temperature, and duration of HCl hydrolysis can influence the staining intensity. The staining protocol must be optimized for a particular tissue. For general visualization of DNA, the bisulfite water washes can be omitted (Bancroft and Gamble 2002). Moreover, proper controls such as staining without the HCl hydrolysis step and deoxyribonuclease treatment need to be performed for critical work.

1. Hydrate slides through the down series of EtOH to water after removing wax with xylene.
2. Rinse sections in the cold (room temperature) 1 N HCl.
3. Place in 1 N HCl at 60 °C for about 12 min. The Coplin jar or staining vessel should be placed in a water bath at 60 °C beforehand.
4. Rinse in cold (room temperature) 1 N HCl and then distilled water.
5. Stain slides in Schiff based for 2 hours at room temperature, preferably in a Coplin jar with a screw cap or a container tightly wrapped with aluminum foil.
6. After staining, in a fume hood, transfer slides into Coplin jars with freshly prepared bisulfite solution wash. Repeat this procedure once before washing the slides in slow-running tap water for 10 min.
7. After color development, dehydrate the sections using the up-series, clear in xylene, and mount.
8. In order to visualize the cell wall, the sections can be counterstained with a Fast Green solution briefly (as in Sect. 3.6.1).

Result: DNA stains purplish red (Fig. 3.1c).

3.6.4 Periodic Acid–Schiff's (PAS) Reaction for Insoluble Polysaccharides

Cell wall characteristics are essential markers for the identification of cell types. The PAS reaction marks the walls of plant cells. The reaction is reliable. Similar to the Fuelgen reaction for DNA, the PAS procedure needs to be optimized. Several variables, that is, periodic acid concentration and treatment duration, and section thickness need to be optimized. Again, the sodium bisulfite washes can be omitted. Moreover, to confirm PAS-positive materials, proper controls are required.

1. Bring slides through the down-series of EtOH to water after removal of wax using xylene.
2. Place the slides in an 0.5% periodic acid solution in distilled water at room temperature for 5–30 min. For root tips, 15 min work well.
3. Wash in running water for 10 min,
4. Stain in Schiff's reagent for 30 min.
5. Rinse sections in water, and place them in bisulfite solution for 1–2 min.
6. Wash in running tap water for 5–10 min.

7. Dehydrate in the up-series of EtOH, clear, and mount. If needed, the amido black 10B can be used as a counterstain for the PAS reaction.

Results: Insoluble carbohydrates such as cell walls and starch grains stain intense purplish-red (Fig. 3.1d).

3.7 Mounting and Drying

Ensure the necessary supplies and tools, that is, the mounting medium, coverslips of proper dimension and thickness, needles, and forceps are available before making permanent mounts. Since the slides are stored in xylene, excess xylene needs to be drained from the surface by touching the paper towel to remove it from the slide. This is followed at once by the addition of a mountant and a coverslip. The entire process must be done quickly; this is to prevent the evaporation of the residual xylene from the surface of the slide. If xylene is allowed to evaporate completely from the slide surface, air will be trapped by cell walls, generating air bubbles within the preparation. The amount of mountant added should be sufficient and not be excessive; extra-mountant oozing onto the slides' surface is hard to clean. It is best to predetermine how many "drops" of the mounting medium are sufficient for a particular size of a coverslip. If the mountant becomes too viscous, a small amount of xylene or toluene can be added to thin the solution. Whenever possible, perform the entire process in a fume hood to avoid inhaling xylene vapor. It is an excellent practice to place prepared slides in a horizontal position for several days on a slide warmer before placing them inside a slide box for storage. A mountant usually takes several days to become firm, bonding together the coverslip and the slide.

Slide Mounting Protocol

1. Remove a slide from the xylene solution, gently touch the slide on an absorbing paper or paper towel, removing excess xylene from slides.
2. Depending on the size of the coverslip use, place one to four drops of mounting medium over the sections. Apply a coverslip gently onto the mounting medium, allowing the medium to spread and cover the underside of the coverslip. Use a needle to help in lowering the coverslip onto the mounting medium. The coverslip is thin, and it can break easily, handle with care. Be sure to use only a single coverslip per slide and place it on the correct side with sections.
3. Place slides horizontally on a slide warmer. A slide warmer is preferred as any trapped air bubbles will move to the edge of the slide, not interfering with the subsequent examination. Allow the mounting medium to firm up for several days before further processing. Once the slides are dried, if necessary, relabel the slides properly before viewing and storage.

3.8 Concluding Remarks

This article has detailed the PE method to obtain thin sections of tissues for histological studies. The procedures are relatively simple and less expensive when compare to other embedding methods. With practice, the technique can be learned and mastered in a short time. To be successful with this and any other histological methods, we need to aim for perfection in our preparations. There are limitations to all current embedding methods. Moreover, to further our understanding and improvements of different histological techniques, continual trial and error and comparative studies will be key to our success.

References

Bancroft JD, Gamble M (2002) Theory and practice of histological techniques. Churchill Livingstone, London, pp 63–84

Berlyn GP, Miksche JP (1976) Botanical microtechnique and cytochemistry. Iowa State University Press, Ames, pp 1–18

Goodbody KC, Lloyd CW (1994) Immunofluorescence techniques for analysis of the cytoskeleton. In: Harris N, Oparka KJ (eds) Plant cell biology. IRL Press, Oxford, pp 221–243

Pappas PW (1971) The use of a chrome alum–gelatin (subbing) solution as a general adhesive for paraffin sections. Stain Technol 46:121–124

Paulraj S, Lopez-Villalobos A, Yeung EC (2014) Abscisic acid promotes shoot regeneration in *Arabidopsis* zygotic embryo explants. In Vitro Cell Dev Biol Plant 50:627–637

Paulraj S, Lopez-Villalobos A, Yeung EC (2015) Shoot apical meristem ontogeny in *Arabidopsis* embryo explants treated with abscisic acid. Botany 93:445–452

Ruzin SE (1999) Plant microtechnique and microscopy. Oxford University Press, New York, pp 57–128

Sanderson JB (1994) Biological microtechnique. BIOS Scientific Publishers Ltd, Oxford, pp 37–83

Stasolla C, Yeung EC (2015) Paraffin and polyester waxes. In: Yeung EC, Stasolla C, Sumner MJ, Huang BQ (eds) Plant microtechniques and protocols. Springer, New York, pp 45–66

Tung HT, Yeung EC, Cuong LK, Nhut DT (2019) The paraffin embedding technique in the study of plant histology. J Biotechnol 17:197–212

Yeung EC (2012) The study of *in vitro* development in plants: general approaches and photography. In: Loyola-Vargas VM, Ochoa-Alejo N (eds) Plant cell culture protocols. Humana Press, New York, pp 95–108

Yeung EC (2015) A guide to the study of plant structure with emphasis on living specimens. In: Yeung EC, Stasolla C, Sumner MJ, Huang BQ (eds) Plant microtechniques and protocols. Springer, New York, pp 3–21

A Simple Guide to the Use of Compresstome in Plant Research

4

Mohamed M. Mira, Edward C. Yeung, and Claudio Stasolla

Abstract

Microscopic examinations of plant tissue are essential to address biological questions, and protocols to enhance the quality and integrity of the specimens have been developed. These include pretreatments such as fixation and dehydration, which are essential for preserving the structural integrity of the tissue but are often not compatible with the detection of more sensitive and labile molecules or antigens, requiring fresh tissue. Free-hand sectioning and, more recently, the use of vibratomes have facilitated analyses of fresh tissue, although both techniques pose some limitations. The inconsistency in the section thickness and the difficulties in dealing with minute specimens limit the use of free-hand sectioning, while chatter marks and tissue displacement are common drawbacks of vibratomes. These limitations can be overcome by using compresstomes, inexpensive and versatile instruments able to produce good quality sections from a variety of materials and for a wide range of applications. Despite the many advantages of the compresstome over the vibratome in animal research, the use of this instrument in plant research is very limited. This chapter examines the benefits of the compresstome and, most importantly, provides a step-by-step procedure that inexperienced users can utilize.

Keywords

Compresstome · Fresh tissue sectioning · GFP · Nitric oxide · Vibratome

M. M. Mira · C. Stasolla (✉)
Department of Plant Science, University of Manitoba, Winnipeg, MB, Canada
e-mail: Claudio.Stasolla@umanitoba.ca

E. C. Yeung
Department of Biological Sciences, University of Calgary, Calgary, AB, Canada

D. T. Nhut et al. (eds.), *Plant Tissue Culture: New Techniques and Application in Horticultural Species of Tropical Region*,
https://doi.org/10.1007/978-981-16-6498-4_4

4.1 Introduction

Methods and techniques for studying animal and plant tissue often rely on pretreatments, such as dehydration, fixation, and embedding preceding histochemical and/or immunohistochemical analyses. Despite the considerable improvements in the selection of chemicals and protocols minimizing artifacts, these procedures can result in the loss of cellular constituents. They might not be suitable for detecting molecules or antigens sensitive to harsh changes in chemical or physical environments. Therefore, the utilization of fresh and untreated tissue is often required to answer biological questions using histological and immunological analyses targeting specific cell, tissue, and organ types.

There is plenty of literature dealing with studies using free-hand sectioning, a technique that, if performed correctly, produces sections comparable to those obtained with precision instruments, such as microtomes (Yeung 2015). Applied also to specimens requiring hardening through ethanol treatments, free-hand sectioning with razor blades (Fig. 4.1a) can be significantly enhanced by using methods to support the specimen during sectioning better. This can be achieved using fresh carrot tissue, which can be split longitudinally to accommodate the proper orientation of the specimen. The use of hand microtomes (Fig. 4.1b) can further facilitate the procedure as the carrot cylinder supporting the specimen can be bound to the well of the microtome, previously lubricated with low melting point wax or grease. Minute adjustments allow the support block and specimen to advance, producing sections as thin as a few microns.

While inexpensive and accessible to all users, proper free-hand sectioning with or without the assistance of a hand microtome often cannot be applied successfully to minute specimens difficult to immobilize, support, and orient during sectioning. The technique also requires a significant amount of practice and optimization, which can hamper the processing of specimens not readily available or requiring labor-intensive dissections. This is often the case of tissue culture explants, such as seed embryos, which are prone to damage and difficult to extract from the embedded maternal tissue. Furthermore, even when properly mastered, hand sectioning has some major drawbacks, including tissue shattering and shearing, uneven sectioning, and/or imprecise cutting angles. These limitations can be a deterrent to many, especially those new to plant anatomy.

Vibratomes can be a successful alternative to hand sectioning; routinely used in animal and plant research (Steele-Perkins et al. 2005; Rocha et al. 2014; Hunziker et al. 2019), they can generate good quality thin sections without the need to freeze or fix.

Common vibratomes use disposable razor blades, which vibrate while moving horizontally over the tissue imparting a cutting pattern resembling a sine wave. The tissue, which can be fresh or partially fixed, is trimmed and glued onto the vibratome's support stub. Desirable results with the vibratome have been obtained for small specimens, including plant embryos (Lentini et al. 2020), and even for tissue preparation to be used in gene expression analyses (Olsen and Krause 2019). This latter study showed that the quality of vibratome-produced sections was

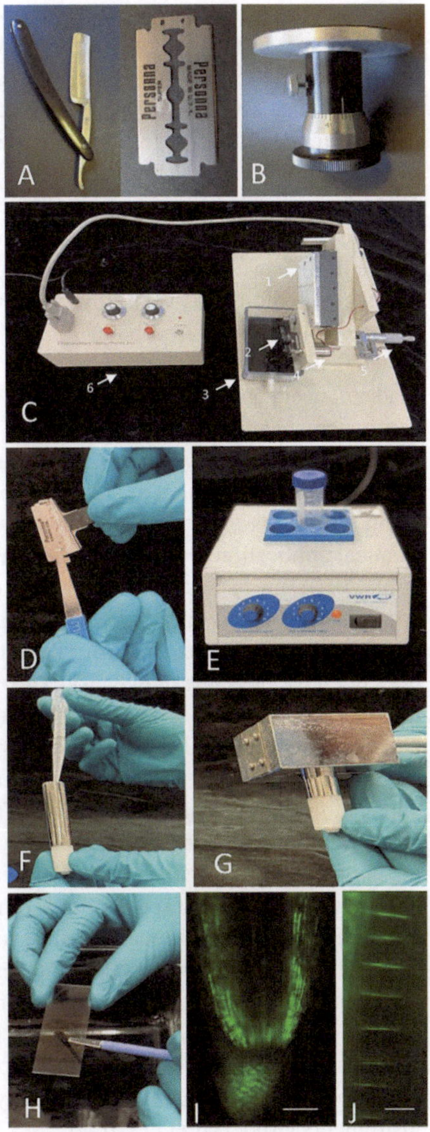

Fig. 4.1 Razor blades are commonly used for free-hand sectioning. (**a**) Hand microtome; (**b**) Compresstome composed of a vibration head (1), blade holder (2), reservoir tank (3), specimen tube (4), manual dial (5), and operation box (6); (**c**) Razor blades are placed on the holder using glue; (**d**) Before embedding the sample, the agarose solution should be kept warm using a dry bath; (**e**) Tissue is embedded by filling the space created between the plunger and the sleeve with the agarose solution, ensuring that the specimen is fully submerged; (**f**) To rapidly cool down the agarose, a cold chilling block is placed over the specimen tube and held for about 1 min; (**g**) Sections are placed on a microscope slide for further treatments using a small brush; (**h**) Visualization of nitric oxide (NO); (**i**) PIN1: GFP signal; (**j**) In corn roots. Scale bar = 300 nm (i) and 10 nm (j)

suitable for laser microdissection and RNA extraction, thus replacing the laborious fixing and embedding procedures for tissue preparation.

The use of the vibratome is convenient; it does not require special blades. Most importantly, it is effective on fresh tissues without pretreatment, such as fixation and dehydration that might cause artifacts typical of cryo- or paraffin-embedded sections. Vibratome sections preserve good structural and ultrastructural characteristics, and adjustments to the vibrating razor blades in terms of speed and vibration amplitude permit the user to optimize sectioning on diverse types of specimens. Furthermore, sectioning can be performed at low temperatures by filling the reservoir tank with ice and a suitable buffer designed for a specific application. These features, not compatible with free-hand and hand microtome sectioning, are crucial when dealing with liable antigens sensitive to suboptimal temperature and chemical environments.

Despite these advantages, however, vibratomes' use has several major drawbacks that need to be considered and addressed when embarking on a new project. Vibratome-generated sections are often characterized by chatter marks (also referred to as vibratome lines), which can be more or less conspicuous depending on the nature of the specimens. Furthermore, compared to microtomes, vibratome sectioning can be tedious and very slow as the movement of the vibrating blade is almost undetectable; this is a problem for projects requiring large numbers of sectioned material. Finally, a very common drawback when using the vibratome is that its blade often slides on soft tissues while "catches" on hard or fibrous tissues displacing components and compromising the integrity of the specimens. This has been demonstrated in several studies, including during the analyses of lymph node and spleen tissue (Abdelaal et al. 2015). These limitations can be partially overcome by using a compresstome, an automated instrument designed for rapid sectioning with frequent applications in animal tissues (White et al. 1997; Skinner et al. 2000; Yang et al. 2010) but far less in plant tissues (Mira et al. 2020). Like the vibratome, the compresstome is versatile for generating free-floating sections (from a few microns to several hundred microns) from fresh or fixed material but, unlike the vibratome, does not have the same drawbacks. This was clearly documented in a study comparing the compresstome VF-300 (Precisionary Instruments Inc., San Jose, California) and the vibratome 3000 (Technical Products International, St. Louis, MO) for sectioning lymphoid and genital tissues of primates for immunofluorescence applications (Abdelaal et al. 2015). The authors observed that while the time required for sample preparation was very similar for two instruments, both allowing the ability to control temperature during sections, the compresstome offered several advantages. The cutting speed of the compresstome was significantly faster than that of the vibratome, a difference ascribed to the larger size of the block supporting the specimens. These characteristics also resulted in larger sections obtained with the same instrument. Unlike vibratome sections, compresstome sections were devoid of chatter marks and, most importantly, retained better structural integrity as the compresstome blade did not cause displacement of tissue. No differences in the quality of the sections for microscopy analyses were noted between the two instruments (Abdelaal et al. 2015).

Based on the advantages and wide range of applications of the compresstome described above, it is surprising to see a very limited utilization in plant research. The purpose of this chapter is to illustrate the step-by-step procedure for the utilization of a compresstome VF-300 (Precisionary Instruments Inc., San Jose, California) and provide case studies for localization of nitric oxide (NO) and GFP signal in corn roots.

4.2 Overview of the Compresstome VF-300

The instrument (Fig. 4.1c) is composed of the following components: a vibration head connected to a blade holder, a reservoir tank, a specimen tube comprising an external metal sleeve with an internal plunger, and a manual dial. The blade holder can accommodate diverse types of razor blades, ranging from ceramic blades, tungsten carbide blades to double edge razor blades that can be glued directly onto the holder. The reservoir tank holds the buffer solution for the free-floating sections. The selection of the buffer and the ability to control its temperature enhance the preservation of the antigen or molecule of interest and the structural integrity of the tissue. With an external metal sleeve securing the tube to the reservoir tank, the specimen tube houses an internal plunger to which the specimen is glued. Advancement of the plunger during sectioning is regulated by either a manual dial equipped with a scaled crank or, in newer models, by a controlled box with an automated advancement controlled system. An operation box with a stop and start function regulates the degree of oscillation of the blade and its advancement speed during sectioning.

4.2.1 Tissue Preparation

The compresstome can be used to section fresh or fixed tissue. For fresh tissue preparation, it is highly recommended to cut the specimen in iced cold Petri plates with either a scalpel or a double edge razor blade. For fixed specimens, the most suitable fixation buffer should be selected depending on the application. It is recommended to wash the tissue with a buffer after fixation. Fresh or fixed tissue should be trimmed to the size of the base of the plunger of the specimen tube, making sure that the height of the tissue does not exceed its width. This precaution will prevent the specimen from wobbling during sectioning. Once trimmed, the specimen can be sectioned directly without any further treatment, or embedded in low melting point agarose, a process yielding uniform sections and highly recommended when dealing with minute specimens that cannot be easily oriented. From our experience, we found embedding in agarose be essential when sectioning Arabidopsis embryos used as explants for somatic embryogenesis and to generate cross-sections of roots. The procedure for agarose embedding is presented in Sect. 4.2.3.

4.2.2 Mounting of the Blade

Depending on the nature of the specimens, different blades can be used, although double edge razor blades are the most commonly used, being relatively inexpensive and disposable. After removing the blade holder from the vibration head using an Allen key, squeeze a small amount of glue (in our hands, Krazy glue or Gorilla glue gives the best results) on a petri dish. With a narrow spatula, scoop some glue and gently apply it to the blade holder, making sure to cover the whole surface housing the blade. Avoid using more glue than necessary as this will interfere with the positioning of the blade and sectioning, especially if the glue comes in contact with the sharp side of the blade edge. To cut the blade, hold one half of the blade with a pair of pliers, and with another pair of pliers, gently twist the other half. The twisting motion will sever the two halves, each of which can be used as a cutting blade. Gently place the cutting blade on the holder and apply pressure with the forceps to ensure the whole surface of the blade adheres uniformly to the glue (Fig. 4.1d). Avoid touching the sharp side of the blade as this might cause scratches on the sections, and most importantly, do not allow the glue to stick on the side of the blade to prevent tissue damage during sectioning. Leave the blade to dry for at least 5 min before mounting the blade holder back onto the vibration head of the compresstome.

It is imperative to clean the blade holder thoroughly before mounting a new blade, as this might interfere with the adhesion of the blade to the mount. Do so by detaching the old blade from the mount using a pair of forceps and then immerse the blade older in acetone to remove residual hardened glue. Wipe the mount with a clean paper towel and repeat the wash in acetone if needed. The blade holder's surface needs to be completely smooth and devoid of any debris before mounting a new blade. Acetone-resistant containers should be used for the washes.

4.2.3 Embedding the Specimens

When possible and depending on the sensitivity of the antigen or molecule of interest, it is recommended to embed the tissue in a low melting point agarose. While some instruction manuals suggest using agarose tablets for consistency in results, embedding can be performed using any powdered low melting point agarose. Generally, a 2–3% agarose solution is recommended, although, from our experience, the agarose gel's consistency should be empirically determined. The best results are obtained when the consistency of the gel is close to the consistency of the specimen, as this will prevent the tissue from detaching or lifting during sectioning.

When using tablets, the instruction manual will provide details on the number of tablets and the volume of water to be used for the desired percentage of agarose. Furthermore, when possible, replacing water with a buffer can further stabilize the antigen or molecule of interest. Using a 50 mL Falcon tube, dissolve the tablets in water (or a buffer solution) by swirling the tube, and then microwave for a few seconds several times until the tablets are fully dissolved. This procedure also applies when using powered agarose. The agarose solution can then be placed in a hot water bath, or a dry bath (Fig. 4.1e) set anywhere between 32 and 35 °C, depending on the melting point of the agarose used, for several minutes. It is crucial to stabilize the agarose solution's temperature before starting the embedding process, making sure it does not congeal as this might have negative effects on the embedding process. It is therefore recommended to check the temperature of the water bath frequently with a hand thermometer.

To help solidify the agarose solution uniformly, compresstomes are equipped with a metallic chilling block that needs to be precooled in the fridge (or ice) for at least 30 min, or, alternatively, −20 °C for about 15 min. While the block is chilling, the tissue can be glued to the specimen tube. After retracting the metallic sleeve of the specimen tube, a drop of glue can be placed on the base of the white plunger; with a pair of forceps, gently press the specimen against the glue in the desired orientation. The glue will be cured in about 1–2 min, although more time might be required. It is very important not to use too much glue. Ensure the glue does not leak down along the plunger's surface and sides, as this might glue the metallic sleeve to the plunger. Tissue paper moistened in acetone can be used to remove excess glue. Glue can also interfere with the visualization of antigens or molecules of interest; a problem often encountered with minute samples, such as plant embryos. For small specimens common in tissue culture applications, it is suggested to squeeze a drop of glue on a petri dish, and while holding the specimen with a pair of forceps, gently touch the glue with the side of the tissue to be secured to the specimen tube. With a gentle motion, press the specimen onto the base of the plunger. While the glue is still liquid, the position of the tissue can be adjusted to reach the desired orientation. If not properly glued, the tissue will be wobbling during sectioning and, in some cases, can detach from the specimen tube. To prevent this from happening, it is important to ensure that the base of the plunger is clean before gluing the tissue (acetone can be used to remove old glue) and that the glue covers the whole surface area of the specimen. To increase the contact area between the tissue and the base of the plunger, it is recommended, when possible, to cut a flat surface on the specimen before applying the glue.

Once the specimen is secured to the plunger and the glue has cured, retract the plunger while holding the metallic sleeve until the full height of the specimen enters the sleeve. Using a pipette, fill the space created between the plunger and the sleeve with the agarose solution, ensuring that the specimen is fully submerged (Fig. 4.1f). It is important to check the temperature and the consistency of the agarose solution before this procedure to ensure the agarose has not been partially solidified. It is also crucial to dispense the agarose solution very slowly to minimize air bubbles forming that might interfere with sectioning.

Place the cold chilling block over the specimen tube and hold for about 1 min to ensure a rapid and uniform agarose solidification (Fig. 4.1g). Before this, it is recommended to remove any condensation that might have formed on the chilling block when in ice or at −20 °C. Once the agarose solution is fully congealed, remove any residual agarose that might have solidified outside the metal sleeve using a spatula or pair of forceps.

4.2.4 Sectioning

Before starting sectioning, the blade mount, housing a new blade, needs to be reconnected to the vibration head with the Allen key. Clean the reservoir tray with a wet paper towel (do not use acetone unless the tray is acetone-proof). Make sure not to disturb the agarose block, insert the plunger tube into the reservoir tray opening by gently twisting the tube as you advance. Push the tube as far as it goes into the tray; a small knob in the tray will prevent the tube from sliding through. Align the manual dial with the specimen tube, and by turning the knob on the dial, advance the micrometer until it touches the plunger tube. Fill the reservoir tank with the desired buffer making sure the agarose block is fully submerged. In the operation box, select the desired speed and oscillation of the blade. Although some instruction manuals suggest optimal values for both speed and oscillation depending on the tissue to be sectioned, from our experience, these parameters should be selected empirically based on the characteristics of the specimen. For better results, and if the dimensions of the tissue allow, it is recommended to start sectioning several microns above the desired thickness and then gradually reduce the thickness while sectioning until the desired microns are reached. As the blade advances, the sections will be released into the buffer within the reservoir tank. Using a small paintbrush, gently extract the sections from the solution and place them on a microscope slide for further treatments (Fig. 4.1h). To obtain a serial sequence, the sections can be arranged in order on labeled microscope slides. For hard-to-section samples, it might be necessary to replace the blade several times during sectioning.

4.2.5 Staining

Staining can be performed directly on the slides using the most appropriate dye, depending on the applications. For general histological analyses, toluidine blue O (TBO) is routinely used due to its polychromatic characteristic. TBO binds to anionic groups imparting unique colors to diverse cellular components, ranging from pink for carboxylated polysaccharides, green-blue for phenolics, and purple for nucleic acids (O'Brien et al. 1964). After incubation in 0.1% TBO for 1 min, use a small filter paper to remove the dye. Wash the sections with water several times, add a drop of water, and place a coverslip on top of the section for examination. More detailed information on dyes and staining procedures is available elsewhere (Chaps. 2 and 3).

4.2.6 Troubleshooting

Sectioning with a compresstome seldom results in immediate successful results. Optimizations in the procedure, depending on the characteristics of the specimens and the overall objective of the analyses, are often necessary. Thus, the user needs to have a proper understanding of the function of the equipment and be able to address possible pitfalls throughout the procedure, from the preparation of the tissue to its visualization. The table below outlines simple suggestions that should be considered when initiating a new project.

1	Uneven sectioning	Mechanical problems with the vibrating head of the microtome or loose blade. Tighten the blade mount to the unit and ensure the blade is glued firmly to the mount The concentration of the agarose might be too low, thus not providing enough support during sectioning. The firmness of the agarose block should be as close as possible to that of the tissue. Empirical adjustments in concentration are required
2	Sample and agarose block wobble during sectioning	Agarose might not have polymerized enough. Allow the chilling block to solidify the agarose solution for a longer period of time Agarose was partially congealed before it was poured onto the sample. Make sure the agarose is fully dissolved during microwaving and check the temperature of the water bath. Reduce the time needed to pipette the agarose on the specimen. The glue might be old or of poor quality The height of the block and/or specimen exceeds the width. Trim the specimen, ensure a larger area of contact with the plunger, and reduce the block's height as much as possible during its preparation
3	Marks on sections	The razor blade might be scratched, or glue residues might be present on the blade. Replace the blade making sure to apply the minimum amount of glue needed Residual glue on the specimen. This is a very common problem when gluing minute specimens that can be overcome by applying a drop of glue on a petri dish and then gently touching the glue with the specimen before securing it onto the plunger of the specimen tube Improper selection and speed and/or oscillation parameters. Start reducing both and then increase gradually. Speed ad oscillation parameters need to be determined empirically Air bubbles in the block. This is due to poor polymerization of the agarose or agarose solution that might be partially solidified when it is poured onto the sample
4	Artifacts when visualizing the sections	Residual glue on the sections interferes with many visualization procedures. Specimens need to be reglued to the specimen tube; use less glue Debris in the reservoir tank. Make sure the tank is cleaned properly as debris can get trapped within the section
5	Agarose cannot be removed from the sectioned specimen	Some procedures require the removal of agarose after sectioning. To facilitate this process, reduce the concentration of agarose when preparing the block

4.3 Equipment and Supplies

Compresstome VF-300 (Precisionary Instruments Inc., San Jose, California).
Microwave oven.
Water bath.
Low melting point agarose.
Petri dishes.
Spatula.
Forceps.
Double edge razor blades.
Pipette.

4.4 Case Studies

4.4.1 Visualization of Nitric Oxide

Nitric oxide (NO) is an endogenous signaling molecule participating in several biological processes related to plant development or response to stress conditions. The brief description below outlines a simple procedure to localize NO in corn roots.

Corn seeds can be germinated on moist filter paper for 5 days at a high relative humidity (95%). Visualization of NO was performed using the fluorescent stain diaminofluorescein (DAF) (Mira et al. 2020). The whole root was incubated in a buffer (50 mM Tris and 50 mM KCl, pH 7.2) containing 1% (v/v) Triton X-100 and 50 μM DAF-FM DA for 1 h at 37 °C with gentle agitation in the dark. Roots were washed twice in the same buffer devoid of DFA and cut in segments of 0.5 cm thick. Following the procedure outlined in Sects. 4.2 and 3.4, root segments were then oriented vertically and glued onto the plunger, and sectioned (10 μm) using the compresstome VF-200. Nitric oxide visualization was performed with confocal microscopy (excitation 495 nm; emission 515 nm) (Fig. 4.1i).

4.4.2 Visualization of PIN1

The localization of the Zea mays PIN1 (ZmPIN1), an auxin efflux transporter, was performed using a PIN1:GFP reporter line (Mira et al. 2020). Germination of corn and tissue collection and processing were identical to those described in Sect. 4.4.1, excluding the use of the buffer and washes. Dissected root segments were directly glued onto the plunger without any treatment. GFP signal was localized by confocal microscopy (excitation 530 nm; emission 580 nm) (Fig. 4.1j).

4.5 Concluding Remarks

The analysis of fresh tissue is crucial in plant research to help to address fundamental biological questions. The conventional use of free-hand sectioning or hand microtomes poses some serious limitations to those new to sectioning procedures, as well as more experienced users dealing with minute specimens often used in tissue culture practices. While still underutilized in plant research, the use of the compresstome represents a better alternative as it can generate uniform sections suitable for many downstream applications. Compresstomes can section fresh or fixed material of various dimensions, and compared to vibratomes, they offer several advantages, including a faster cutting speed, the ability to section larger specimens without chatter marks, and improved section quality to a reduction in tissue displacement during sectioning. The ability to immobilize the tissue in agarose is another desirable feature for processing minute samples or difficult-to-handle specimens. Like any other procedure, the use of compresstomes requires some practice and optimizations that need to be geared towards the characteristics of the sample and the final application.

References

Abdelaal HM, Kim HO, Wagstaff R, Sawahata R, Southern PJ, Skinner PJ (2015) Comparison of vibratome and compresstome sectioning of fresh primate lymphoid and genital tissues for in situ MHC-tetramer and immunofluorescence staining. Biol Proc Online 17:2

Hunziker P, Halkier BA, Schulz A (2019) Arabidopsis glucosinolate storage cells transform into phloem fibres at late stages of development. J Exp Bot 16:4305–4317

Lentini Z, Tabares E, Buitrago ME (2020) Vibratome sectioning and clearing for easing studies of cassava embryo formation. Front Plant Sci 11:1180

Mira M, El-Khateeb E, Gaafar RM, Igamberdiev AU, Hill RD, Stasolla C (2020) Stem cell fate in hypoxic root apical meristems is influenced by phytoglobin expression. J Exp Bot 71:1350–1362

O'Brien TP, Feder N, McCully ME (1964) Polychromatic staining of plant cell walls by toluidine blue O. Protoplasma 59:367–373

Olsen S, Krause K (2019) A rapid preparation procedure for laser microdissection-mediated harvest of plant tissue for gene expression analysis. Plant Methods 15(1):1–10

Rocha S, Monjardino P, Mendonça D, da Câmara Machado A, Fernandes R, Sampaio P, Salema R (2014) Lignification of developing maize (*Zea mays* L.) endosperm transfer cells and starchy endosperm cells. Front Plant Sci 5:102

Skinner PJ, Daniels MA, Schmidt CS, Jameson SC, Haase AT (2000) Cutting edge: *in situ* tetramer staining of antigen-specific T cells in tissues. J Immunol 165:613–617

Steele-Perkins G, Plachez C, Butz KG, Yang G, Bachurski CJ, Kinsman SL, Litwack D, Richards L, Gronostaisky RM (2005) The transcription factor gene Nfib is essential for both lung maturation and brain development. Mol Cell Biol 25(2):685–698

White HD, Yeaman GR, Givan AL, Wira CR (1997) Mucosal immunity in the human female reproductive tract: cytotoxic T lymphocyte function in the cervix and vagina of premenopausal and postmenopausal women. Am J Reprod 37:30–38

Yang J, Sanderson NS, Wawrowsky K, Puntel M, Castro MG, Lowenstein PR (2010) Kupfer- type immunological synapse characteristics do not predict anti-brain tumor cytolytic T-cell function *in vivo*. Proc Natl Acad Sci U S A 107:4716–4721

Yeung EC (2015) A guide to the study of plant structure with emphasis on living specimens. In: Yeung EC, Stasolla C, Sumner MJ, Huang BQ (eds) Plant microtechniques and protocols. Springer, New York, pp 3–21

Establishment of Nylon Bag Culture System in Regeneration and Micropropagation

5

Duong Tan Nhut, Ha Thi My Ngan, Truong Hoai Phong, and Hoang Thanh Tung

Abstract

In this chapter, a new culture system has been established and applied to improve the efficiency of plant micropropagation. Research on using nylon bag culture system with some advantages such as ventilation, lightweight and low cost (0.003–0.005 USD/bag) in micropropagation of *Sinningia* spp. This culture system enables the plantlets to grow better than in a closed culture system. The plantlets derived from in vitro culture in a nylon bag system with sucrose concentrations reduced to zero (photoautotrophic growth conditions) gave good growth and development after 7 or 14 weeks under nursery. Plants had higher stem height, number of leaves, fresh weight and tuber diameter compared to plants grown under heterotrophic conditions (culture medium supplemented with sucrose).

Keywords

Glass bottles · Nylon bag culture system · *Sinningia* spp. · Photoautotrophic growth

5.1 Introduction

The plantlets cultured in vitro under a photoautotrophic condition with a controlled rate of photosynthesis have significantly improved adaptability and survival at the nursery stage. The plantlets were cultured in a well-ventilated system, the more CO_2 supplied, the reduced moisture content and the sucrose content were the solutions to

D. T. Nhut (✉) · H. T. M. Ngan · T. H. Phong · H. T. Tung
Tay Nguyen Institute for Scientific Research, VAST, Dalat City, Vietnam

D. T. Nhut et al. (eds.), *Plant Tissue Culture: New Techniques and Application in Horticultural Species of Tropical Region*,
https://doi.org/10.1007/978-981-16-6498-4_5

this problem. Kozai and Iwanami (1988) have successfully developed a culture system using flasks with the following material: tetrafluoroethylene perfluoro-alkyl vinyl ether copolymer or tetrafluoroethylene hexafluoropropylene copolymer, simultaneously with increasing CO_2 concentration and light intensity. However, this system proved ineffective because it was difficult to apply in commercial laboratories.

The nutritional content of the culture medium and the composition of the atmosphere have a direct influence on the growth and development of cultured plants (Jackson et al. 1987; Blazková et al. 1989). Subsequent studies have shown that high humidity in closed or poorly ventilated flasks causes the plantlets cultivated in these flasks to open the stomata to maintain equilibrium with their surroundings (Losch and Tenhanen 1981; Shackel et al. 1990) will lead to stomata abnormalities (stomata malformations and damage, stomata open even when plantlets are under sudden environmental changes) (Fabbri et al. 1986; Sutter et al. 1992). Besides, the low CO_2 concentration in the closed culture flasks and the high sucrose content in the medium reduced the plant's photosynthetic capacity. These phenomena not only affect the growth and development of in vitro plantlets, but also ex vitro plantlets because they need to be kept under conditions of relatively high humidity to prevent water loss through the stomata until they restore photosynthesis (Shackel et al. 1990; Preece and Sutter 1991; Kirdmanee et al. 1995).

Up to now, the improvement, modification, and optimization of the technology system have always been the main purpose of commercial micropropagation to create a large number of genetically and physiologically homogeneous plantlets in the shortest time; the plantlets have high photosynthetic or photoautotrophic potential and well adapted to ex vitro conditions (Jeong et al. 1995). Therefore, in this study, we used polyethylene (PE), commonly known as nylon, to design a micropropagation system on *Sinningia* (Gesneriaceae)—an ornamental of economic value and loved gesneriads. PE with good ventilation, different sizes and shapes, high melting point (106–124 °C), resistance to pressure during autoclaving and cheap prices (0.003–0.005 USD/bag) so that it can be used in commercial micropropagation to increase plant growth and development in comparison with traditional micropropagation systems (Tanaka et al. 1988).

5.2 Photoautotrophic Micropropagation and Plant Growth and Development

In the late 1980s at the University of Chiba, Japan, intensive researches on photoautotrophic micropropagation methods began to be carried out with several experiments on environmental factors in vitro plants in small vessels, such as CO_2 concentration, photosynthetic photon flux density. These studies have shown that most in vitro plantlets can grow photoautotrophically. Furthermore, the low net photosynthesis rate of cultured plants is due to poor photosynthetic capacity and low CO_2 concentrations in airtight culture vessels (Kozai et al. 2005). Studies have demonstrated that increasing the CO_2 concentration and light intensity in culture vessels simultaneously with decreasing relative humidity; in addition, the use of fibrous or porous support materials to replace gelling agents (agar) or using a highly ventilated culture system that enhances the photoautotrophic growth of in vitro plants (Kozai 2010; Xiao et al. 2011). These studies have shown that to enhance in vitro plant photosynthesis as well as to increase survival under nursery conditions, it is necessary to understand the state of the in vitro environment inside the culture vessels and facilitate optimal environmental conditions for maximizing in vitro plant growth.

The natural ventilation methods improve the exchange of air (CO_2 and water vapor) between the inside and outside of the culture vessel. A simple way to increase the air exchange capacity of the culture vessel is to use gas-permeable membranes or culture systems with improved ventilation properties (such as nylon bags). On the other hand, by increasing the culture vessel ventilation, the ethylene concentration in the culture vessels can also be reduced, thereby avoiding the harmful effects of ethylene on shoot regeneration, growth of plants, and the aging process of flowers and leaves (Jackson et al. 1991; Biddington 1992).

5.3 Advantages and Disadvantages of Photoautotrophic Micropropagation on In Vitro Plant Growth and Development

Micropropagation underventilated and photoautotrophic conditions have many advantages over traditional micropropagation in the closed vessel using the sugar-containing medium in both biological and technical aspects. Biological advantages include: (a) promoting plant growth and photosynthesis in vitro; (b) shortening the propagation cycle; (c) reducing microbiological contamination; (d) prevention of morphological and physiological abnormalities; and (e) increasing in vitro plant survival when transferred to ex vitro condition. Technical advantages include: (1) simplification of the culture system; (2) increase productivity and reduce labor costs; and (3) easily automated culture system. However, this culture system also has disadvantages such as the higher cost of using gas-permeable membranes to increase air exchange, the cost for lighting and cooling equipment; requires knowledge of the physical environment in vitro; applied plant species restriction (photosynthesis plants following C4 or CAM).

5.4 General Information on How to Establish Nylon Bag Culture System

Commercial nylon bags (Tan Saigon Company, Ho Chi Minh City, Vietnam,) measuring 21 × 18 cm were used in these studies. The nylon bags culture system (NS) is prepared as shown in Fig. 5.1. The folded nylon bags were then wrapped in paper and autoclaved at 121 °C, 15 psi for 20 min.

Then, 50 mL of sterile culture medium was poured into bags under aseptic conditions (Fig. 5.2). Bags with the culture medium inside were sealed with paper

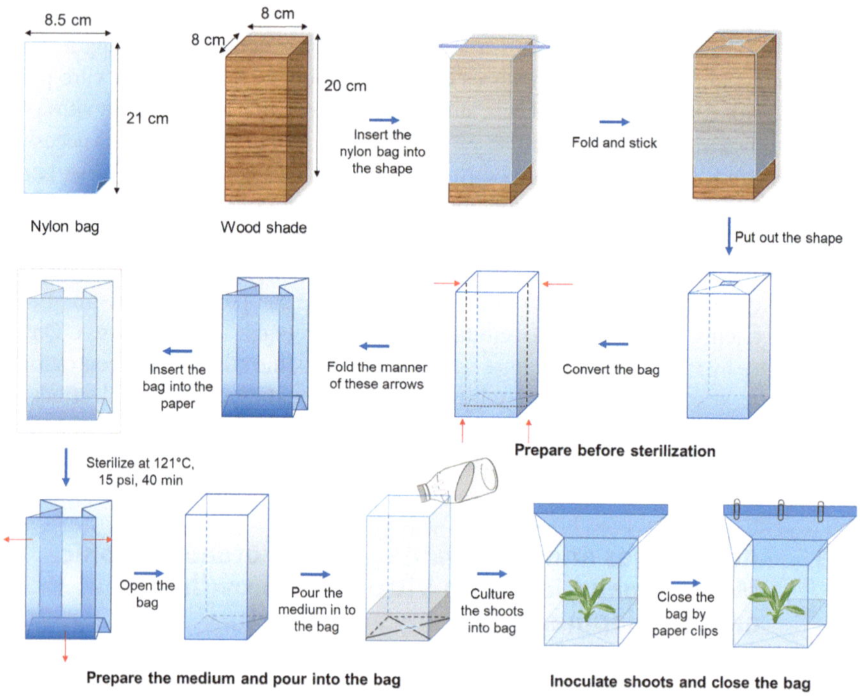

Fig. 5.1 The process of nylon bag culture system preparation

Fig. 5.2 Pour the liquid medium into a nylon bag

Fig. 5.3 Nylon bag with solid medium after autoclaved

clips and let the culture medium solidify at room temperature (Fig. 5.3). A traditional system (glass bottle, 500 mL) was used as the control.

5.5 A Case Study with *Gloxinia*

Gloxinia is a flower of economic value and very diverse (from large flowering hybrids of *Sinningia speciosa* to small flower *Sinningia latexilla* varieties), originating from the tropical forests of South America. It has been cultivated for over 100 years, with many different varieties produced year-round. In recent years, *Gloxinia* has become a popular ornamental plant in Vietnam, but large-scale propagation using tissue culture technology has not been widely applied. Therefore, in this study, we propagate in the new in vitro culture system (the nylon bag systems).

5.5.1 Materials and Methods

Gloxinia shoots (*Sinningia speciosa* × *Sinningia maxima*) 20 days old were cultured on MS medium (Murashige and Skoog 1962) containing 0.1 mg/L N6-benzyladenine (BA), 30 g/L sucrose, and 8 g/L agar was used as the initial plant material source. The healthy and uniform shoots with 1.5–2 cm in height (4 leaves) were used as initial explants for the rooting experiment (Fig. 5.4).

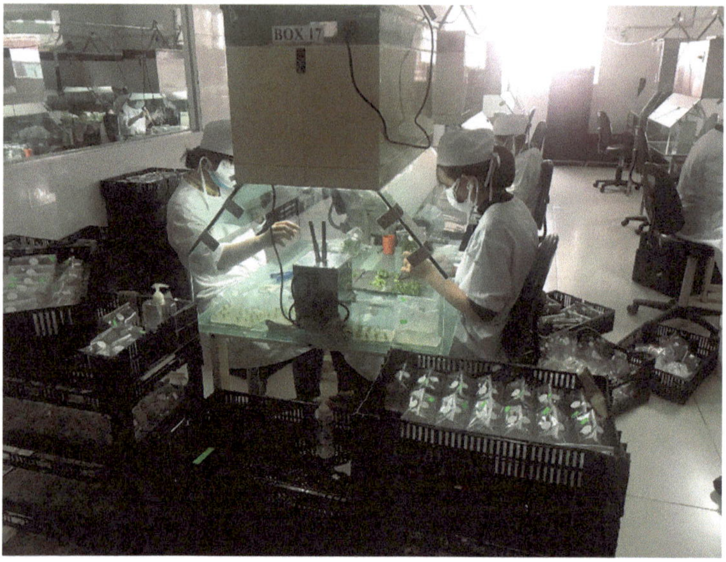

Fig. 5.4 Culture the shoots into a nylon bag

5.5.1.1 Culture Medium

The rooting medium is MS medium containing 0.5 mg/L α-naphthaleneacetic acid (NAA), sucrose (0, 20, 30 g/L), and 8 g/L agar. For control, 30 g/L sucrose was used. The pH of the medium was adjusted to 5.8 before autoclaving at 121 °C for 20 min.

5.5.1.2 Culture Conditions

All cultures were placed at 25 ± 2 °C under a fluorescent lamp (Cool White) in a 10 h/day photoperiod, with an intensity of light of 30 $\mu mol.m^{-2}.s^{-1}$ (Figs. 5.5, 5.6, and 5.7).

For subsequent growth in the nursery stage, plantlets from the in vitro rooting experiment were collected, washed agar, transplanted into plastic pots containing a soilless mixture (Metro-Mix® 350, Scotts, Marysville, Ohio), and placed in the greenhouse. Plants are watered twice a day in the first week, then watered once a day in the early morning. Survival rate and growth indicators were recorded to assess the quality of in vitro plants derived from the nylon bag culture system.

5.5.1.3 Growth Parameters

Monitor the growth and development of explants and check for microbial contamination under a magnifying glass with light to ensure that the plantlets with good quality and disease-free (Figs. 5.8 and 5.9).

The number of leaves, leaf length and width, plant height and fresh weight, tuber diameter of plantlets in in vitro and ex vitro conditions were recorded after 3 (in vitro), 7 or 14 weeks (ex vitro).

Fig. 5.5 Nylon bag with shoots was put in the culture room

Fig. 5.6 Apply nylon bag in shoot proliferation

Fig. 5.7 Apply nylon bag in rooting stage

5.5.1.4 Statistical Analysis

Each treatment was repeated tree times and data were recorded at 3, 7, or 14 weeks of culture. The experiments were arranged in a randomized complete block with 9 shoots per treatment. The data were analyzed by analyzing variance with mean separation using Duncan's multiple range test (Ducan 1995).

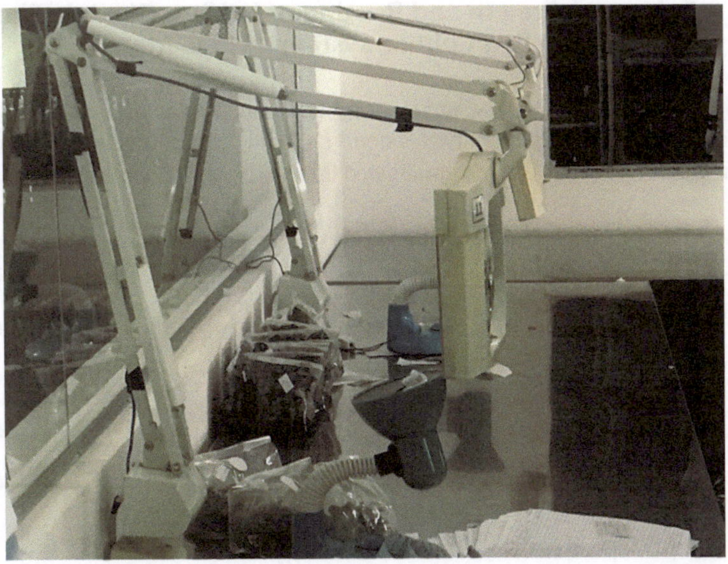

Fig. 5.8 Using large magnifying glass with light for microbial and fungi detection

Fig. 5.9 Microbial and fungi detection were checked by a large magnifying glass with light

5.5.2 Enhanced Growth and Rooting of In Vitro Gloxinia Shoots Using Nylon Bag Culture System

Plant height, number of leaves, and fresh weight of in vitro plantlets after 3 weeks of culturing on rooting medium in two systems (NS and glass bottle) were recorded and presented in Fig. 5.10a. The results showed that the NS system improved the growth

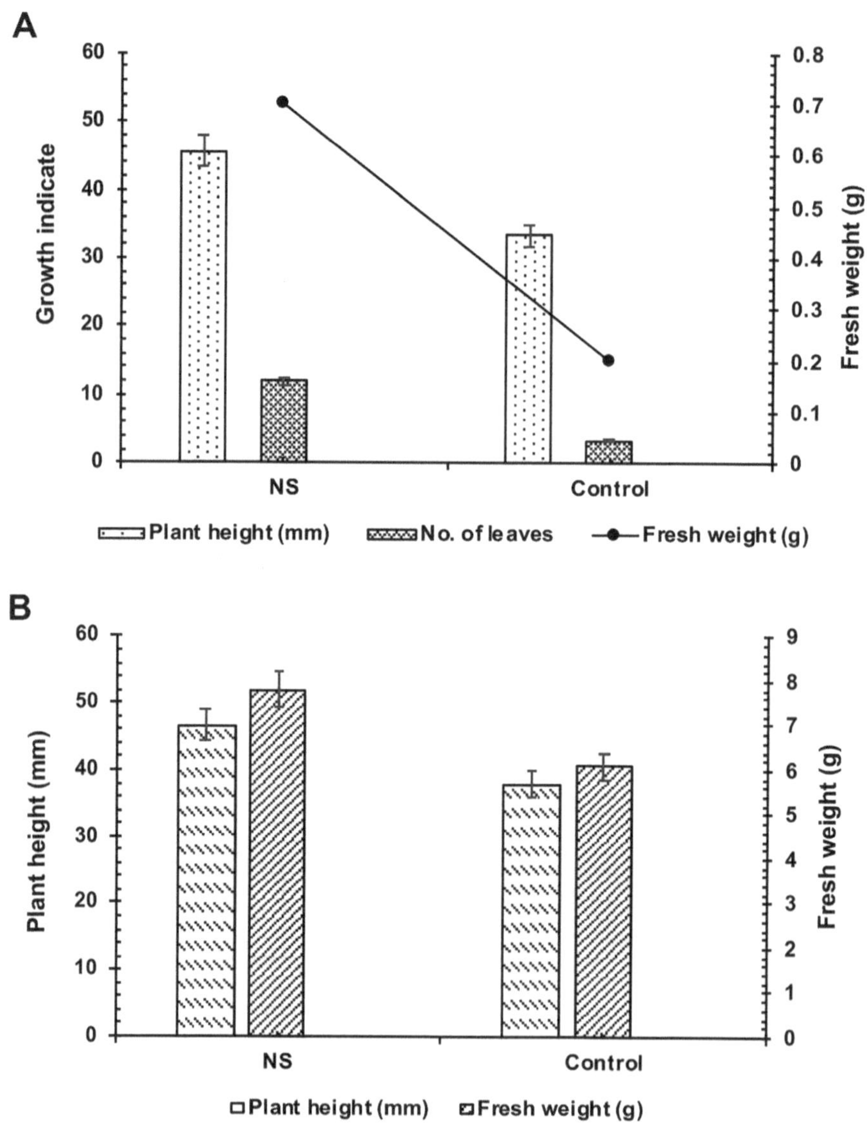

Fig. 5.10 The effect of in vitro culture systems on the growth and development of *Gloxinia* plantlets after 3 weeks. (**a**) In vitro; (**b**) Ex vitro

and development of *Gloxinia* plantlets better than those in the glass bottle (stem height and the number of leaves were three times higher, fresh weight of the plantlet is four times higher). The NS system with increased ventilation capacity has a positive effect on the growth and development of plantlets in vitro.

Several studies reported that the growth and development at the rooting stage of some crops have been significantly improved by using a novel film culture system (Neoflon® film, Daikin Industries, Japan) with thermal stability, high light transmittance, and air permeability (Tanaka et al. 1988, 1996; Tanaka 1991; Nhut 2002; Nhut et al. 2003); however, the price is still too high for use in commercial tissue culture. The important role of ventilation capacity in micropropagation has been observed in a variety of plant species (Jackson et al. 1987; Kozai and Iwanami 1988; Blazková et al. 1989; Kozai et al. 1992; Nguyen et al. 1999). The main advantages of this system are increased photosynthesis of in vitro plantlets due to stabilization of stomata structure and increased chlorophyll content. Cassells and Walsh (1994) report that good ventilation increases evapotranspiration resulting in increased calcium uptake into leaves and improved stomatal function. This indicates that in vitro plantlets in NS with higher concentrations of CO_2 and low humidity increased their adaptability to ex vitro conditions, resulting in a higher survival rate than those in control treatment after 3 weeks in a greenhouse (Fig. 5.10b).

5.5.3 The Growth and Development of In Vitro Gloxinia at Different Sucrose Concentrations and Subsequent Growth of Plantlets in the NS System

One of the advantages of the NS system is photoautotrophic enables of the in vitro plantlets, so sucrose in the medium is presumably no longer the main carbohydrate source for the plantlets. Therefore, in this study, we have investigated the effect of the concentration of sucrose on the growth and development of plants in NS.

Growth and development of plantlets cultured at different concentrations of sucrose are shown in Fig. 5.11a. The results showed that plantlets from sucrose-free medium were shorter, with lower fresh weight than those on 20 or 30 g/L sucrose supplemented. This suggests that, although NS is well-ventilated, the CO_2 content supply is insufficient to sustain the plantlet's photosynthesis on a low sucrose medium. Therefore, the plantlets grown on a sucrose-free medium could not grow as well as those on a 20 or 30 g/L sucrose medium. There have been some studies showing that upstream inhibition of photosynthesis is often due to the high in vitro sucrose content provided (Langford and Wainwright 1987; Desjardins et al. 1995), so the plantlets on medium with lower sucrose level better photoautotrophic and have a major impact on growth, development, and survival of ex vitro plantlets.

Subsequent growth of plantlets derived from in vitro cultures on different sugar concentrations was observed after 7 and 14 weeks in greenhouse conditions (Fig. 5.11b,c). Results after 7 weeks at nursery stage showed that stem height, number of leaves, fresh weight, and tuber diameter of plants from sucrose-free in vitro medium performed better than plants from sucrose-containing medium

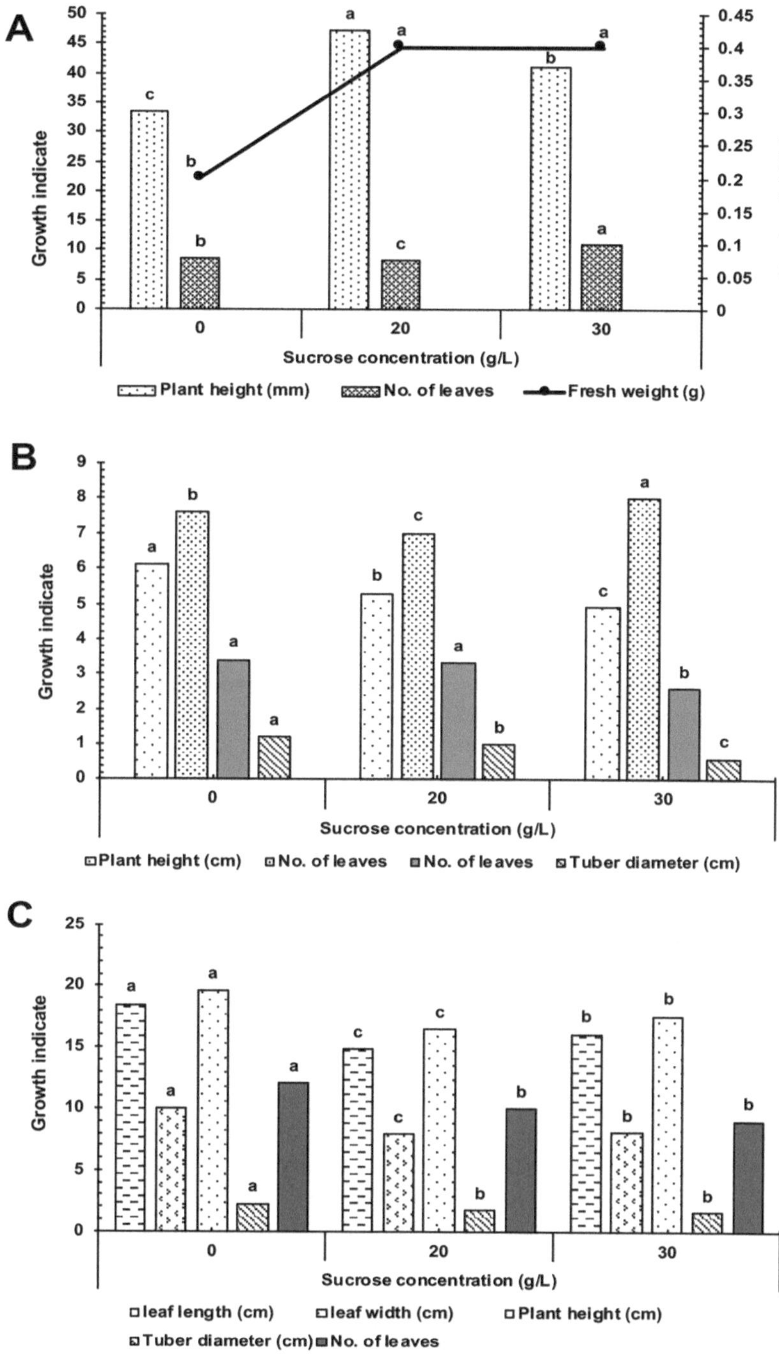

Fig. 5.11 The effect of sucrose concentration on the growth and development of *Gloxinia* plantlets. (**a**) In vitro micropropagation in NS system after 3 week of culture; (**b**) Subsequent growth in greenhouse (7 weeks); (**c**) Subsequent growth in greenhouse (14 weeks)

(particularly stem height and tuber diameter), demonstrating that in vitro plantlet's photoautotrophic capacity is the main factor affecting the quality, adaptability, growth, and development of plantlets ex vitro (Fig. 5.11b). Nguyen et al. (1999) demonstrated that the enhanced photoautotrophic of coffee plantlets when cultured on the sucrose-free medium was similar to the results of Hdider and Desjardins (1994) for that sucrose supply in micropropagation induces inhibition of Rubisco, which in turn leads to low P_n in in vitro plantlets.

After 14 weeks, ex vitro plants begin to flower and the dome plant's characteristics at this stage are recorded in Fig. 5.11c. During the flowering stage, observing the leaf area, stem height, tuber diameter, and the number of leaves, it can be concluded that plants grown from a sucrose-free medium obtained the best traits, showing a positive effect of NS for the growth and development of *Gloxinia* plantlets in vitro and ex vitro conditions.

5.6 Conclusion

The present study showed that the NS was a suitable system for the tissue culture of Gloxinia, because of its aeration, ease in transportation, and low price. Plants originated from NS grew faster and healthier than those from conventional culture system (glass bottle). Furthermore, the sucrose concentration in the medium can be decreased when using the NS. The decrease in sucrose concentrations (even 0) leads to a consequent decrease in contamination and in the price of in vitro plantlets.

In the future, this system can be applied at a large scale in order to save space and energy and to decrease the price of the plant in vitro. Because of the flexible property of nylon, NS can be used in various shapes and sizes depending on the purpose (Fig. 5.12).

Fig. 5.12 Production of plantlets by using nylon bag in Dalat companies (Vietnam)

References

Biddington NL (1992) The influence of ethylene in plant tissue culture. Plant Growth Regul 11: 173–187

Blazková AJ, Ullmann J, Josefusova Z, Machackova I, Krekule J (1989) The influence of gaseous phase on the growth of plants *in vitro*: the effect of different types of stoppers. Acta Hortic 251: 209–214

Cassells AC, Walsh C (1994) The influence of gas permeability of the culture lid on calcium uptake and stomatal function in *Dianthus* microplants. Plant Cell Tissue Organ Cult 37:171–178

Desjardins Y, Hdider C, Deriek J (1995) Carbon nutrition *in vitro*: regulation and manipulation of carbon assimilation in micropropagated system. In: Aitken-Christie J, Kozai T, Smith MAL (eds) Automation and environmental control in plant tissue culture. Kluwer Academic Publishers, Dordecht, pp 441–471

Ducan DB (1995) Multiple ranges and multiple F test. Biometrics 11:1–42

Fabbri A, Sutler E, Dunston SK (1986) Anatomical changes in persistent leaves of tissue cultured strawberry plants after removal from culture. Sci Hortic 28:331–337

Hdider C, Desjardins Y (1994) Effects of sucrose on P_n and phosphonolpyrovate carboxylase activity of *in vitro* cultured strawberry plantlets. Plant Cell Tissue Organ Cult 36:27–33

Jackson MB, Abbott AJ, Belcher AR, Hall KC (1987) Gas exchange in plant tissue cultures. In: Jackson MB, Mantell S, Blake J (eds) Advances in the chemical manipulation of plant tissue cultures, Monograph 16. British Plant Growth Regulation Group, Bristol, pp 57–71

Jackson MB, Abbott AJ, Belcher AR, Hall KC, Butler R, Camerson J (1991) Ventilation in plant tissue cultures and effects of poor aeration on ethylene and carbon dioxide accumulation, oxygen depletion and explant development. Ann Bot 67:229–237

Jeong BR, Fujiwara K, Kozai T (1995) Environmental control and photoautotrophic micropropagation. Hortic Rev 17:12–171

Kirdmanee C, Kitaya Y, Kozai T (1995) Rapid acclimatization of *Eucalyptus* plantlets by controlling photosynthetic photon flux density and relative humidity. Environ Control Biol 33:123–132

Kozai T (2010) Photoautotrophic micropropagation: environmental control for promoting photosynthesis. Propag Ornam Plants 10(4):188–204

Kozai T, Iwanami Y (1988) Effects of CO_2 enrichment and sucrose concentration under high photon fluxes on plantlet growth of carnation (*Dianthus caryophyllus* L.) in tissue culture during the propagation stage. J Jpn Soc Hortic Sci 57:279–288

Kozai T, Fujiwara K, Hayashi M, Aitken-christie J (1992) The *in vitro* environment and its control in micropropagation. In: Kurata K, Kozai T (eds) Transplant production systems. Kluwer Academic Publishers, Dordecht, pp 247–282

Kozai T, Afreen F, Zobayed SMA (2005) Photoautotrophic (sugar-free medium) micropropagation as a new propagation and transplant production system. Springer, Dordrecht, pp 226–308

Langford PJ, Wainwright H (1987) Effects of sucrose concentration on photosynthetic ability of rose shoots *in vitro*. Ann Bot 60:633–640

Losch R, Tenhanen JD (1981) Stomatal response to humidity-phenomenon and mechanism, in stomatal physiology. In: Jarvis PG, Mansfield TA (eds) Stomatal physiology. Cambridge University Press, New York, pp 137–161

Murashige T, Skoog F (1962) A revised medium for rapid growth and bioassays with tobacco tissue cultures. Plant Physiol 15:473–477

Nguyen QT, Kozai T, Nguyen KL, Nguyen UV (1999) Photoautotrophic micropropagation of tropical plants. In: Altman A, Ziv M, Izhar S (eds) Plant biotechnology and *in vitro* biology in the 21st century. Kluwer Academic Publishers, Dordrecht, pp 659–662

Nhut DT (2002) *In vitro* growth and physiological aspects of some horticultural plantlets cultured under red and blue light-emitting diodes (LEDs). Ph.D thesis, Kagawa University, Japan, pp 1–196

Nhut DT, Takamura T, Watanabe H, Okamoto K, Tanaka M (2003) Responses of strawberry plantlets cultured *in vitro* under superbright red and blue light-emitting diodes (LEDs). Plant Cell Tissue Organ Cult 73:43–52

Preece JE, Sutter EG (1991) Acclimatization of micropropagated plants to the greenhouse and field. In: Debergh PC, Zimmermann RH (eds) Micropropagation: technology and application. Kluwer Academic Publishers, Dordrecht, pp 71–91

Shackel KA, Novello V, Sutler EG (1990) Stomatal function and cuticular conductance in whole tissue cultured apple plants. J Am Soc Hortic Sci 115:468–472

Sutter EG, Shackel K, Diaz JC (1992) Acclimatization of tissue cultured plants. Acta Hortic 314: 115–119

Tanaka M (1991) Disposable film culture vessels, In: Bajaj YPS (Ed.) Biotechnology in agriculture and forestry. Springer, Berlin 17(1), 212–228

Tanaka M, Goi M, Higashiura T (1988) A novel disposable culture vessel made of fluorocarbon polymer films for micropropagation. Acta Hortic 226:663–670

Tanaka M, Nagae S, Takamura T, Kusanagi N, Ujike M, Goi M (1996) Efficiency and application of film culture systems in the *in vitro* production of plantlets in some horticultural plants. J Agric Sci Technol 8:280–285

Xiao Y, Niu G, Kozai T (2011) Development and application of photoautotrophic micropropagation systems. Plant Cell Tissue Organ Cult 105:149–158

Wounding Manipulation and Shoot Tip Removal Methods in the Micropropagation of *Paphiopedilum callosum* 6

Duong Tan Nhut, Vu Quoc Luan, and Hoang Thanh Tung

Abstract

Paphiopedilum spp. are high-value horticultural crops due to their wild populations are under the threat of extinction as a result of overcollection and loss of suitable habitats. They are commercially propagated via seed germination in vitro, *Paphiopedilum* is considered to be difficult to propagate in vitro, especially by plant regeneration from tissue culture. Therefore, it is necessary to find out adequate solutions to improve the culture techniques of these recalcitrant species. Furthermore, there have been few researches on breeding of *Paphiopedilum* spp. This chapter aimed to improve the shoot regeneration rate of *Paphiopedilum callosum* orchids by applying the wounding manipulation and shoot tip removal methods. The results of the study were successful in using wounding manipulation or shoot tip removal methods in combination with plant growth regulators in increasing the efficiency of shoot regeneration as well as improving the quality of *Paphiopedilum callosum* plants on the in vitro rooting medium supplemented with organic and increasing plant growth on peat moss substrate in the greenhouse.

Keywords

Wounding manipulation · Shoot tip removal · *Paphiopedilum callosum*

D. T. Nhut (✉) · V. Q. Luan · H. T. Tung
Tay Nguyen Institute for Scientific Research, VAST, Dalat City, Vietnam

D. T. Nhut et al. (eds.), *Plant Tissue Culture: New Techniques and Application in Horticultural Species of Tropical Region*,
https://doi.org/10.1007/978-981-16-6498-4_6

6.1 Introduction

Paphiopedilum is a terrestrial orchid genus, which grows from the Himalayas in Southeast Asia to Papua New Guinea (Teoh 2005). Species of this genus have been favorite potted plants for centuries due to their attractive colors and distinctive shapes. However, conventional propagation methods for low multiplication rates as well as high market demand led to these orchids being endangered. Many *Paphiopedilum* species are endangered species, recognized, and protected by the Convention on International Trade of Endangered Species of Wild Flora and Fauna (CITES). Consequently, plant cell tissue culture through which a large number of plantlets can be obtained within a short period has become an ideal solution for preserving this genus from extinction. Moreover, the success of *Paphiopedilum* micropropagation is relatively limited due to the difficult bacterial and fungal decontamination of ex vitro derived explants and the poor development and survival of explants under in vitro conditions (Stewart and Button 1977; Huang 1988).

To date, there has been some success in the micropropagation of *Paphiopedilum* species. The main methods used in micropropagation of *Paphiopedilum* are in vitro seedling germination, callus, and protocorm-like body formation, shoot regeneration using different explant types. Zeng et al. (2016) reported 58 protocols for in vitro seed germination studies which occupied 80.6% of all *Paphiopedilum* in vitro studies. Some achievements on the micropropagation of *Paphiopedilum* are shown in (Table 6.1).

Although commercial scale in vitro production of *Paphiopedilum* species has been achieved for several species (Liao and Chen 2006; Zeng et al. 2006, 2010, 2016; Liao et al. 2011; Chen et al. 2015), strategies for enhancing mass productivity of in vitro shoots especially for the endangered orchids are a challenge for researchers. In previous studies, we demonstrated a high shoot regeneration rate of *Paphiopedilum delenatii* by the wounding method and stem node culture method (Nhut et al. 2005, 2007). The present chapter aims to provide protocols that could improve the shoot regeneration rate through shoot apex decapitation and wounding methods and to establish suitable in vitro and ex vitro conditions for plant development using *Paphiopedilum callosum* as an example.

6.2 Technical Information on the Micropropagation of *Paphiopedilum* Species by Wounding Manipulation and the Shoot Tip Removal Methods

6.2.1 Plant Materials

The initial plant materials of *Paphiopedilum callosum* were collected from wild populations and planted in the greenhouse of Tay Nguyen Institute for Scientific Research, Vietnam. The 3-month-old shoots (3–7 cm in length) with 3–4 leaves were collected and cleaned the surface with liquid soap (Unilever Co., Ltd., Vietnam), and

Table 6.1 Some achievements on micropropagation of *Paphiopedilum*

No.	Methods	Plant materials	References
1	In vitro seedling germination	Seeds	Dennis and Stimart Peter (1981) Pierik (1997) Tay et al. (1988) Songjun et al. (2012, 2013)
2	Callus and PLB	Young plant without leaves	Stewart and Button (1975)
		Seedling embryos	Cho et al. (1987)
		Seedling protocorm	Lin et al. (2000)
		Seedling plant	Patcharawadee et al. (2011)
			Chyuam and Saleh (2011)
		In vitro young shoot	Luan et al. (2012)
3	Direct shoot		Chen et al. (2004)
4	Axillary bud	Seedling plant	Chyuam et al. (2010)
		In vitro young shoot	Aree et al. (2013)
5	Wounding	Seedling plant	Nhut et al. (2005)
6	Elongated stem	In vitro plant	Nhut et al. (2007) Waraporn et al. (2012)
		Ex vitro plant	Huy et al. (2019a)
7	Shoot regeneration	Flower stalk	Liao et al. (2011)
8	In vitro polyploidy	Shoot	Huy et al. (2019b)

placed under running tap water for 2–3 h. The shoots were soaked with Streptomycin (Pharmaceutical Co., Ltd., China) 1.0 ppm for 30 min. Then, the shoots of *Paphiopedilum* orchids were dipped into 70% (v/v) ethanol 30 s, followed by agitation for 10 min in sodium hypochlorite and rinsed five times with sterilized distilled water. The shoots were cultured on the Schenk and Hildebrandt (1972) (SH) medium, with 1.0 mg/L thidiazuron (TDZ), 0.3 mg/L naphthaleneacetic acid (NAA), 30 g/L sucrose, and 9 g/L agar for 90 days. The medium was adjusted pH to 5.8 before autoclaving at 121 °C, 15 psi for 30 min. The culture condition was set up under 16 hours photoperiod with light intensity 15–20 μmol m^{-2} s^{-1} of cool white fluorescent tubes at 25 ± 1 °C.

Vigorous and uniform adventitious shoots formed after 90 days of culture were used for shoot apex decapitation and wounding method.

6.2.2 Wounding Manipulation and Shoot Tip Removal Procedures, and Selections of Plant Growth Regulators

Three-month-old adventitious shoots with three leaves approximately 5 cm in length were excised from the original explant. For wounding manipulation, each excised shoot was pierced three to four times at its base using a sterile sharp needle. The puncture wounds were approximately 0.3 mm. For shoot tip removal, the shoot tip of

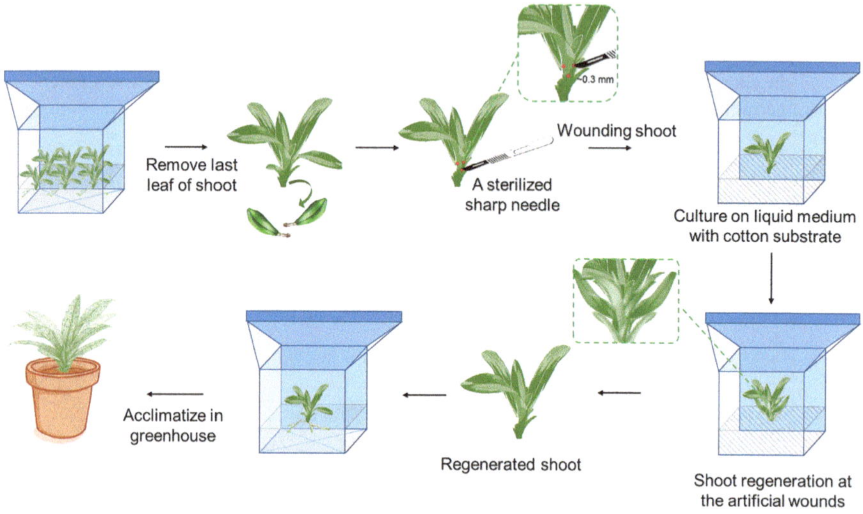

Fig. 6.1 Diagram depicting shoot regeneration of *P. callosum* via wounding manipulation and shoot tip removal methods

adventitious shoots was removed by the forceps. Diagram depicting shoot regeneration of *P. callosum* via wounding manipulation and shoot tip removal methods was shown in Fig. 6.1.

After the manipulation, shoots were cultured on the liquid SH medium supplemented with difference concentrations of 0.2–1.0 mg/L TDZ or 0.5–2.5 mg/L BA, 30 g/L sucrose for shoot regeneration. These shoots were placed on a 250 mL bottle with 40 mL of medium. The absorbent cotton (7 × 7 cm in size and 2.00 g in weight) (Bao Thach JSC., Vietnam) was used as the substrate in the culture for shoot regeneration.

6.2.3 Rooting of Regenerated Shoots

During shoot growth, organic compounds (coconut water; potato; peptone; banana, etc.) were added to the culture medium to increase shoot growth and in vitro rooting of some orchids. Depending on the specific crop, the type and concentration of organic compounds are added accordingly. In this study, different concentrations of organic compounds were added to the rooting medium in vitro to evaluate shoot growth.

For rooting of regenerated shoots, shoots were transferred to SH medium supplemented with 0.5 g/L BA, 0.5 g/L NAA, 30 g/L sucrose, 9 g/L agar, 1 g/L AC, and difference concentration of organic compounds such as coconut water (100, 200, 300, 400, and 500 mL/L); potato (50, 100, 150, 200, and 250 g/L); peptone (1, 2, 3, 4, and 5 g/L); banana (50, 100, 150, 200, and 250 g/L). Adventitious roots began to form 90 days after culture and the rooted plants were then transferred to greenhouse for acclimatization.

6.2.4 Acclimatization Procedures

In *Paphiopedilum* micropropagation, transferring plants from in vitro to acclimatization was an important issue to evaluate the effectiveness of the whole micropropagation process; in which, the growth of plantlets depends on the selection of suitable substrates for each type of plant. This will help increase the survival rate as well as subsequent growth and development at the acclimatization stage.

For *P. callosum* acclimatization, the plantlets were transplanted into plastic pots (9 cm diameter) containing rice hush ash (RHA), coconut fiber (CF) (Eco Source Co. Ltd., Vietnam), and two kinds of fern fibers including Cibotium barometz fiber (CFF), grind fern (GFF), which is native to Vietnam and peat moss (PM) (Eco Source Co. Ltd., Vietnam). Transplanted plantlets were grown in greenhouse at 16–25 °C ambient temperature, 60–90% relative humidity, and natural sunlight with PPFD less than 200 μmol m^{-2} s^{-1}. Watering was done once a day from November to April (dry season) and twice a day from May to October (rainy season). The plants were fertilized once per month with a slow-release Hi-control 13-11-11+ME (Arysta Health and Nutrition Science Co., Ltd., Japan).

6.3 Shoot Regeneration of *P. callosum* by Wounding Manipulation and the Shoot Tip Removal Methods

Shoots grown on shoot regeneration medium without PGRs (TDZ or BA) showed low shoot regeneration for wounding manipulation method (1.22 shoots per explant) and shoot tip removal method (1.52 shoots per explant) (Table 6.2). The addition of PGRs to the culture medium increased the shoot regeneration as well as shoot quality of *P. callosum* than those in control (without PGRs) (Table 6.2).

The combination of wounding manipulation method and TDZ or BA obtained positive affected on shoot regeneration of *P. callosum* (Table 6.2). The higher number of shoots per explant (4.24–4.48 shoots) were found at SH liquid medium supplemented with 0.4–1 mg/L TDZ. Shoots obtained on liquid SH supplemented with 0.4–0.6 mg/L were large, strong, normal shaped; while the shoots obtained on liquid SH supplemented with 0.8–1.0 mg/L were morphological abnormalities such as short, thick leaves, not forming pigment.

The shoots of *P. callosum*, where shoot apex was decapitated, and cultured on SH liquid medium supplemented with 0.4 mg/L TDZ showed the highest regeneration frequency (5.61 shoots per explant) than BA (Table 6.2).

The results of this study show that wounding manipulation method combined with 0.4–0.6 mg/L TDZ (Fig. 6.2) or shoot tip removal method combined with 0.6 mg/L TDZ (Fig. 6.3) gives about twofold more shoot regeneration efficiency

Table 6.2 Combination of wounding manipulation or shoot tip removal methods and PGRs (TDZ and BA) on shoot multiplication of *P. callosum* after 90 days of culture

PGRs (mg/L)		No. of shoots per explant	
TDZ	BA	Wounding manipulation method	Shoot tip removal method
0	0	1.22^{d}	1.52^{h}
0.2	–	2.47^{c}	3.75^{f}
0.4	–	4.48^{a}	5.61^{a}
0.6	–	4.48^{a}	5.17^{b}
0.8	–	$4.24^{a,b}$	$4.87^{b,c}$
1.0	–	$3.97^{a,b}$	$4.68^{c,d}$
–	0.5	2.47^{c}	2.57^{g}
–	1.0	$3.23^{b,c}$	3.52^{f}
–	1.5	3.71^{b}	3.50^{f}
–	2.0	3.71^{b}	4.25^{e}
–	2.5	3.48^{b}	$4.37^{d,e}$

Note: Different letters shown in the same column represent significant differences at $p < 0.05$ *in Duncan's test*

Fig. 6.2 Shoot regeneration of *P. callosum* by the wounding manipulation method combined with 0.4 mg/L or 0.6 mg/L TDZ (left to right) after 90 days of culture

than the shoot regeneration method on medium supplemented medium 2.25 mg/L BA (Patcharawadee et al. 2011).

The shoot regeneration is associated with the type and amounts of cytokinin used (Amoo et al. 2011). In this study, TDZ was observed to be relatively more effective than BA because TDZ was shown to be 50 to 100 times more effective than other cytokinins (Genkov and Ivanova 1995).

Fig. 6.3 Shoot regeneration of *P. callosum* by the shoot tip removal method combined with 0.4 mg/L TDZ after 90 days of culture

Shoot regeneration of *P. callosum* at optimal PGR concentrations by shoot tip removal method (5.61 shoots per explant) is as effective as the wounding manipulation method (4.48 shoots per explant). In addition, the morphology and quality of shoots were no different in both regeneration methods.

6.4 In Vitro Rooting of *P. callosum*

Three-month-old shoot with three leaves of *P. callosum* was cultured on SH medium supplemented with difference concentration of organic compounds such as coconut water, potato, peptone, and banana showed significant differences in growth and plant morphology compared with the control medium (without supplementation) (Tables 6.3, 6.4, 6.5, and 6.6 and Fig. 6.4). Shoot growth on control medium was lower than on medium supplemented with organic compounds at optimal concentrations.

The culture medium supplemented with 200 mL/L coconut water gave better plant growth with leaf length (5.10 cm) and fresh weight (1.67 g) compared with other concentrations of coconut water and without coconut water (Table 6.3 and Fig. 6.4a). Moreover, coconut water content higher than 200 mL/L reduced plant growth, light green leaves, leaf length, and fresh weight decreased.

Table 6.3 In vitro rooting of *P. callosum* on medium supplemented with coconut water after 90 days of culture

Coconut water (mL/L)	Leaf length (cm)	No. of leaves	No. of roots	Root length (cm)	Fresh weight (g)
0	3.40^{c}	$3.73^{a,b}$	3.92^{a}	3.22^{b}	1.10^{d}
100	4.37^{b}	4.48^{a}	4.10^{a}	3.32^{b}	1.42^{b}
200	5.10^{a}	4.40^{a}	3.90^{a}	3.67^{a}	1.67^{a}
300	4.42^{b}	4.48^{a}	3.80^{a}	3.77^{a}	1.45^{b}
400	3.57^{c}	3.23^{b}	2.92^{b}	3.12^{b}	$1.30^{b,c}$
500	2.55^{d}	3.25^{b}	1.85^{c}	1.55^{c}	0.77^{d}

Note: Different letters shown in the same column represent significant differences at $p < 0.05$ *in Duncan's test*

Table 6.4 In vitro rooting of *P. callosum* on medium supplemented with potato extract after 90 days of culture

Potato extract (g/L)	Leaf length (cm)	No. of leaves	No. of roots	Root length (cm)	Fresh weight (g)
0	3.40^{c}	3.75^{c}	$3.92^{b,c}$	3.22^{d}	1.10^{c}
50	4.42^{b}	$4.75^{a,b}$	$4.22^{a,b}$	$4.02^{b,c}$	$1.30^{b,c}$
100	5.05^{a}	$4.75^{a,b}$	$4.20^{a,b}$	4.10^{b}	$1.45^{a,b}$
150	5.10^{a}	$5.00^{a,b}$	$3.97^{b,c}$	$4.00^{b,c}$	$1.52^{a,b}$
200	5.45^{a}	5.50^{a}	4.32^{a}	4.60^{a}	1.65^{a}
250	4.02^{b}	$4.25^{b,c}$	3.77^{c}	3.75^{c}	$1.27^{b,c}$

Note: Different letters shown in the same column represent significant differences at $p < 0.05$ *in Duncan's test*

Table 6.5 In vitro rooting of *P. callosum* on medium supplemented with peptone after 90 days of culture

Peptone (g/L)	Leaf length (cm)	No. of leaves	No. of roots	Root length (cm)	Fresh weight (g)
0	3.40^{c}	3.73^{b}	3.92^{a}	3.22^{c}	1.10^{c}
1	4.97^{a}	4.74^{a}	4.27^{a}	4.57^{a}	1.52^{a}
2	$4.67^{a,b}$	4.74^{a}	3.42^{a}	3.95^{b}	$1.37^{a,b}$
3	4.20^{b}	4.74^{a}	3.00^{a}	3.20^{c}	1.30^{b}
4	2.80^{d}	3.48^{b}	2.12^{b}	1.27^{d}	$1.22^{b,c}$
5	2.15^{e}	3.48^{b}	0.00^{c}	0.00^{e}	0.52^{d}

Note: Different letters shown in the same column represent significant differences at $p < 0.05$ *in Duncan's test*

For potato extract, the addition of 100–200 g/L potato extract to the culture medium resulted in best plant growth in leaf length, number of leaves, number of roots, root length, and fresh weight than 0, 50, 250 g/L potato extract after 90 days of culture (Table 6.4 and Fig. 6.4b).

For peptone or banana extract, plant growth on culture medium supplemented with 1 g/L peptone or 100 g/L banana extract was optimal compared with other

Table 6.6 In vitro rooting of *P. callosum* on medium supplemented with banana after 90 days of culture

Banana extract (g/L)	Leaf length (cm)	No. of leaves	No. of roots	Root length (cm)	Fresh weight (g)
0	$3.40^{b,c}$	$3.75^{b,c,d}$	3.92^{a}	3.22^{b}	1.10^{c}
50	3.67^{b}	4.50^{a}	4.20^{a}	4.15^{a}	$1.35^{a,b}$
100	4.60^{a}	$4.25^{a,b}$	4.10^{a}	4.37^{a}	1.50^{a}
150	3.22^{c}	$4.00^{a,b,c}$	2.70^{b}	3.12^{b}	$1.22^{b,c}$
200	3.22^{c}	$3.50^{c,d}$	2.47^{b}	2.52^{c}	$1.17^{b,c}$
250	2.37^{d}	3.25^{d}	1.75^{c}	1.62^{d}	0.75^{d}

Note: Different letters shown in the same column represent significant differences at $p < 0.05$ in Duncan's test

Fig. 6.4 In vitro rooting of *P. callosum* on medium supplemented. (**a**) Coconut water; (**b**) Potato extract; (**c**) Peptone; (**d**) BN extract after 90 days of culture

concentrations of peptone or banana extract after 90 days of culture (Tables 6.4 and 6.5 and Figs. 6.4c, d).

The addition of some extracts or compounds to the rooting medium has been carried out on some plant species such as coconut water on *Calanthe* hybrids (Abdullahil et al. 2011), potato on *Celosia* sp. (Norhayati et al. 2011), peptone on *Paphiopedilum rothschildianum* (Chyuam and Saleh 2011), *Dianthus* (Ramage and Williams 2001), banana on *Dendrobium* (Aktar et al. 2008), *Cymbidium* (Saranjeet and Bhutani 2012). The results of these studies show that adding some organic compounds with appropriate concentrations helps to increase in vitro rooting as well as plant growth.

6.5 Acclimatization of *Paphiopedilum* sp. on Various Types of substratessubstrates

All plantlets transplanted into plastic pots containing RHA, CF, CFF, GFF, and PM gave a 100% survival rate. However, the growth of 24-month-old plants on various types of substrates in the greenhouse was significantly different (Fig. 6.5).

Plants grown on PM substrate showed the highest leaf width, leaf length, root length, and fresh weight among the five substrates (Figs. 6.5 and 6.6).

The results of this study showed that plants grown on PM substrate had higher survival and growth rates than reported on *Paphiopedilum P. villosum* var *densissinum* (Bo et al. 2010) or *P. rothschildianum* (Chyuam et al. 2010).

Currently, *Paphiopedilum* plants on the market are mainly seeded in vitro of hybrid species; however, plants derived from seeding were variable and take a long time to mature before flowering (Liao et al. 2011; Zeng et al. 2016). In addition,

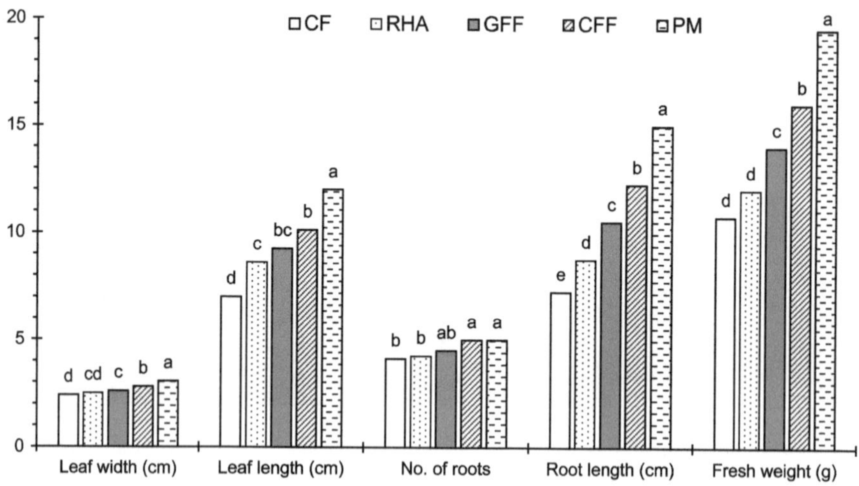

Fig. 6.5 The growth of *Paphiopedilum callosum* on various types of substrates after 24 months in the greenhouse

Fig. 6.6 The growth of *P. callosum* plant on PM substrate after 24 months in the greenhouse

plantlets are expected to have the same genetic characteristics as their parent plants, capable of producing in large quantities. The results of this study indicated that high shoot regeneration efficiency was induced by applying wounding manipulation and shoot tip removal, and survival rate and acclimatization, and good growth of *P. callosum*.

6.6 Conclusion

The results showed that *P. callosum* shoots were treated by wounding manipulation method combined with 0.4–0.6 mg/L TDZ or shoot tip removal method combined with 0.6 mg/L TDZ gives high shoot regeneration efficiency. Moreover, *P. callosum* shoots were cultured on in vitro rooting medium supplemented with organic compounds increased plant growth. The plantlets transplanted into plastic pots containing PM were well developed after 24 months in the greenhouse.

References

Abdullahil BM, Shin YK, Elshmari T, Lee EJ, Paek KY (2011) Effect of light quality, sucrose and coconut water concentration on the micropropagation of *Calanthe* hybrids. Australian J Crop Sci 5(10):1247–1254

Aktar S, Nasiruddin KM, Hossain K (2008) Effects of different media and organic additives interaction on *in vitro* regeneration of *Dendrobium* orchid. J Agric Rural Dev 6:69–74

Amoo SO, Finnie JF, Van-Staden J (2011) The role of metatopolins in alleviating micropropagation problems. Plant Growth Regul 63(2):197–206

Aree T, Ekasit N, Chockpisit T (2013) Multiple shoot formation of *P. "Delrosi"*. World Acad Sci Eng Tech 78

Bo L, Alex X, Niemiera ZC, Chun L (2010) *In vitro* propagation of four threatened *Paphiopedilum* species (Orchidaceae). Plant Cell Tissue Organ Cult 101:151–162

Chen TY, Cirier I, Chang WC (2004) Plant regeneration through direct shoot bud formation from leaf culture of *Paphiopedilum* orchids. Plant Cell Tissue Organ Cult 76:11–15

Chen Y, Goodale UM, Fan XL, Gao JY (2015) Asymbiotic seed germination and *in vitro* seedling development of *Paphiopedilum spicerianum*: an orchid with an extremely small population in China. Global Ecol Conserv 3:367–378

Cho MS, Helen L, Valmayor (1987) Effects of culture media on the growth of seedlings derived from embryo culture in *P. philippinense* (tropical orchid). Korean J Plant Tiss 15(2):103–110

Chyuam YN, Saleh NM (2011) *In vitro* propagation of *Paphiopedilum* orchid through formation of protocorm-like bodies. Plant Cell Tissue Organ Cult 105:193–202

Chyuam YN, Saleh NM, Zaman FQ (2010) *In vitro* multiplication of the rare and endangered slipper orchid, *P. rothschildianum* (Orchidaceae) Afr J Biotech 9(14):2062–2068

Dennis P, Stimart Peter DA (1981) *In vitro* germination of *Paphiopedilum* seed on a completely defined medium. Sci Hortic 14:165–170

Genkov T, Ivanova I (1995) Effect of cytokinin-active phenylurea derivatives on shoot multiplication, peroxidase and superoxide dismutase activities of *in vitro* cultured carnation. Bulg J Plant Physiol 21(1):73–83

Huang LC (1988) A procedure for asexual multiplication of Paphiopedilum in vitro. Am Orchid Soc Bull 57(3):274–278

Huy NP, Luan VQ, Cuong LK, Nam NB, Tung HT, Hien VT, Le TD, Paek KY, Nhut DT (2019a) Strategies for the regeneration of *Paphiopedilum callosum* through internode tissue cultures using dark light cycles. HortSci 54(5):920–925

Huy NP, Tam DTT, Luan VQ, Tung HT, Hien VT, Ngan HTM, Duy PN, Nhut DT (2019b) *In vitro* polyploid induction of *Paphiopedilum villosum* using colchicine. Sci Hortic 252: 283–290

Liao YZ, Chen JJ (2006) Asymbiotic seed germination of *Paphiopedilum*. *Paphiopedilum* in Taiwan IV. Taiwan Paphiopedilum Society, pp 11–14

Liao YJ, Tsai YC, Sun YW, Lin RS, Wu FS (2011) *In vitro* shoot induction and plant regeneration from flower buds in *Paphiopedilum* orchids. In Vitro Cell Dev Biol Plant (47):702–709

Lin YH, Chen C, Chang WC (2000) Plant regeneration from calus culture of a *Paphiopedilum* hybrid. Plant Cell Tissue Organ Cult 62:21–25

Luan LQ, Uyen NHP, Ha VTT (2012) *In vitro* mutation breeding of *Paphiopedilum* by ionization radiation. Sci Hortic 144:1–9

Nhut DT, Trang PTT, Vu NH, Thuy DTT, Khiem DV, Binh NV, Tran Thanh Van K (2005) A wounding method and liquid culture in *P. delenatii* propagation. Propag Ornam Plants 5(3):158–163

Nhut DT, Thuy DTT, Don NT, Luan VQ, Hai NT, Tran Thanh Van K, Chinnappa CC (2007) Stem elongation of *P. delenatii* Guillaumin and shoot regeneration via stem node culture. Propag Ornam Plants 7(1):29–36

Norhayati D, Rosna MT, Nor NMN, Hasimah A (2011) Provision of low cost media options for *in vitro* culture of *Celosia* sp. Afr J Biotech 10(80):18349–18355

Patcharawadee W, Eric B, Kongkanda C, Sureeya T (2011) Effect of cytokinins (BAP and TDZ) and auxin (2,4-D) on growth and development of *P. callosum*. Kasetsart J Nat Sci 45:12–19

Pierik RLM (1997) In vitro culture of higher plants. Kluwer Academic Publishers, Dordrecht, p 348p

Ramage CM, Williams RR (2001) Mineral nutrition and plant morphogenesis. In Vitro Cell Dev Biol Plant 38:116–124

Saranjeet K, Bhutani KK (2012) Organic growth supplement stimulants for *in vitro* multiplication of *Cymbidium pendulum*. Hortic Sci 9(1):47–52

Schenk RU, Hildebrandt AC (1972) Medium and techniques for induction and growth of monocotyledonous and dicotyle-donous plant cell cultures. Can J Bot 50:199–204

Songjun Z, Kunlin W, Jaime ATDS, Jianxia Z, Zhilin C, Nianhe X, Jun D (2012) Asymbiotic seed germination, seedling development and reintroduction of *Paphiopedilum wardii* Sumerh, an endangered terrestrial orchid. Sci Hortic 138:198–209

Songjun Z, Jia W, Kunlin W, Jaime ATDS, Jianxia Z, Jun D (2013) *In vitro* propagation of Paphiopedilum hangianum Perner & Gruss. Sci Hortic 151:147–156

Stewart J, Button J (1975) Tissue culture studies in *Paphiopedilum*. Am Orchid Soc Bull 44:591–599

Stewart J, Button J (1977) The effect of benzyl adenine on the development of lateral buds of *Paphiopedilum*. Am Orchid Soc Bull 46:415–418

Tay LJ, Kiyotoshi T, Yataka H (1988) Culture conditions suitable for *in vitro* seed germination and development of seedlings in *Paphiopedilum*. J Japanese So Hortic Sci 57(2):243–249

Teoh ES (2005) Orchids of Asia. Marshall Cavendish, Singapore, 368p

Waraporn U, Wen PJ, Chin SW, Chen FC (2012) Shoot multiplication of *Paphiopedilum* orchid through *in vitro* cutting methods. Afr J Biotech 11(76):14077–14082

Zeng SJ, Chen Z, Duan J (2006) Asepsis sowing and *in vitro* propagation of *Paphiopedilum hirsutissimum* Pfitz. Plant Physiol Commun 42:247

Zeng SJ, Chen ZL, Wu KL, Zhang JX, Duan J (2010) *In vitro* propagation of *Paphiopedilum henryanum* Bream. Plant Physiol Commun 46: 471–472

Zeng SJ, Huang W, Wu K, Zhang J, Teixeira da Silva JA, Duan J (2016) *In vitro* propagation of *Paphiopedilum* orchids. Crit Rev Biotechnol 36:521–534

Stem Elongation for Plant Micropropagation

7

Hoang Thanh Tung, Vu Quoc Luan, Le Thi Van Anh, and Duong Tan Nhut

Abstract

Nodal and internode explant culture is a simple and effective method in micropropagation. However, some plants have very short and not well-defined internodes, such as *Paphiopedilum* and *Nepenthes*. As a result, defined nodal and internodal explants are difficult to obtain for micropropagation purposes. Furthermore, the close clustering of leaves makes surface decontamination of explants difficult. Red LED, dark conditions, and gibberellic acid (GA_3) have been reported to stimulate stem elongation under ex vitro and in vitro conditions. In this chapter, the effects of LED lights, different blue to red LED ratios, dark conditions, and GA_3 are used to study the stem elongation of *Paphiopedilum*, *Anthurium,* and *Nepenthes*. The results of this study could increase the propagation efficiency in these plants.

Keywords

LED · GA_3 · Stem elongation · Paphiopedilum · Anthurium · Nepenthes

7.1 Introduction

Stem tissues are favorable materials used in plant micropropagation. The nodal and internodal sections have good regeneration potential. The stem pieces are readily excised from shoots, and surface decontamination is easy to perform. Moreover, in

H. T. Tung · V. Q. Luan · D. T. Nhut (✉)
Tay Nguyen Institute for Scientific Research, VAST, Dalat City, Vietnam

L. T. Van Anh
Graduate University of Science and Technology, VAST, Hanoi City, Vietnam

D. T. Nhut et al. (eds.), *Plant Tissue Culture: New Techniques and Application in Horticultural Species of Tropical Region*,
https://doi.org/10.1007/978-981-16-6498-4_7

some plants, leaves are formed close to one another, and the internodes can be very short, making precise excision impossible. In *Paphiopedilum*, for example, there are many challenges in plant micropropagation. One of the difficulties is that the nodes are few in number, and they are very close to one another. As a result, it is difficult to obtain a large quantity of stem tissues for shoot regeneration. In order to generate more stem tissues, the best strategy is to induce stem elongation using gibberellins and/or altering the light intensity, photoperiod, and light quality using specific light-emitting diodes (LEDs). In pineapple, etiolation generates more stem tissue and nodes for successful regeneration (Kiss et al. 1995).

Sachs (1965) summarized earlier information on stem elongation in his review. It was recognized that gibberellins and light quality, and photoperiod are key to the stem elongation process. Gibberellin (GA) application effectively stimulates stem elongation by promoting cell elongation and cell division (Huttly and Phillips 1995). There is clear evidence that some light effects are mediated by the modification of active GA levels (García-Martinez and Gil 2001). Thomas et al. (2005) summarized the mechanisms related to GA actions on elongation. The current model of GA signaling is discussed by Schwechheimer (2008) and Hirano et al. (2008). Recent studies indicate that light can regulate gibberellin biosynthesis. Low light intensity increases stem elongation and active GA content in pea and *Brassica* (García-Martinez and Gil 2001). Red (R) LED illumination is more effective in promoting seedlings growth than Blue (B) LED light in tomato seedlings (Matsuo et al. 2019). It was subsequently determined that the level of bioactive GA_1 and GA_4 was significantly higher in seedlings grown under R LEDs (Matsuo et al. 2019). Through manipulating light quality and quantity, and growth regulator levels, one can stimulate stem elongation, generating more stem tissues for micropropagation purposes.

In this chapter, information is detailed to demonstrate the effects of LED lights on stem elongation in *Paphiopedilum* species, *Anthurium,* and *Nepenthes* and the successful use of internodal and nodal explants in the micropropagation of these species (Fig. 7.17).

7.2 Light-Emitting Diodes (LEDs) in Stem Elongation

Light affects many aspects of plant growth and development; such as stem elongation. Some observations on the effects of LEDs on in vitro plant growth and morphogenesis are summarized in Gupta and Jatothu (2013) review. Plantlets whose stems are elongated under the R LED are usually slender, with yellowish leaves; the chlorophyll content, photosynthetic rate, and fresh weight of stems and roots were lower than those grown under a combination of red and B LEDs (Nhut 2002).

Some studies showed that under R LEDs, the shoots and stem length give the highest elongation as in Chrysanthemum (Kim et al. 2004) and potato (Jiang et al. 2019). However, in other studies, R LEDs were not suitable for stem elongation. Hahn et al. (2000) indicated that R LEDs were ineffective in inducing *Rehmannia glutinosa* stem elongation. The plants had shorter stems and lower fresh weights and photosynthesis rate than under B LED conditions. These and other results indicate that the response to each type of light will be different, depending on the plant species (Hirai et al. 2006).

Plantlets cultured under R with B LED combinations have different responses. Different species react to the ratio of red and blue lights differently. One needs to determine the proper ratio of blue/red light for optimum response. Many studies have attempted to determine the role of combining red and blue lights but have yet to clarify their effects in detail; although, the individual effects of each type of light have been well studied.

The role of light quality in plant morphogenesis, in general, is not clearly understood. Its effects depend on the emitted wavelength, the type of crop being studied, the stage of plant growth, and culture conditions, such as light intensity (Kurilčik et al. 2008), media composition (Schuerger et al. 1997), or aeration conditions (Hahn et al. 2000). In this chapter, the effects of R and B LEDs in stem elongation on *Paphiopedilum* species, *Anthurium,* and *Nepenthes* are detailed.

The lighting equipment used in our studies is as follows. Fluorescent lamps (FL) with a wavelength of 320–800 nm (voltage of 220 V) and a length of 1.2 m (40 W/T10) (Rang Dong Company, Hanoi, Vietnam) were used in the experiments as controls. B LEDs with a wavelength of 450–470 nm (voltage of 3 V) and R LEDs with wavelengths 650–665 nm (voltage of 2 V) (Super Bright LEDs Inc., St. Louis, Missouri, USA) were used as LED light sources. The power per LED was 0.1 W with a resistor of 330 Ω for R LEDs and a 220 Ω resistor for B LEDs (TQCOM Joint Stock Company—Hanoi). The power supply was 12 V with voltage 5 A (AXT 450—Golden Field Firm). Each circuit board (10 × 50 cm) had 480 LEDs divided into 10 LED bars. Each bar was made up of 48 bulbs divided into 16 small circuits, and each circuit consisted of 3 LED chips and 1 resistor connected in series. The small circuits were connected in parallel with the power supply (12 V). The proportional mix will be based on the number of R LEDs or B LEDs on the board. The combined ratio of B and R LEDs depends on the number of bulbs used between them, according to Nhut (2002).

7.3 Stem Elongation in *Paphiopedilum* Species

7.3.1 Achievements in Research on Regeneration and Micropropagation of Paphiopedilum Species

Paphiopedilum species are highly valued ornamental plants. The pioneering work in the study of *Paphiopedilum* micropropagation was carried out by Bubeck (1973), while Morel (1974) recorded the first successful tissue culture propagation protocol using *Paphiopedilum* stem apices. *Paphiopedilum* species are generally propagated via axillary buds by divisions from the mother plant. However, this method is very inefficient and requires a long time before each subdivision (Ng et al. 2010; Ng and Saleh 2011). Natural seed germinates relatively slowly because of the absence of an endosperm (Lee et al. 2006). In *Paphiopedilum,* in vitro seed germination protocols have been described (Pierik et al. 1988; Chen et al. 2004b; Ding et al. 2004; Nhut et al. 2005; Kauth et al. 2008; Long et al. 2010; da Silva 2013). However, the germination rate for many *Paphiopedilum* species/cultivars is extremely low and is often affected by many unknown factors (Pierik et al. 1988; Arditti 2008).

The morphogenesis of orchids often depends on the explants used in micropropagation (Chugh et al. 2009; Hossain et al. 2013). Until now, only a few reports documenting the use of ex vitro explants for *Paphiopedilum* micropropagation (Stewart and Button 1975; Huang 1988; Liao et al. 2011; Luan et al. 2015; Huy et al. 2019). Seeds, protocorm-like bodies (PLBs), and in vitro seedlings were often used as initial explants for *Paphiopedilum* micropropagation. Other explants include shoot apices (Huang et al. 2001; Ng et al. 2010), internodes (Nhut et al. 2007; Ng et al. 2010; Ng and Saleh 2011; Huy et al. 2019), and leaves (Chen et al. 2004a; Lin et al. 2000).

The media used in the micropropagation of *Paphiopedilum* species include Heller medium (Stewart and Button 1975), Murashige and Skoog (1962), MS medium (Huang 1988; Nhut et al. 2005; Nhut et al. 2007), modified MS medium (Chen et al. 2002; Chen et al. 2004a; Liao et al. 2011), and ½ MS medium (Lin et al. 2000; Hong et al. 2008; Ng et al. 2010; Ng and Saleh 2011). Moreover, at present, comparative studies on the effects of basal media *Paphiopedilum* regeneration are not readily.

The plant growth regulators (PGRs) used in *Paphiopedilum* micropropagation include auxin (2,4-D and NAA), and cytokinin (BA, TDZ, zeatin and kinetin). The types and concentrations of PGRs required for *Paphiopedilum* micropropagation are species-specific (Morel 1974; Stewart and Button 1975; Huang 1988; Lin et al. 2000; Huang et al. 2001; Chen et al. 2002; Nhut et al. 2005, 2007; Hong et al. 2008; Long et al. 2010; Ng et al. 2010; Liao et al. 2011).

There has been limited success in inducing callus in *Paphiopedilum* due to the initial difficulty in the induction process itself. Subsequent slow growth rate, low regeneration capacity, and eventual browning of calluses often occur (Arditti 2008; Lin et al. 2000; Long et al. 2010; Ng and Saleh 2011). Research on shoot multiplication has also been reported (Huang et al. 2001; Chen et al. 2002; Nhut et al. 2007; Long et al. 2010; Zeng et al. 2013).

Micropropagation through the use of ex vitro explants has been relatively limited. This is caused by bacterial and fungal contaminations and the poor development of explants under in vitro conditions (Stewart and Button 1975; Huang 1988). Moreover, recently, we reported successes of using the in vitro propagation of *Paphiopedilum* species using stem-elongated ex vitro explants as the source under dark-light cycles for plant regeneration through internode tissue cultures (Huy et al. 2019). The general approach is detailed below.

7.3.2 Ex Vitro Stem Elongation in Paphiopedilum sp. Under Lighting Condition

Ex vitro 1-month-old young shoot tips of *Paphiopedilum* species (Fig. 7.1) were placed under different lighting conditions (natural lighting, 50B:50R, 90B:10R,

Fig. 7.1 Ex vitro 1-month-old young shoot tips of *Paphiopedilum* sp. (**a**) *P. callosum*; (**b**) *P. delenatii*; (**c**) *P. gratrixianum*

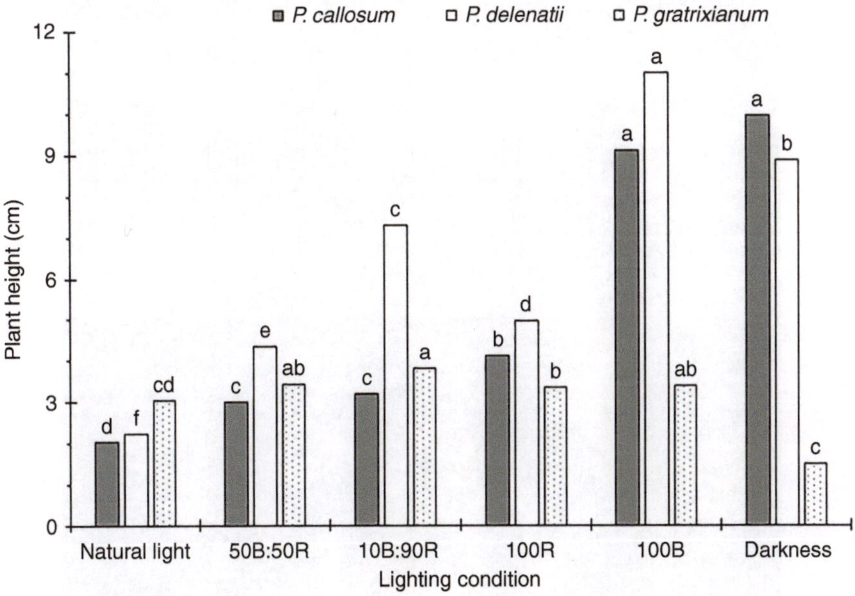

Fig. 7.2 Plant height of ex vitro stem elongation in *Paphiopedilum* sp. after 4 months placed under lighting conditions

100R, 100B, darkness). Young plants of *P. callosum*, *P. delenatii*, and *P. gratrixianum* showed a difference in height after 4 months of culture (Fig. 7.2).

Optimal plant height of *P. callosum* was noted under 100B (9.11 cm) and darkness (9.97 cm) (Fig. 7.2), while plant height of *P. delenatii* was optimal at 100B (11 cm) (Fig. 7.2). However, *P. gratrixianum* young plants placed under LEDs from natural light and darkness, little changes in stem elongation could be found as seen on *P. callosum* and *P. delenatii* (Figs. 7.2 and 7.3). In addition, *P. gratrixianum* young plant placed under darkness had browning leaves and showed necrosis. *P. callosum* and *P. delenatii* placed under 100B had a clear stem elongation and the best plant height, whereas *P. gratrixianum* had an unclear stem elongation (Fig. 7.3).

Young shoots obtained from ex vitro stem elongation in *Paphiopedilum* sp. under lighting condition were defoliated, cleaned with 1% Sunlight (Unilever Vietnam,

Fig. 7.3 The morphology of ex vitro stem elongation in *Paphiopedilum* sp. after 4 months placed under lighting conditions (natural lighting, 50B:50R, 90B:10R, 100R, 100B, darkness; from left to right). (**a**) *P. callosum*; (**b**) *P. gratrixianum*

Vietnam), and washed under running water for 2–3 h. Then, the shoots soaked in 1 mg/mL streptomycin for 30 min and washed with 70% ethanol for 30 s. Next, soak with sterilized water three times. Finally, the shoots were sterilized with 0.1% $HgCl_2$ for 6 min and washed with sterilized water three times.

Table 7.1 Shoot regeneration of ex vitro stem elongation in *Paphiopedilum* sp. under lighting condition after 1.5 months of culture

Lighting condition	*P. callosum*	*P. delenatii*	*P. gratrixianum*
Natural light	16.65^{e*}	18.65^{c}	19.85^{a}
50B:50R	20.00^{d}	18.55^{c}	19.85^{a}
10B:90R	21.65^{c}	24.85^{b,c}	21.20^{a}
100R	25.00^{b}	26.05^{b}	17.40^{a}
100B	31.65^{a}	34.40^{a}	22.20^{a}
Darkness	11.65^{f}	9.70^{d}	9.70^{b}

*Note: *Different letters shown in the same column represent significant differences at $p < 0.05$ in Duncan's test*

Fig. 7.4 Shoot regeneration of ex vitro stem elongation in *P. gratrixianum*, *P. delenatii*, and *P. callosum* (left to right) under lighting condition after 1.5 months of culture

After sterilization, the shoots were collected and cut into nodes, which were cultured into culture flasks. The in vitro young shoots of *P. callosum* and *P. delenatii* grown under different light conditions showed significantly higher shoot regeneration rates than the natural light or darkness conditions after 1.5 months of culture (Table 7.1 and Fig. 7.4).

Shoot regeneration rates were optimal on *P. callosum* (31.65%) and *P. delenatii* (34.40%) reaching the highest in treatment 100B (Table 7.1). Meanwhile, shoot regeneration rates at different lighting treatments (except for dark conditions) on *P. gratrixianum* plants did not show any significant difference (Table 7.1). The results of this study show that ex vitro stem elongation in *Paphiopedilum* sp. under lighting condition helps to increase shoot growth and propagation efficiency through increasing shoot regeneration efficiency.

7.3.3 In Vitro Stem Elongation in Paphiopedilum sp. Under Different Lighting Conditions

In vitro 1-month-old young plants (0.5 cm) with 4 leaves of *Paphiopedilum* sp. were cultured on SH medium supplemented with 0.5 mg/L BA, 0.5 mg/L NAA, 30 g/L sucrose, 9 g/L agar, 1 g/L AC and placed under different lighting conditions (FL, 50B:50R, 70B:30R, 70R:30B,100B, 100R, 100G, 100Y, 100W and darkness) after 4 months showed difference in plant morphology (Fig. 7.5).

For *P. callosum*, the plants were elongated stem under 100W, 100Y, 100G, 30B: 70R, and darkness; plant height (8.55 cm) was optimal under dark conditions (Fig. 7.5). Meanwhile, *P. delenatii* plants were elongated stem under 100G, 100Y, 100W, and darkness; plant height (7.85 cm) was optimal under 100B (Fig. 7.5).

For *P. gratrixianum*, plants placed under different lighting conditions did not show a clear difference in stem elongation and plant height (Fig. 7.5). Plant height (2.1 cm) was optimal in darkness conditions (Fig. 7.5). Therefore, depending on the *Paphiopedilum* species, there are differences in in vitro elongation stem.

7.3.4 In Vitro Stem Elongation in Paphiopedilum sp. by GA_3

In vitro 1-month-old young plants (0.5 cm) with 4 leaves of *Paphiopedilum* sp. were cultured on SH medium supplemented with 30 g/L sucrose, 9 g/L agar, and 1 g/L AC and different concentrations of GA_3 (0, 1, 2, 3, 4, 5 mg/L) after 4 months showed a difference in plant morphology (Figs. 7.6 and 7.7). Stem elongation was significantly different when *P. callosum* and *P. delenatii* were cultured in different concentrations of GA_3 (Fig. 7.6). Plant height (4.65 cm) was optimal in 3 mg/L

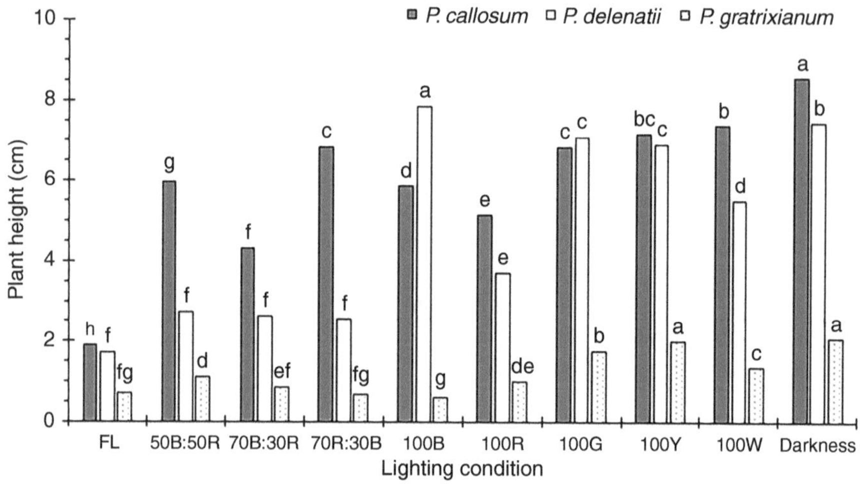

Fig. 7.5 Plant height of in vitro stem elongation in *Paphiopedilum* sp. after 4 months placed under lighting conditions

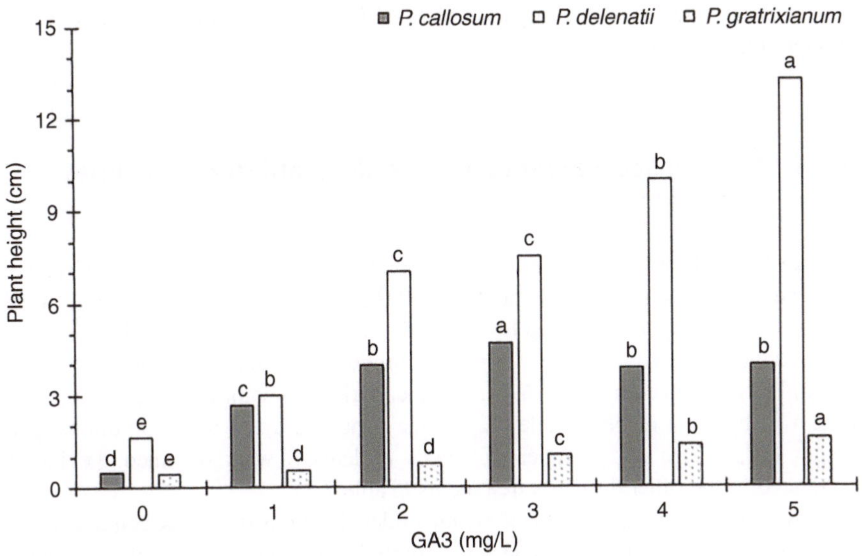

Fig. 7.6 Plant height of in vitro stem elongation in *Paphiopedilum* sp. by GA_3 after 4 months of culture

Fig. 7.7 Growth of stem node derived *P. delenatii* plantlet after transferring to plant developmental medium

GA_3 treatment of *P. callosum* (Fig. 7.6); meanwhile, the optimal plant height (13.25 cm) in 5 mg/L GA_3 treatment for *P. delenatii* (Fig. 7.6). For *P. gratrixianum* stem elongation is also different but not as clear as *P. callosum* and *P. delenatii*. The addition of GA_3 to culture medium shows stem elongation ability on many crops such as *Cuscuta chinensis*, Potato, Pepper, Olive, Soybean, *Mucuna pruriens*, Red oak, Ashwagandha, Date palm, Chili pepper, *Paphiopedilum*. In this study, 3 mg/L

GA_3 was suitable for stem elongation of *P. callosum*; meanwhile, 5 mg/L GA_3 was suitable for *P. delenatii*.

7.3.5 In Vitro Stem Elongation in P. delenatii Under Darkness Condition

In vitro 1-month-old young plants (0.5 cm) with 4 leaves of *Paphiopedilum* sp. were cultured on SH medium supplemented with 0.5 mg/L BA, 0.5 mg/L NAA, 30 g/L sucrose, 9 g/L agar, and 1 g/L AC and placed under dark conditions (1, 2, 3, 4 months) showed difference in plant morphology (Table 7.2). Stem elongation was different when cultured *P. delenatii* under darkness conditions (1, 2, 3, 4 months) as shown by plant height (Table 7.2). After 4 months under dark conditions, plant height (10.50 cm) and number of leaves (5 leaves) were obtained the highest compared to the other darkness treatments (Table 7.2).

However, leaf length and leaf width under 4-month darkness condition were lower than other conditions (Table 7.2). Young leaves were nearly white due to lack of chlorophyll, and primary leaves turned brown or necrotic. To increase growth and quality of *P. delenatii*, 4-month-old plantlets were transferred into FL condition for 2 months and 6-month-old plantlets used as a material for propagation by stem nodes method and planting on the Taiwanese bush.

Stem node positions influence on the percentage of survival and growth of 12-month-old *P. delenatii* plant at the greenhouse (Table 7.3). The first stem node position (corresponding to the shoot tip) for the survival rate (100%) and growth were optimal. Through a stem node culture, the plants are able to reproduce and grow well and depending on the stem node position, the propagation efficiency is different. With stem elongation that can produce many stem nodes and each stem node can regenerate into separate plantlets, the propagation efficiency is higher than using only 1 original mother plant transferred into to the acclimatization stage.

Table 7.2 In vitro stem elongation in *P. delenatii* placed under darkness conditions

Lighting condition	Plant height (cm)	No. of leaves	Leaf length (cm)	Leaf width (cm)	Plant morphology
FL	4.25^{d*}	3.00^{c}	4.95^{a}	3.37^{a}	Dark green plant
1-month darkness	4.37^{d}	1.00^{e}	3.80^{b}	1.65^{b}	Young leaves are light green
2-month darkness	5.65^{c}	2.42^{d}	3.00^{c}	$1.45^{b,c}$	The original leaves are yellow, the young leaves are greenish-white
3-month darkness	7.12^{b}	3.75^{b}	2.22^{d}	1.15^{c}	The original leaves are yellow, the young leaves are greenish-white
4-month darkness	10.50^{a}	5.00^{a}	2.00^{d}	1.12^{c}	The original leaves are brown, the young leaves are greenish-white

*Note: *Different letters shown in the same column represent significant differences at $p < 0.05$ in Duncan's test*

Table 7.3 Effects of stem node position on the percentage of survival and growth of 12-month-old *P. delenatii* plant at the greenhouse

Stem node position	The percentage of survival (%)	No. of new plant	No. of leaves	Leaf width (cm)	Plant morphology
1	**100.00^{a}**	**1.00^{a}**	**3.87^{a}**	**3.07^{a}**	Large and strong plant
2	23.75^{d}	0.00^{d}	0.00^{d}	0.00^{d}	–
3	33.75^{c}	0.75^{b}	2.50^{c}	1.42^{c}	Weak and light green plant
4	32.50^{c}	0.72^{b}	3.12^{b}	1.45^{c}	Weak and light green plant
5	56.25^{b}	0.53^{c}	4.12^{a}	2.12^{b}	Large and dark green plant

Note: Different letters shown in the same column represent significant differences at $p < 0.05$ in Duncan's test

7.4 Stem Elongation in *Anthurium andreanum* "Tropical"

7.4.1 Achievements in Research on Regeneration and Micro-Propagation of A. andreanum "Tropical"

Anthurium, a perennial herbaceous plant, is loved by many people for its charm, elegance, colorful, and long-lasting flowers. Therefore, this flower quickly dominates the world market and has economic value (Chen et al. 2011). Currently, in vitro propagation is applied for the purpose of large-scale production. However, the in vitro propagation efficiency in *Anthurium* was not high due to the long growth time, low propagation coefficient, slow growth and development, poor rooting leading to reduced survival rate at the acclimatization stage; therefore, plantlets are not in sufficient supply and prices increase (Elibox and Umaharan 2008). Methods of seeding and separation of shrubs used in the propagation of Anthurium; however, these methods cause high heterozygosity and prolonged growth time, reduced flower quality, etc. (Viégas et al. 2007).

Organogenesis and embryogenesis micropropagation protocols of *Anthurium* were successful depending on media used and culture conditions (Te-Chato et al. 2006; Viégas et al. 2007; Yu et al. 2009; Liendo and Mogollón 2009; Oropeza et al. 2010; Islam et al. 2010; Atak and Çelik 2009; Farsi et al. 2012). The explants, including lateral shoots, shoot tips, leaves, stems, and petioles were used in micropropagation (Martin et al. 2003; Nhut et al. 2006; Gantait et al. 2008; Jahan et al. 2009).

Stem elongation is an essential component of a successful regeneration protocol. Methods on stem elongation studies were conducted on many crop species such as strawberries, *Eucalyptus*, *Cymbidium*, *Phalaenopsis*, *Paphiopedilum*, and banana, under light-emitting diode conditions (Nhut 2002; Nhut et al. 2003, 2005, 2007; Nanya et al. 2012). However, little information is available on in vitro stem elongation studies on *A. andraeanum* "Tropical."

7.4.2 Stem Elongation Under LEDs

Single shoots (2 cm in height) with 2 leaves were cultured on MS medium supplemented with 30 g/L sucrose and 8 g/L agar and placed under different lighting conditions: FL, 100R, 50R:50B, 60R:40B, 70R:30B, 80R:20B, 90R:10B. The shoot height (cm), fresh weight (mg), dry weight (mg), stem nodes/shoot, number of leaves/shoot and SPAD (SPAD-502, Minolta Co., Ltd., Osaka, Japan) (nmol/cm^2) were obtained after 2 months of culture. The results showed that different lighting conditions also had different effects on the regeneration of *A. andreanum* after 2 months of culture (Figs. 7.12, 7.13, and 7.14). Under LED lighting, the shoot height and stem nodes/shoot were higher than those in FL condition (Figs. 7.8 and 7.9). Shoot height, and the number of stem nodes/shoot were highest under 70R:30B condition (Figs. 7.8, 7.9, 7.10 and 7.11). It is important to note that under R condition, plant height was highest (Fig. 7.8); however, the number of stem nodes per plant was only 4 stem nodes with shoots of unequal height.

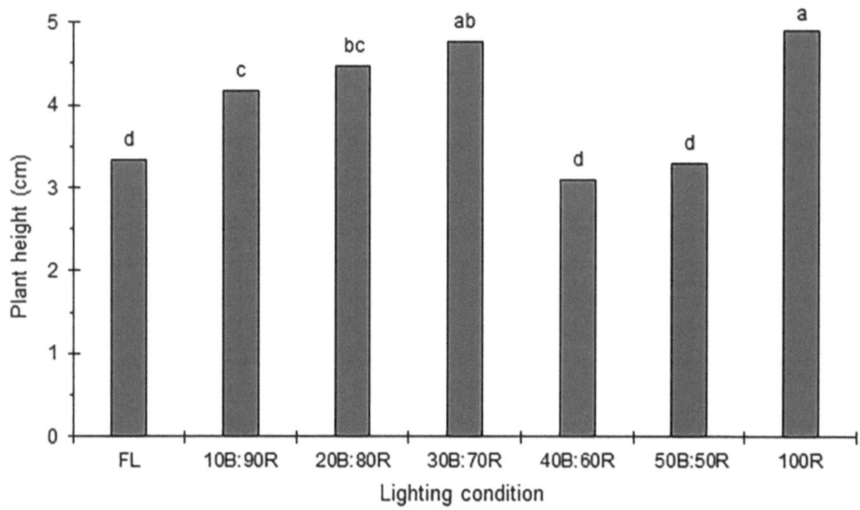

Fig. 7.8 Plant height of in vitro stem elongation in *A. andreanum* "Tropical" after 2 months placed under lighting conditions

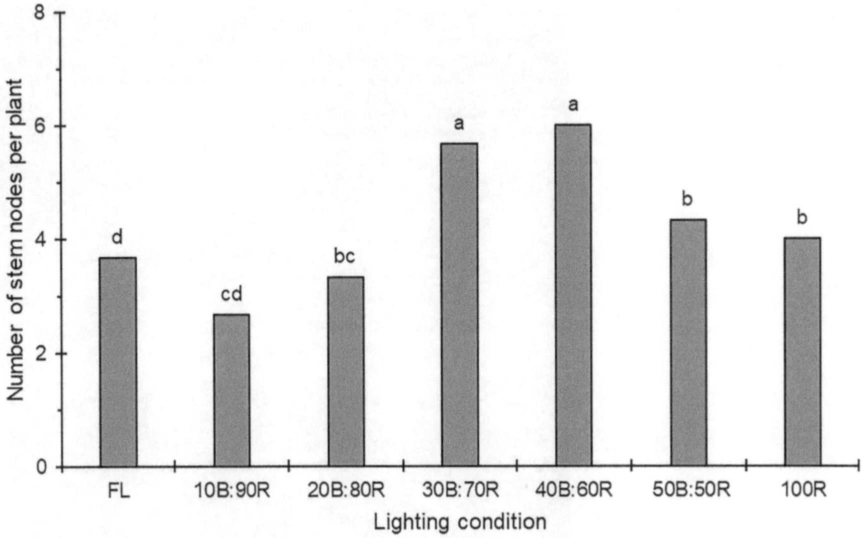

Fig. 7.9 Number of stem nodes per plant of in vitro stem elongation in *A. andreanum* "Tropical" after 2 months placed under lighting conditions

7.4.3 Shoot Multiplication Via Stem Node Culture

Stem nodes (0.5 cm in height) derived from elongated stem under lighting conditions in the above experiment were cultured on the MSC medium supplemented with 30 g/L sucrose and 8 g/L agar. Apical shoot (0.5 cm in height) was used for control. The shoot height (cm), fresh weight (mg), dry weight (mg), number of shoots, number of leaves/shoots, leaf width (cm), and leaf length (cm) were obtained results after 6 weeks of culture.

The shoots formed from stem node culture were similar to that of the control (shoot tip culture) after 8 weeks of culture. The shoot height, fresh and dry weight, leaf length, number of leaves per shoot, and leaf length and leaf width did not differ significantly between the two treatments. The results indicate that shoots derived from stem node culture are similar to those from direct shoot tip culture. Moreover, the total number of shoots was significantly different between stem node culture and the control (shoot tip culture). The total number of shoots (number of shoot/explants of different node positions) obtained from stem node (1, 2, 3, 4, 5, and 6) culture was 12.00 shoots per explant versus one shoot per explant in the control treatment. Until now, no study has been performed on the propagation in *Anthurium* sp. This study showed that shoot multiplication through stem node culture is potentially useful in propagating *Anthurium* sp. after 8 weeks of culture (Fig. 7.12).

Fig. 7.10 Stem elongation of *A. andreanum* "Tropical" after 2 months placed under 30 blue:70 red

7.5 In Vitro Stem Elongation in *Nepenthes mirabilis*

7.5.1 Achievements in Research on Regeneration and Micropropagation of N. mirabilis

Nepenthes mirabilis, a carnivorous plant in the family Nepenthaceae, is distributed mainly in China, Vietnam, Thailand, Indonesia, Philippines, Peninsular Malaysia, Sulawesi, Maluku, New Guinea, Palau Island, and Australia (Hua and Li 2005). In the wild, this plant lives in an acidic and nitrogen-deficient environment, so its leaves have elongated veins and develop into pitchers, which trap insects to compensate for the nitrogen deficiency (Kitching and Schofield 1986). *N. mirabilis* plant was commonly used as an ornamental and has high commercial value because of its distinctive leaf shape (Plachno 2007).

In vitro seed germination method is commonly used in the propagation of *Nepenthes* (Bahadur et al. 2008; Nongrum et al. 2009). Previous studies on *Nepenthes* micropropagation focused only on callus induction, shoot induction, shoot proliferation, and improved shoot quality (Bahadur et al. 2008; Nongrum et al. 2009; Muangkroot 2015; Budisantoso et al. 2018). However, the shoot multiplication and regeneration efficiency were not high.

Fig. 7.11 The morphology of in vitro stem elongation in *A. andreanum* "Tropical" after 2 months placed under 30 blue:70 red and FL

7.5.2 Stem Elongation in N. mirabilis by LED

The *Nepenthes* shoots (2 cm) with 3 leaves were cultured on Schenk and Hildebrandt medium (1972) supplemented with 1 mg/L 6-benzylaminopurine combined with 0.5 mg/L α-naphthalene acetic acid (NAA) under different LEDs lighting conditions (100R, 100B, 80R: 20B, 70R: 30B, 60R: 40B, 50R: 50B).

Shoots placed under FL were used as the control treatment. Shoots cultured under FL grew well. However, these plants did not show clear stem elongation. Whereas plants grown under LEDs had more clearly defined plant height and stem elongation and optimally recorded under 80R:20B and 60R:40B. In particular, the indicators of plant height (8.00 cm), and the number of stem nodes (6.33) is highest in plants grown under 60R:40B (Figs. 7.13 and 7.14).

Plant height and number of stem nodes are the lowest below 50R:50B (Fig. 7.13). Based on this result, it can be seen that the LED light is suitable for stem elongation and increased number of stem nodes than in conventional culture under fluorescent light (the number of burns is three times higher) (Figs. 7.13 and 7.14). Thus, 60R:40B is a suitable light source for the growth and stem elongation of *Nepenthes* plants.

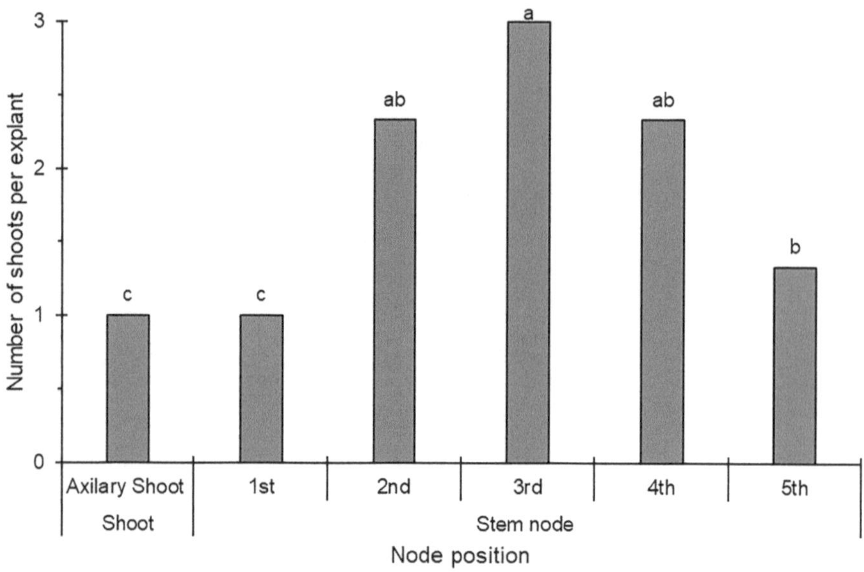

Fig. 7.12 The number of shoots per explant of *A. andraeanum* "Tropical" plantlets derived from stem node and axillary shoot after 8 weeks of culture

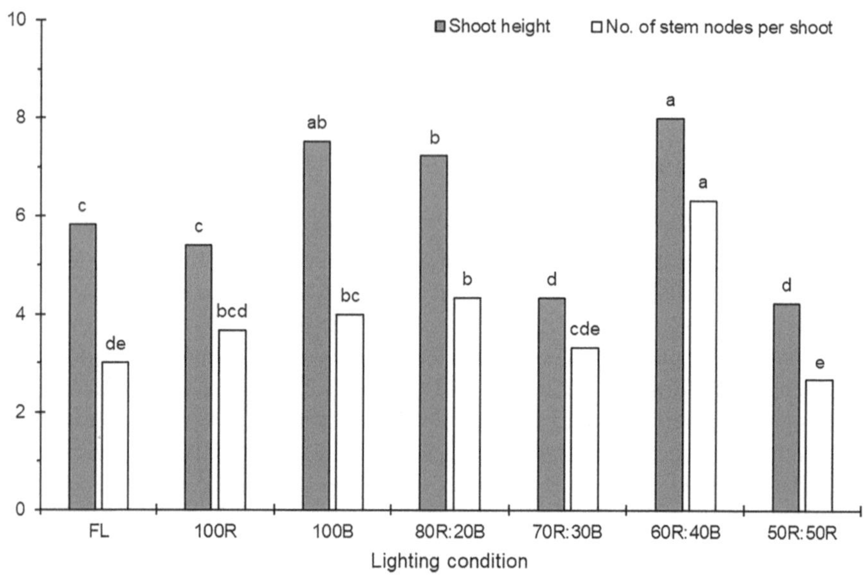

Fig. 7.13 Effect of lighting conditions on shoot height and number of stem nodes per shoot of *Nepenthes* after 2 months of culture

Fig. 7.14 Stem elongation of *Nepenthes* sp. under 60R:40B (lest) and FL (right) after 2 months of culture

7.5.3 In Vitro Stem Elongation in N. mirabilis by GA_3

The shoots cultured on medium supplemented with GA_3 at different concentrations (0.3, 0.5, 0.7, and 1 mg/L) had significant differences on in vitro shoot elongation, shoot growth, and development compared with shoots cultured on without GA_3 and placed under 60R:40B condition after 60 days of culture (Table 7.4). The shoots cultured on the medium supplemented with 0.3 mg/L GA_3 were highest elongated and grown such as shoot height (6.67 cm), number of leaves per shoot (8.33 leaves), leaf length (5.73 cm), leaf width (1.43 cm) (Table 7.4).

Table 7.4 Effect of GA_3 on in vitro stem elongation of *N. mirabilis* after 60 days of culture

Treatment	Concentration (mg/L)	Shoot height (cm)	No. of stem nodes	No. of leaves/ shoot	Leaf length (cm)	Leaf width (cm)
GA_3	0	3.67^{e}	2.33^{b}	5.33^{d}	3.07^{d}	0.73^{a}
	0.3	$\mathbf{6.67^{a}}$	$\mathbf{5.67^{a}}$	$\mathbf{8.33^{a}}$	$\mathbf{5.73^{a}}$	$\mathbf{1.43^{a}}$
	0.5	4.83^{b}	3.33^{b}	8.67^{a}	3.74	1.20^{b}
	0.7	4.03^{d}	$2.67^{bc,d}$	7.00^{b}	3.40^{c}	$1.10^{b,c}$
	1.0	3.13^{f}	2.00^{e}	$6.67^{b,c}$	2.83^{e}	1.07^{b}
60R:40B		4.53^{c}	$3.00^{b,c}$	$5.67^{c,d}$	3.13^{d}	0.63

Note: Different letters shown in the same column represent significant differences at $p < 0.05$ *in Duncan's test*

Fig. 7.15 In vitro stem elongation of *N. mirabilis* shoot cultured on medium supplemented without/with 0.3 mg/L GA_3 after 60 days of culture (*Bars: 1 cm*). (**a**) Without GA_3; (**b**) 0.3 mg/L GA_3

The height of the shoots (3.13 cm) obtained on the medium supplemented with 1 mg/L GA_3 was the lowest, yellow leaves, weak stems; meanwhile, shoots obtained on the culture medium supplemented with 0.3 mg/L GA_3 had clear segmentation, strong, uniform leaves and dark green color (Fig. 7.15b). This result shows that the role of GA_3 in plant growth and development is complex, varying by species and growth stage. Increasing the concentration of GA_3 in the culture medium may increase shoot elongation in this species but may also inhibit shoot elongation in some other species. The results of this study showed that, adding 0.3 mg/L GA_3 to the culture medium was suitable for growth, stem elongation, and increased member of stem nodes of *N. mirabilis* compared with shoots cultured under the 60R:40B condition.

This result could be explained that, low concentration of GA_3 can help increase shoot height by promoting cell division and elongation in the sub-foliation of *N. mirabilis* shoots, promoting growth, biomass production, and xylem chain elongation (Eriksson et al. 2000). Anandan et al. (2011) showed that regeneration of *Carica papaya* through indirect organogenesis, shoots cultured on medium supplemented with 0.1 mg/L GA_3 gave the highest grown and development.

GA_3 has been shown to be effective in in vitro stem elongation studies on some plants such as *Vigna mungo* (Muruganantham et al. 2005), *J. curcas* (Purkayastha et al. 2010), *P. dactylifera* (Rasmia et al. 2011), *D. latifolia* (Boga et al. 2012). The results of this study once again demonstrate the effectiveness of GA_3 in in vitro stem elongation and as a source of material for *N. mirabilis* micropropagation.

7.5.4 Growth and Development of iln Vitro N. mirabilis Plant Derived from Different Explants

Stem nodes (1.0 cm) derived from in vitro stem elongation (GA_3 and 60R:40B) of *N. mirabilis* were cultured on SH medium supplemented with 1.0 mg/L BA, 0.5 mg/L NAA, 30 g/L sucrose, 8 g/L agar, and 1 g/L AC. Growth and development of in vitro *N. mirabilis* plant derived from different explants was significantly different after 60 days of culture (Table 7.5 and Fig. 7.16).

Plants derived from in vitro stem elongation by GA_3 had shoot height (4.96 cm), number of nodes/plant (3.67), number of leaves/plant (8.00 leaves), leaf width (1.97 cm), leaf length (5.24 cm) were higher than that of the control and 60R:40B condition (Table 7.5 and Figs. 7.16 and 7.17). In addition, these plants were uniform, with green, long leaves, and fatter stems than the shoots in the other treatments.

Table 7.5 Growth and development of in vitro *N. mirabilis* plant derived from different explants after 60 days of culture

Explant	Shoot height (cm)	No. of stem nodes	No. of leaves/ plant	Leaf length (cm)	Leaf width (cm)
Control	3.60^c	1.67^c	6.67^b	3.96^c	1.58^b
60R:40B	4.43^b	2.33^b	7.66^a	4.50^b	1.86^a
0.3 mg/L GA_3	**4.96^a**	**3.67^a**	**8.00^a**	**5.24^a**	**1.97^a**

Note: Different letters shown in the same column represent significant differences at $p < 0.05$ in Duncan's test

Fig. 7.16 The in vitro *N. mirabilis* plant derived from different explants after 60 days of culture (*Bars: 1 cm*). (**a**) Control; (**b**) 60R:40B; (**c**) 0.3 mg/L GA_3

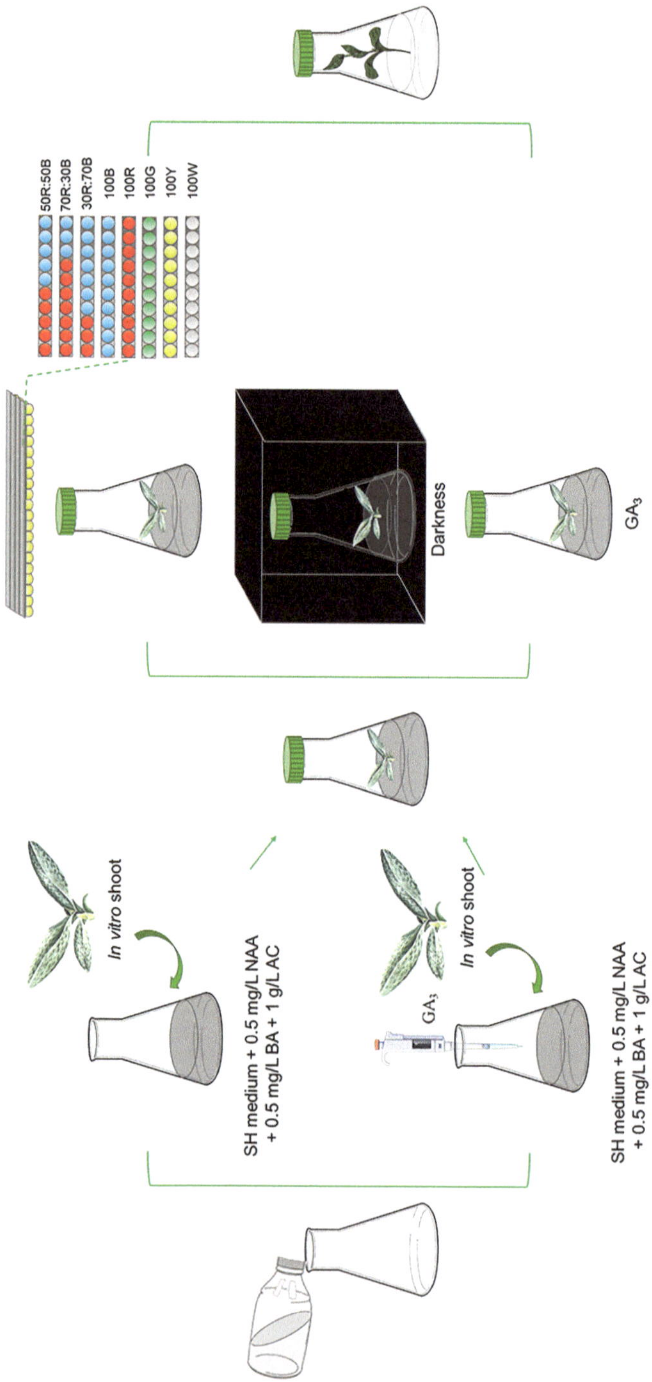

Fig. 7.17 Stem elongation for plant micropropagation

7.6 Conclusion

Stem elongation in both ex vitro and in vitro stages is an effective method in propagating some of today's difficult micropropagation plants. This method can help explants rejuvenate and facilitate in vitro morphogenesis. In this chapter, the effects of GA_3 and LEDs on stem elongation ability were shown in some plants such as *Paphiopedilum*, *Anthurium*, and *Nepenthes*. The results of the study are applicable to many other crops.

References

Anandan R, Thirugnanakumar S, Sudhakar D, Balasubramanian P (2011) *In vitro* organogenesis and plantlet regeneration of (*Carica papaya* L.). J Agric Tech 7(5):1339–1348

Arditti J (2008) Micropropagation of orchids, 2nd edn. Blackwell Publishing Ltd, Maiden, 1560p

Atak Ç, Çelik Ö (2009) Micropropagation of *Anthurium andraeanum* from leaf explants. Pak J Bot 41:1155–1161

Bahadur V, Kirad KS, Mathew A, Singh DB (2008) Tissue culture studies in *Nepenthes khasiana*. Acta Hortic 786:287–293

Boga A, Ram B, Reddy GRS (2012) Effect of benzyl amino purine and gibberellic acid on *in vitro* shoot multiplication and elongation of *Dalbergia latifolia* Roxb: an important multipurpose tree. Biotechnol Bioinformatics Bioeng 2(1):597–602

Bubeck SK (1973) A study of *Paphiopedilum* meristem culture. Ph.D. thesis, Rutgers University. University Microfilm International, Ann Arbor, pp 109

Budisantoso I, Indriani M, Kamsinah K (2018) Effect of BAP (6-benzyl amino purine) concentration on growth microcutting of *Nepenthes ampullaria*. Biosaintifika 10(3):678–683

Chen TY, Chen JT, Chang WC (2002) Multiple shoot formation and plant regeneration from stem nodal explants of *Paphiopedilum* orchids. In Vitro Cell Dev Biol Plant 38:595–597

Chen TY, Chen JT, Chang WC (2004a) Plant regeneration through direct shoot bud formation from leaf cultures of *Paphiopedilum* orchids. Plant Cell Tissue Organ Cult 76:11–15

Chen ZL, Ye XL, Liang CY, Duan J (2004b) Seed germination *in vitro* of *Paphiopedilum armeniacum* and *P. micranthum*. Acta Hortic Sinica 31:540–542

Chen C, Hou X, Zhang H, Wang G, Tian L (2011) Induction of *Anthurium andraeanum* "Arizona" tetraploid by colchicine *in vitro*. Euphytica 181(2):137–145

Chugh S, Guha S, Rao IU (2009) Micropropagation of orchids: a review on the potential of different explants. Sci Hortic 122:507–520

da Silva JAT (2013) Orchids: advances in tissue culture, genetics, phytochemistry and transgenic biotechnology. Floricult Orn Biotech 7:1–52

Ding CC, Wu H, Liu FY (2004) Factors affecting the germination of *Paphiopedilum armeniacum*. Acta Bot Yunn 26:673–677

Elibox W, Umaharan P (2008) Morphophysiological characteristics associated with vase life of cut flowers of *Anthurium*. HortScience 43(3):825–831

Eriksson ME, Israelsson M, Olsson O, Moritz T (2000) Increased gibberellin biosynthesis in transgenic trees promotes growth, biomass production and xylem fiber length. Nat Biotech 18(7):784–788

Farsi M, Taghavizadeh MEY, Qasemiomran V (2012) Micropropagation of *Anthurium andreanum* cv. Terra. Afr J Biotechnol 11:13162–13166

Gantait S, Mandal N, Bhattacharyya S, Das PK (2008) *In vitro* mass multiplication with pure genetic identity in *Anthurium andreanum* Lind. Plant Tissue Cult Biotechnol 18(2):113–122

García-Martinez JL, Gil J (2001) Light regulation of gibberellin biosynthesis and mode of action. J Plant Growth Regul 20(4):354–368

Gupta SD, Jatothu B (2013) Fundamentals and applications of light-emitting diodes (LEDs) in *in vitro* plant growth and morphogenesis. Plant Biotech Rep 7(3):211–220

Hahn EJ, Kozai T, Paek KY (2000) Blue and red light-emitting diodes with or without sucrose and ventilation affects *in vitro* growth of *Rehmannia* glutinose plantlets. J Plant Biol 43:247–250

Hirai T, Amaki W, Watanabe H (2006) Effects of monochromatic light irradiation by LED on the internodal stem elongation of seedlings in eggplant, leaf lettuce and sunflower. J Soc High Technol Agric 18(2):160–166

Hirano K, Ueguchi-Tanaka M, Matsuoka M (2008) GID1-mediated gibberellin signaling in plants. Trends Plant Sci 13:192–199

Hong PI, Chen JT, Chang WC (2008) Plant regeneration via protocorm-like body formation and shoot multiplication from seed derived callus of a Maudiae type slipper orchid. Acta Physiol Plant 30:755–759

Hossain M, Kant R, Van PT, Winarto B, Zeng SJ, da Silva JAT (2013) The application of biotechnology to orchids. Crit Rev Plant Sci 32:69–139

Hua Y, Li H (2005) Food web and fluid in pitchers of *Nepenthes mirabilis* in Zhuhai, China. Acta Bot Gallica 152(2):165–175

Huang LC (1988) A procedure for asexual multiplication of *Paphiopedilum in vitro*. Am Orchid Soc Bull 57:274–278

Huang LC, Lin CJ, Kuo CI, Huang BL, Murashige T (2001) *Paphiopedilum* cloning *in vitro*. Sci Hort 91:111–121

Huttly AK, Phillips AL (1995) Gibberellin-regulated plant genes. Physiol Plant 95:310–317

Huy NP, Luan VQ, Cuong LK, Nam NB, Tung HT, Hien VT, Le TD, Paek KY, Nhut DT (2019) Strategies for the regeneration of *Paphiopedilum callosum* through internode tissue cultures using sark-light cycles. Hort Sci 54:920–925

Islam SA, Dewan MMR, Mukul MHR, Hossain MA, Khatun F (2010) *In vitro* regeneration of *Anthurium andreanum* cv. Nitta. Bangl J Agril Res 35:217–226

Jahan MT, Islam MR, Khan R, Mamun ANK, Ahmed G, Hakim L (2009) *In vitro* clonal propagation of anthurium (*Anthurium andraeanum* L.) using callus culture. Plant Tissue Cult Biotechnol 19(1):61–69

Jiang L, Wang Z, Jin G, Lu D, Li X (2019) Responses of favorita potato plantlets cultured in vitro under fluorescent and light-emitting diode (LED) light sources. Am J Potato Res 96(4):396–402

Kauth PJ, Dutra D, Johnson TR, Stewart SL, Kane ME, Vendrame W (2008) Techniques and applications of *in vitro* orchid seed germination. In: da Silva JAT (ed) Floriculture, ornamental and plant biotechnology: advances and topical issues, vol V, 1st edn. Global Science Books, Isleworth, pp 375–391

Kim SJ, Hahn EJ, Heo JW, Paek KY (2004) Effects of LEDs on net photosynthetic rate, growth and leaf stomata of *Chrysanthemum* plantlets *in vitro*. Sci Hortic 101:143–151

Kiss E, Kiss J, Gyulai G, Heszky LH (1995) A novel method for rapid micropropgation of pineapple. HortSci 30:127–129

Kitching R, Schofield C (1986) Every pitcher tells a story. New Sci 23:48–51

Kurilčik A, Miklušytė-Čanova R, Dapkūnienė S, Žilinskaitė S, Kurilčik G, Tamulaitis G, Duchovskis P, Žukauskas A (2008) *In vitro* culture of Chrysanthemum plantlets using light-emitting diodes. Cent Eur J Biol 3:161–167

Lee YI, Yeung EC, Lee N, Chung MC (2006) Embryo development in the lady's slipper orchid, *Paphiopedilum delenatii*, with emphasis on the ultrastructure of the suspensor. Ann Bot 98: 1311–1319

Liao YJ, Tsai YC, Sun YW, Lin RS, Wu FS (2011) *In vitro* shoot induction and plant regeneration from flower buds in *Paphiopedilum* orchids. In Vitro Cell Dev Biol Plant 47:702–709

Liendo M, Mogollón N (2009) Multiplicación clonal *in vitro* del Anturio (*Anthurium andreanum* Lind. cv. Nicoya). Bioagro 21:179–182

Lin YH, Chang C, Chang WC (2000) Plant regeneration from callus culture of a *Paphiopedilum* hybrid. Plant Cell Tissue Organ Cult 62:21–25

Long B, Niemiera AX, Cheng ZY, Long CL (2010) *In vitro* propagation of four threatened *Paphiopedilum* species (Orchidaceae). Plant Cell Tissue Organ Cult 101:151–162

Luan VQ, Huy NP, Nam NB, Huong TT, Hien VT, Hien NTT, Nhut DT (2015) *Ex vitro* and *in vitro* *Paphiopedilum delenatii* Guillaumin stem elongation under light-emitting diodes and shoot regeneration via stem node culture. Acta Physiol Plant 37(7):1–11

Martin KP, Joseph D, Madassery J, Philip VJ (2003) Direct shoot regeneration from lamina explant of two commercial cultivars of *Anthurium andreanum* Hort. In Vitro Cell Dev Biol Plant 39: 500–504

Matsuo S, Nanya K, Imanishi S, Honda I, Goto E (2019) Effects of blue and red lights on gibberellin metabolism in tomato seedlings. Hort J 88(1):76–82

Morel G (1974) Clonal multiplication of orchid. In: Withner LJ (ed) The orchid. Scientific studies. Wiley, New York, p 222

Muangkroot A (2015) Appropriate medium for micropropagation of *Nepenthes lowii* Hook. f. Mahidol J 2:27–34

Murashige T, Skoog F (1962) A revised medium for rapid growth and bio-assays with tobacco tissue cultures. Plant Physiol 1:473–497

Muruganantham M, Ganapathi A, Amutha S, Vengadesan G, Selvaraj N (2005) Shoot regeneration from immature cotyledonary nodes in black gram [*Vigna mungo* (L.) Hepper]. Indian J Biotechnol 4:551–555

Nanya K, Ishigami Y, Hikosaka S, Goto E (2012) Effects of blue and red light on stem elongation and flowering of tomato seedlings. Acta Hortic 956:261–266

Ng CY, Saleh NM (2011) *In vitro* propagation of *Paphiopedilum* orchid through formation of protocorm-like bodies. Plant Cell Tissue Organ Cult 105:193–202

Ng CY, Saleh M, Zaman FQ (2010) *In vitro* multiplication of the rare and endangered slipper orchid, *Paphiopedilum rothschildianum* (Orchidaceae). Afr J Biotechnol 9:2062–2068

Nhut DT (2002) *In vitro* growth and physiological aspects of some horticultural plantlets cultured under red and blue light-emitting diodes (LEDs). Doctoral thesis, Kagawa University, Japan, pp 1–196

Nhut DT, Takamura T, Watanabe H, Okamoto K, Tanaka M (2003) Responses of strawberry plantlets cultured *in vitro* under superbright red and blue light-emitting diodes (LED). Plant Cell Tissue Organ Cult 73:43–52

Nhut DT, Trang PTT, Vu NH, Thuy DTT, Khiem DV, Binh NV, Van KTT (2005) A wounding method and liquid culture in *Paphiopedilum delenatii* propagation. Prop Orna Plants 5:158–163

Nhut DT, Duy N, Vy NNH, Khue CD, Khiem DV, Vinh DN (2006) Impact of *Antihurium* spp. genotype on callus induction derived from leaf explants, and shoot and root regeneration capacity from callus. J Appl Hortic 8:135–137

Nhut DT, Thuy DTT, Don NT, Luan VQ, Hai NT, Van KTT, Chinnappa CC (2007) *In vitro* stem elongation of *Paphiopedilum delenatii* Guillaumin and shoot regeneration via stem node culture. Prop Orna Plants 7:29–36

Nongrum I, Kumaria S, Tandon P (2009) Multiplication through *in vitro* seed germination and pitcher development in *Nepenthes khasiana* Hook. f., a unique insectivorous plant of India. J Hortic Sci Biotechnol 84(3):329–332

Oropeza M, Alexander M, Vargas TE (2010) Micropropagation and organogenesis of *Anthurium andreanum* Lind. cv. 'Rubrun'. Methods Mol Biol 589:3–14

Pierik RLM, Sprenkels PA, Vanderharst B, Vandermeys QG (1988) Seed germination and further development of plantlets of *Paphiopedilum ciliolare* Pfitz *in vitro*. Sci Hort 34:139–153

Plachno BJ (2007) Sweet but dangerous: nectarines in carnivorous plants. Acta Agr 60(2):31–37

Purkayastha J, Sugla T, Paul A, Solleti SK, Mazumdar P, Basu A, Mohommad A, Ahmed Z, Sahoo L (2010) Efficient *in vitro* plant regeneration from shoot apices and gene transfer by particle bombardment in *Jatropha curcas*. Biol Plant 54(1):13–20

Rasmia S, Darwesh Z, Zaid E, Rehab SA (2011) Effect of ammonium nitrate and GA_3 on growth and development of date palm plantlets *in vitro* and acclimatization research. J Agric Biol Sci 7(1):17–22

Sachs RM (1965) Stem elongation. Ann Rev Plant Physiol 16(1):73–96

Schenk RU, Hildebrandt AC (1972) Medium and techniques for induction and growth of monocotyledonous and dicotyledonous plant cell cultures. Can J Bot 50:199–204

Schuerger AC, Brown CS, Stryjewski EC (1997) Anatomical features of pepper plants (*Capsicum annuum L.*) grown under red light-emitting diodes supplemented with blue or far-red light. Ann Bot 79:273–282

Schwechheimer C (2008) Understanding gibberellic acid signaling—are we there yet? Curr Opin Plant Biol 11(1):9–15

Stewart J, Button J (1975) Tissue culture studies in *Paphiopedilum*. Am Orchid Soc Bull 35:88–95

Te-Chato S, Susanon T, Sontikun Y (2006) Cultivar, explant type and culture medium influencing embryogenesis and organogenesis in *Anthurium* spp. Songklanakarin J Sci Technol 28:717–722

Thomas SG, Rieu I, Steber CM (2005) Gibberellin metabolism and signaling. Vitam Horm 72:289–338

Viégas J, Rocha MTR, Ferreira-Moura I, Rosa DL, Souza JA, Correa MGS, Silva JA (2007) *Anthurium andraeanum* (Linden ex André) culture: *in vitro* and *ex vitro*. Floricul Ornam Biotech 1:61–65

Yu YX, Liu L, Liu JX, Wang J (2009) Plant regeneration by callus-mediated protocorm-like body induction of *Anthurium andraeanum*. Hort Agric Sci China 8:572–577

Zeng SJ, Wang J, Wu KL, da Silva JAT, Zhang JX, Duan J (2013) *In vitro* propagation of *Paphiopedilum hangianum* Perner & Gruss. Sci Hort 151:147–156

Somatic Embryo as a Tool for Micropropagating of Some Plants

8

Hoang Thanh Tung, Ha Thi My Ngan, Do Manh Cuong, Vu Thi Hien, Trinh Thi Huong, Bui Van The Vinh, Vu Thi Mo, Truong Thi Lan Anh, Nguyen Van Binh, Le Thi Diem, and Duong Tan Nhut

Abstract

Somatic embryogenesis is derived by several factors in culture medium including plant growth regulators (PGRs), amino acids (proline, serine, threonine, etc.), polyamines (spermidine, spermine), carbohydrate sources, etc. This method has been used for multiplication of plants with high economic values. In this chapter, somatic embryogenesis has been studied on some plants such as *Panax vietnamensis* Ha *et* Grushv.—an endemic species to Vietnam, *Kappaphycus striatus* belongs to the phylum Rhodophyta, distributed mainly in tropical seas, and *Jatropha curcas*, a shrub plant, belongs to the family Euphorbiaceae. The results of this study will further expand the applicability of somatic embryogenesis research on many different plant species.

Keywords

Panax vietnamensis · *Kappaphycus striatus* · *Jatropha curcas* · Somatic embryogenesis

H. T. Tung · H. T. M. Ngan · D. M. Cuong · V. T. Hien · L. T. Diem · D. T. Nhut (✉)
Tay Nguyen Institute for Scientific Research, VAST, Dalat City, Vietnam

T. T. Huong
Ho Chi Minh City University of Food Industry, Hochiminh City, Vietnam

B. V. T. Vinh
Ho Chi Minh City University of Technology—HUTECH, Hochiminh City, Vietnam

V. T. Mo
Nhatrang Institute of Technology Research and Application, VAST, Nha Trang City, Vietnam

T. T. L. Anh · N. Van Binh
University of Dalat, Dalat City, Vietnam

D. T. Nhut et al. (eds.), *Plant Tissue Culture: New Techniques and Application in Horticultural Species of Tropical Region*,
https://doi.org/10.1007/978-981-16-6498-4_8

8.1 Introduction

The somatic embryos contain nutrients similar to zygotic embryos; there are root and shoot sprouts, so they can germinate directly into plants without the stage of shooting and rooting. Somatic tissues and cells in vitro cultured directly create somatic embryos through callus. Callus cells can be divided by multiplier, so they can create a significant number of embryos in a short time. Up to date, over 200 species of plants have been successfully propagated via somatic embryonic technology. The somatic embryos can be preserved for a long time and germinate in the appropriate season. Somatic embryogenesis (SE) technology is still an advanced technology in plant propagation field worldwide (Méndez-Hernández et al. 2019).

The ability of somatic embryogenesis is expressed primarily at the genotype level. This can be easily demonstrated by converting embryogenesis from an embryogenic genotype to a genotype that inhibits embryogenesis through sexual crossbreeding. Although the conditions for inducing embryogenesis have been established for many plant species, there are still a large number of species that have not yet been able to somatic embryo form (Moltrasio et al. 2004). Even within the same species, there is a genotype that is prone to embryogenesis, but there is also a genotype that prevents embryogenesis. It should be emphasized, however, that in many cases, inhibition of embryo formation can be resolved by optimizing the growth conditions of the plant or by selecting the appropriate explants. Thus, genetic trait identification can be used only to determine the location and timing of embryogenesis. The ability of somatic embryogenesis (SE) in soma cells is shown only when these cells are placed in an in vitro culture medium and determined through the plant growth program as well as environmental markers.

The experiments of tissue culture have shown that there exists an embryonic response gradient between different plant organs. Embryonic-derived tissues were most capable of embryogenesis, and this capacity decreased in the lower cotyledons, petiole, leaves, and roots (Neumann 2000). But even if the potential for embryogenesis is lost in the soma cells, it can still be restored. These indirect embryogenesis pathways need to undergo an intermediate callus stage in order to demonstrate the potential for embryogenesis. Obviously, plant cells' ability to produce embryos does not cease to decline during individual development, and it depends on the species. In monocot plants, including the most important cereal species, embryogenesis is mostly limited to cells derived from embryos or meristems including embryos, seeds, and leaf bases (Graminae), shoot tip (Orchidaceae), tuber scale (Liliaceae), and lateral shoot. The embryogenesis of these meristem cells can be maintained if the explant is cultured in media supplemented with 2,4-D to induce callus formation. The transfer of these embryogenic callus cells to medium without plant growth regulators (PGRs) or enriched media at low concentration of auxin can achieve a high rate of asexual embryogenesis (Krishna and Vasil 1995).

Embryogenic cells are characterized by a centrally located nucleus, with prominent microtubules near the nucleus and actin filaments (Šamaj et al. 2003). Besides, they also have a special cell wall structure with cells and are both derived from meristem (or embryogenesis tissue) and can be produced from cells with a large vacuole under certain suitable conditions, for example, after treated with 2,4-D. However, abscisic acid, cytokinin, or other stress can also stimulate the formation of cell types that are capable of producing embryos. Many tissue culture systems used 2,4-D as an effective inducer for SE. Scientists now agree that determination of the state of an embryonic cell occurs in medium with 2,4-D, but at the same time, the development of cells is inhibited. The 2,4-D, used as an agent to transit cell state, has been emphasized by suspension culture experiments of *Medicago* callus in media supplemented with another synthetic auxin is NAA (Dudits et al. 1991). If these cells are transferred to a medium free of growth regulators, they will produce roots with high frequency. If these cells were treated with a high concentration of 2,4-D (100 μM) for a short time (about a few minutes) and then transplanted to a medium without growth regulators, the cells will develop into the somatic embryos. However, it takes 2–3 weeks after treatment before the first embryos can be observed on the surface of the callus. These studies have shown that 2,4-D is essential to initiate the program of embryogenesis. The transfer of cultures to a 2,4-D-free medium has an important role in establishing cell polarity, which is considered one of the first events of embryo development (Feher et al. 2003).

8.2 Somatic Embryogenesis Process

8.2.1 Somatic Embryogenesis

Plant regeneration through the somatic embryogenesis pathway consists of five stages: (1) embryonic culture by selecting a primary inoculated in a medium supplemented with growth regulators, mainly auxin (sometimes cytokinin); (2) growth of embryo culture on solid or liquid medium supplemented with growth regulators similar to step 1; (3) the pre-maturation period of the embryo in an environment without growth regulators, to prevent proliferation and stimulate embryogenesis; (4) embryo maturation stage by culturing embryos in ABA-supplemented medium and/or reducing osmotic pressure; and (5) plantlets growth and development in the absence of plant growth regulators media (Méndez-Hernández et al. 2019).

8.2.2 Application of the Somatic Embryogenesis Process

The somatic embryo formation has brought many practical applications and great commercial potential, especially in plant micropropagation. In theory, an infinite number of embryonic cells could be produced from tissue cultured. The propagation rate from embryos is much higher than those from meristem. Besides, the large quantity of embryos is a significant source of raw materials for other important practical applications such as artificial seed production, improving crop quality (as raw material for cell selection, genetic transformation, somatic hybridization, forming polyploid plants), preservation of germs, elimination of viruses, and production of in vitro metabolites (Carlos and Martinez 1998).

Cell selection is the regeneration of plants from populations of cells combined with the selection of desirable traits of genes. It is most essential to use regenerative systems of plants capable of rebuilding a new plant from a single cell to avoid the occurrence of genetic mutations. The somatic embryogenesis process allows the regeneration of complete plants from single cells, which can then be used in cell-selective programs and can create important agricultural properties such as increased tolerance to salinity, become resistant to herbicides, disease, alum or drought, and may develop other interesting biochemical mutations. For example, cell selection and plant regeneration through the asexual embryo have been used to increase salinity tolerance and disease resistance in citrus plants, sugarcane virus resistance, and coffee plant phytotoxin resistance.

Regeneration of transgenic plants using the somatic embryogenesis pathway: transformed into the culture and then regenerated through direct or indirect embryogenesis; transformed into cells capable of embryogenesis and then regenerated by indirect embryogenesis; transformed into the somatic embryo and regenerated through secondary embryogenesis or indirect embryogenesis; transformed into the somatic embryo, proliferated through the secondary embryo formation pathway, and then regenerated through the organ formation pathway.

The somatic embryogenesis may also create polyploid regenerating plants. Treatment of colchicine or amiprophos-methyl, which suppresses microtubule formation in somatic embryo culture, promotes the formation of polyploid embryos. Triploid plants can also be regenerated if the embryo formed is derived from the endosperm.

The somatic embryo has a highly developed capillary system which is not connected to the vascular system of the explant. This means that somatic embryogenesis can also help to produce virus-free plant lines, which can be combined with thermal shock for better results. The regenerated plants from the somatic embryos showed negative results when tested for the presence of virus and viroid.

8.2.3 Factors Affecting the Process of Asexual Embryogenesis

Plant growth regulators (PGRs): PGRs are used in most somatic embryo culture systems. Synthetic auxins such as 2,4-D are commonly used in primary cultures; other auxins also used such as IBA, picloram; NAA and IAA are used in some culture systems. Auxin is used for inducing embryonic cell formation; in the case of cultures containing embryonic cells, auxin may not be required in the culture medium. Cytokinins are also used for the induction of morphogenesis in dicotyledonous plants. The most commonly used cytokinin in culture media is BAP, but other cytokinins such as kinetin, zeatin, and TDZ also give good results depending on the type of plant. In embryo culture, the concentration of growth regulators is critical for the optimal growth response of the culture. When the concentration is too low, it will not stimulate growth; otherwise, too high content can be toxic to the sample.

The development of carrot embryos included two-stage, each with its medium. Callus tissue was initially cultured and multiplied on a medium with 2,4-D at the concentration range of 0.5–1.0 mg/L. These cells are then transferred to medium with very low auxin concentration (0.01–0.1 mg/L) or without auxin, which will develop into mature embryos. In coffee plants (*Coffea arabica*), the somatic embryos that develop only when the callus formed on 2,4-D media are transferred to a medium free of this hormone. In some plants, other auxins are more effective than 2,4-D, as in the case of pumpkin plants, NAA and IBA are used for embryogenesis. In the embryo culture of *Vitus* plants, embryogenesis occurs in the presence of NAA and BAP (Mullin and Schlegel 1976). Other growth regulators such as GA_3 and ABA at low concentrations also have stimulating effects in the embryogenesis of carrots, *Citrus medica*, *Phaseolus*, GA_3 can promote embryo development to matured stage and development into plantlets.

Nitrogen source: The forms of nitrogen used in culture have an important effect on embryo formation, and each plant type is suitable for the different forms of nitrogen. Inorganic nitrogen such as KNO_3, NH_4, and organic nitrogen including glutamine and asparagine are widely used in embryo culture. Unidentified organic mixtures such as casein and coconut water also have a good effect on embryo formation, as they contain many mineral components and amine acids (Dougall and Verma 1978).

Carbohydrate: Sucrose is commonly used as a carbohydrate source for embryogenesis. Sucrose and auxin interact with each other; the concentration of sucrose or auxin is dependent on each other. Besides sucrose, there are also some double sugars and simple sugars that have been used with success and are more effective in other species. Galactose and lactose outperformed sucrose for the induction of embryo from callus of *Citrus*, while sucrose was superior to glucose and fructose (Kochba et al. 1982). It is sometimes also necessary to combine two different sugars to aid in the formation of the somatic embryo.

pH: Intracellular pH is a factor influencing cell differentiation and division. The changes in pH in the cytoplasm are known to be essential to regulating cell cycle, cell growth, and division. The pH value in the vacuole as well as in chloroplasts can be considered as an indicator of the cell type (capable and incapable to embryogenesis proliferation). Increased cytoplasmic pH is associated with cell division (Pasternak et al. 2002). Low intracellular pH is believed to be involved in the loosening of the cell wall required for direct cell elongation during embryo development.

The light: Somatic embryo can be produced under different light and dark conditions. There are species with a high need for light but there are also species of embryo formation that occur in completely dark conditions. Induction of embryogenesis in *Cauliflower* plants completely requires light (Gupta and Holmstrom 2005). In contrast, induction of embryos in *Podophyllum* plants completely required dark conditions. Red light promotes embryogenesis in date palms while blue light reduces that ability. In carrot somatic embryogenesis, red, green, or dark light have similar effects, while blue or white light have negative effects on embryogenesis (Arumugam and Bhojwani 1990).

Polyamine: Polyamine (PA) are small molecular polycations found in all living organism. In plants, they are described in a series of biological processes including growth, development, and stress responses. They are also involved in protein phosphorylation, improvement after transcription, proper DNA transport, associated with early stage of embryogenesis and plant regeneration (Joshi et al. 2010).

8.3 Somatic Embryogenesis in *Panax vietnamensis* Ha *et* Grushv.: A Case Study

8.3.1 General Introduction

Panax genus (scientific name: *Panax*) is a genus containing about 11 species of very slow-growing rhizome plants of the Araliaceae family. The species of the genus *Panax* has been known and used very early in traditional medicine to restore functional impairment and return the body's activities to normal. Ngoc Linh ginseng (*Panax vietnamensis* Ha *et* Grushv.)—endemic species to Vietnam—is a medicinal plant of the Xe Dang ethnic minority distributing in the high mountains of Kon Tum and Quang Nam provinces (Fig. 8.1); herbaceous plants live interwoven in the diverse flora of cold, rugged mountainous areas, and cloudy almost all year round, with no population.

Ngoc Linh ginseng is a medicinal plant with classified ginsenoside composition among the most numerous species of the genus *Panax* in the world. Through research, scientists have found that Ngoc Linh ginseng not only has the specific pharmacological effects of the genus *Ginseng* but also has typical effects such as anti-stress, anti-depressant, anti-oxidant in vitro and in vivo. The group of substances that have the most decisive role in the pharmacological effects of this ginseng species is triterpenoid saponins, represented by MR_2, G-Rb_1, and G-Rg_1 (Luan 2003). Ngoc Linh ginseng is one of the ginseng species with the highest content of dammaran frame saponins (about 12–15%) and the highest amount of saponins in comparison to other species of the *Panax* genus in the world (Dong et al. 2007).

Traditional Ngoc Linh ginseng propagation methods such as cutting off the shoots have low multiplier and required a long growing time, from 6 to 7 years to harvest. Propagation by sowing method does not give high results for many reasons:

Fig. 8.1 *Panax vietnamensis* Ha *et* Grushv plant. (**a**) Leaf; (**b**) Flower; (**c**) Fruit; (**d**) Rhizome

(1) it is difficult to collect seeds, seeds when sowing stay in the soil after a long time to germinate, so seeds are often affected by animals and insects; (2) the germination rate from seeds is low (only 30–40%). So, application of in vitro propagation method on Ngoc Linh ginseng was performed and got successful by Nhut et al. (2010). Then, Nhut et al. (2012) have also studied successfully the embryogenesis from the rhizome. The propagation through embryogenesis has brought many practical and commercial values. The application of the somatic embryogenesis method in Ngoc Linh ginseng promises to bring many advantages because it can create a large number of good-quality plantlets with rhizome in a short time.

8.3.2 Effect of the Explant Sources on Somatic Embryogenesis

The leaf explants (0.5 × 0.5 cm), petiole (1.0 cm in length), and rhizome (0.5 × 0.5 × 0.1 cm) were cut from 3-month-old in vitro Ngoc Linh ginseng with 1.5 cm in height and a weight of about 0.2 g (Fig. 8.2) and culture on Murashige and Skoog (1962) (MS) medium supplemented with 1.0 mg/L 2,4-D and 1.0 mg/L NAA. After 8 weeks of incubation, the criteria of embryogenesis frequency (%), number of embryos/explant, and weight of explants (mg) were recorded.

The sample source and nutrient composition in the culture medium play an important role in the morphogenesis of the cultured explant (Fleming 2002). In this experiment, the Ngoc Linh ginseng leaf, petiole, and rhizomes explant formed callus after 2 weeks of culture on MS medium supplemented with 1.0 mg/L 2,4-D and 1.0 mg/L NAA. Callus first arising on the rhizomes explant is spongy and formed around the border of the slice of the rhizome. The spongy, clear, succulent calli are also formed from leaf and petiole explants. For leaf culture, callus formed

Fig. 8.2 In vitro *Panax vietnamensis* Ha *et* Grushv. plant (3-month-old)

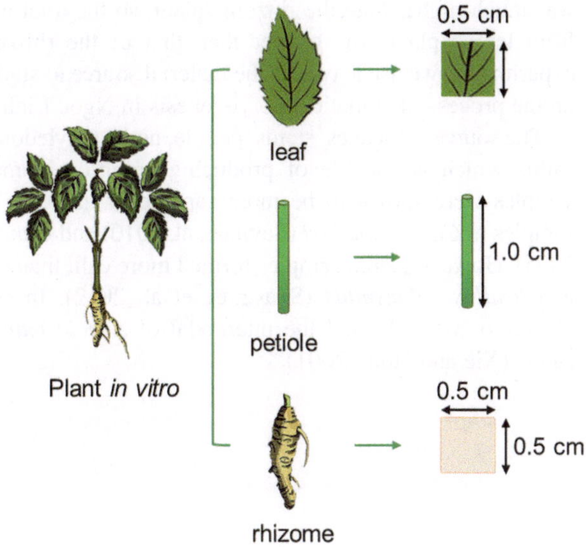

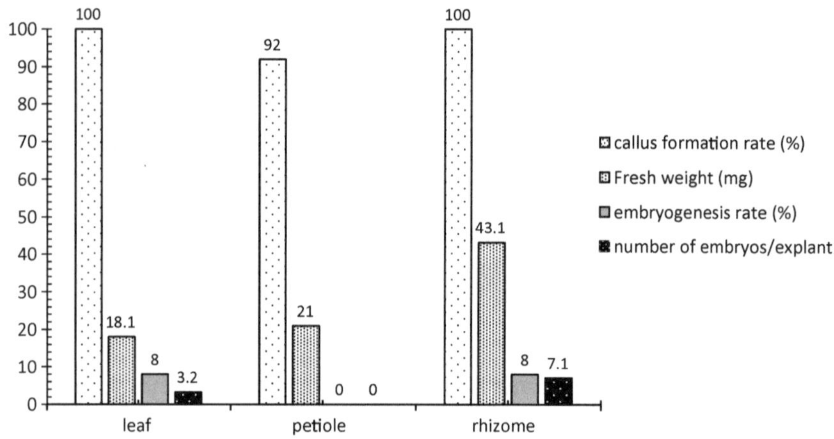

Fig. 8.3 Influence of the explant source on the ability to Ngoc Linh ginseng somatic embryogenesis

first in the cross-section that was exposed to the medium and then on all four sides of the cross-section, whereas, for petiole, callus formed only at the ends where the cut was located (Fig. 8.9a1–a3).

After 8 weeks of culture, callus growth was strongest for leaf and rhizome cultures (Fig. 8.9a1, a3), reaching a rate of 100% (Fig. 8.3). Small purple spots appear on the surface of the spongy callus; most of these structures then develop into adventitious roots with different morphologies (data not shown). Besides, white spherical embryos appeared on the surface of callus derived from leaf and rhizome cultures (Fig. 8.3). The number of embryos from callus derived from rhizomes explants than those from leaf explants; however, the number of leaves per plant was much higher than the rhizome/plant, so the total number of embryos obtained from leaf explant was greater than that of the rhizome. Therefore, in the next experiments, we use leaves as the material source to study the effects of other factors on the process of somatic embryogenesis in Ngoc Linh ginseng.

The sources of leaves, stems, petiole, node, cotyledons, etc. are also used to create callus which is capable of producing embryos in many different species. Leaf samples were shown to be more capable of inducing callus formation than stem samples in *Citrus jambhiri* (Savita et al. 2010) and *Cannabis sativa* L. (Aurelia et al. 2005). However, stem samples formed more calli than cotyledons and young leaves in *Citrullus colocynthis* (Shasthree et al. 2012). In contrast, the root sample of *Cannabis sativa* L. and the internodal of *Acacia mangium* were poorly inducing callus (Xie and Hong 2001).

8.3.3 Effect of PGRs on Somatic Embryogenesis

The callus explant generated from the above experiments was cut into small pieces (0.5 × 0.5 cm) and cultured on MS medium supplemented with 1.0 mg/L 2.4-D combined with NAA and/or KIN at different concentrations (0.1, 0.2, 0.5, 1.0 mg/L) (Fig. 8.9b1). After 12 weeks of incubation, the criteria of embryogenesis frequency (%), number of embryos/explants, and weight of explants (mg) were recorded. The optimal embryo induction (EM) medium in this experiment is used as the base medium for the next experiments.

The results showed that the rate of embryo forming and the number of embryos formed per the explant were different in the experimental treatments. In the control treatment (PGR-free), embryos did not appear and the callus gradually browned and died. The treatments supplemented with 2,4-D combined with NAA or KIN at different concentrations showed the presence of spherical embryonic structures. The combination of NAA and 2,4-D showed better embryo induction through the criteria of embryogenesis ratio and the number of embryos/explants (Fig. 8.4).

In studies of the induction of embryogenesis in many plant species, auxin, especially 2,4-D, is the most commonly used (Umehara and Kamada 2005). Therefore, we have used 1.0 mg/L 2.4-D for all experiments (this is the appropriate concentration on Ngoc Linh ginseng that has been previously studied). However, if only 2,4-D was added individually without the support of other growth regulators such as auxin or cytokinin group, there would be no callus capable of embryogenesis. In this experiment, we combined 2,4-D with another auxin, NAA, which is capable of inducing embryogenesis in some species. NAA is very effective in somatic embryogenesis in soybean and carrot plants when used at a concentration

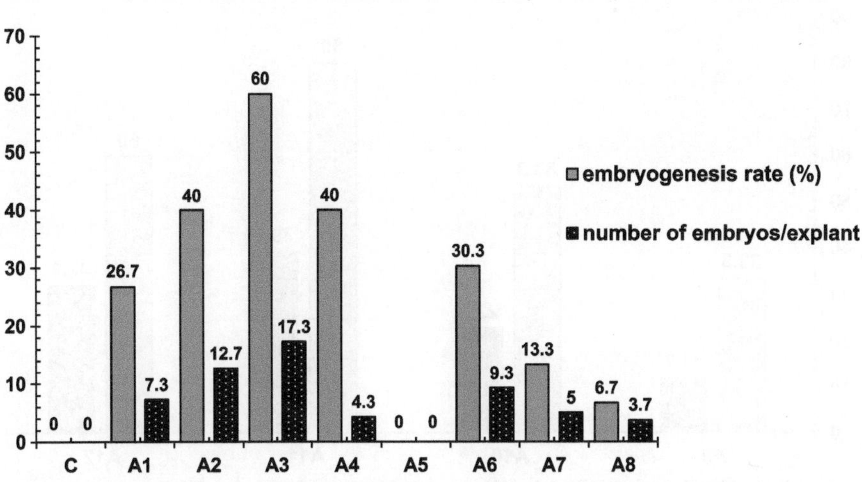

Fig. 8.4 Influence of 2,4-D combined with NAA or KIN on the ability to Ngoc Linh ginseng somatic embryogenesis. (**c**) Control (PGR-free); (**a1**, **a2**, **a3**, **a4**) 2,4-D (1.0 mg/L) combined with NAA (0.1, 0.2, 0.5 and 1.0 mg/L); (**a5**, **a6**, **a7**, **a8**) 2,4-D (1.0 mg/L) in combination with KIN (0.1; 0.2; 0.5 and 1.0 mg/L)

of 1.0 mg/L (Borkird et al. 1986). Meanwhile, in our experiment, the rate of embryogenesis was highest when using 0.5 mg/L NAA combined with 1.0 mg/L 2.4-D (52 embryos/explants) (Fig. 8.4). This result is consistent with the study of embryogenesis in *Momordica charantia* L. (Ananya et al. 2009) and *Panax ginseng* (Shohana et al. 2009).

Cytokinin stimulates the formation of somatic embryos from callus without embryogenesis reported in legumes such as *Phaseolus* (Malik and Saxena 1992) and Peanut (Gill and Saxena 1992). Cytokinin such as KIN is effective in primary culture for the formation of somatic embryos in some woody species. However, the concentration of KIN used also varied according to the plant species. The results in this experiment showed that the 2,4-D and KIN combination was not as effective at the induction of embryogenesis as the combination between 2,4-D and NAA. In treatment A5, there was no sign of embryogenesis; in A6, embryos appeared, not too much, but the embryos were bright white, homogeneous, and larger than those in the 2,4-D and NAA combination treatments. Therefore, we combine these two regulators in the next experiment to evaluate the ability to induce embryogenesis on Ngoc Linh ginseng.

When combining 1.0 mg/L 2,4-D with 0.2 mg/L KIN and NAA at different concentrations, the results obtained were relatively good and uniform at all treatments. Specifically, in treatment A9, the rate of embryo production was 33.3%, treatment A10 was 53.3%, treatment A11 reached the highest rate of 80%, and treatment A12 decreased to 60% (Fig. 8.5). The number of embryos formed on the explant was also higher than those experiments that only combined 2,4-D with NAA or kinetin and was highest in treatment A11. The embryo has spherical, heart-

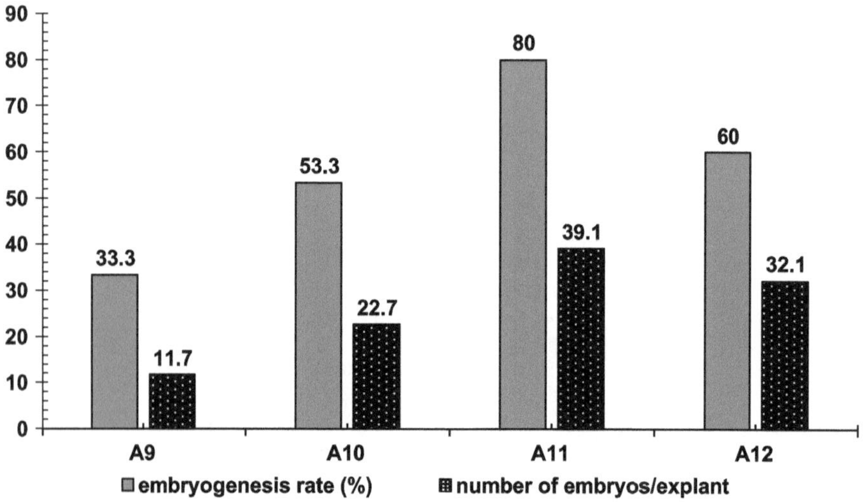

Fig. 8.5 Influence of 2,4-D combined with NAA and KIN on the ability to Ngoc Linh ginseng somatic embryogenesis. (**a9**, **a10**, **a11**, **a12**) 2,4-D (1.0 mg/L) in combination with KIN (0.2 mg/L) and NAA (0.1, 0.2, 0.5 and 1.0 mg/L)

shaped, mine-shaped, two-cotyledons. The embryo is bright yellow, large and uniform, and does not show any abnormalities; some explants have shoot formation.

In plant cell tissue culture, morphogenesis is influenced by different components in the culture medium, especially the concentration of PGR; thus, assessing their effects on plant regeneration is very important. The somatic embryogenesis from a callus in cytokinin-containing medium with auxin has been reported in *Cereals* (Bhaskaran and Smith 1990), *Trifolium pratense* (Haliloglu 2006) have shown the possibility of somatic embryogenesis from the culture of cassava leaf samples in medium combined of auxin and cytokinin. Thus, the combination of 2,4-D or NAA with a cytokinin is believed to be essential to induce embryogenesis in some plants. The majority of somatic embryogenesis is stimulated by the use of auxin and cytokinin alone or in combination in the medium; not all combinations of cytokinin and auxin lead to somatic embryogenesis. In this experiment, we use 2,4-D and NAA in combination with KIN for the highest efficiency of embryogenesis in treatment A11. This result is consistent with those reported in *Pinus taeda* (Vanildo et al. 2004) and Cotton (Hamidou et al. 2005).

8.3.4 Effects of Spermidine on Increasing the Frequency Somatic Embryogenesis

Ngoc Linh ginseng callus was cultured on embryonic induction medium (EM) supplemented with spermidine (polyamine group, concentration 0.01, 0.05, 0.1, 0.2 mM) (Sigma-Aldrich, Germany). Carbohydrate sources (sucrose, glucose, and fructose) were also added to embryonic induction media at different concentrations (10, 20, 30, 40, 50, 60 g/L) for investigating ability increased frequency of somatic embryogenesis in Ngoc Linh ginseng. The criteria such as rate of embryogenesis (%) and number of embryos per explant were recorded after 12 weeks of culture.

A polyamine is a form of amino acid and it is considered to be a growth regulator. Therefore, we used spermidine (a polyamine) in the experiment to investigating the ability of embryogenesis from callus. The results showed that low concentration of spermidine was added to the medium; callus was induced rapidly after 4 weeks of culture. On the surface of the callus, small spherical embryos appeared and the number of embryos increased gradually. After 12 weeks of incubation, the number of embryos appeared much on the surface, having all shapes such as spherical embryo, heart shape, mines, and bi-cotyledon. The embryo was bright white and did not appear currently mutated embryo. As the spermidine concentration gradually increased, the indicators were also increased and spermidine supplementation with 0.1 mM concentration showed the highest efficiency on embryogenesis rate (93.3%) and the number of embryos formed on the explant (Figs. 8.6 and 8.9b2, f1, g1). Continuing to increase spermidine concentration to 0.2 mM, the rate of embryo formation and the number of embryos per explant decreased. This result is similar to that of Kevers et al. (2000) on *Panax ginseng* species. But on coffee trees, the

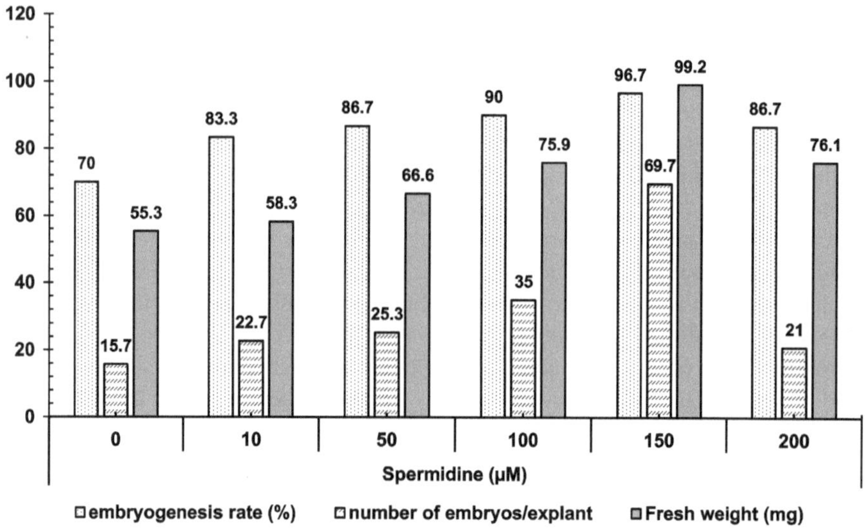

Fig. 8.6 Influence of spermidine (Spd) on the ability to Ngoc Linh ginseng somatic embryogenesis

appropriate concentration is 0.05 mM, and the appropriate concentration on *Momordica charantia* L. is 0.001 mM (Ananya et al. 2009).

Polyamine plays an important role in cell differentiation for embryo formation (Rajam 1997). However, its mechanism is not clearly understood and is still being studied. In high concentrations, polyamines help cells grow and divide. During the induction stage of embryos, polyamine helps cells to divide rapidly and embryos to form. A decreased concentration of polyamine can increase callus but decrease embryogenesis. Therefore, the use of polyamine in combination with other growth regulators in the culture medium is effective and meaningful in the embryogenesis stage as well as the complete plant formation stages (Kevers et al. 2000).

Spermidine is a specific polyamine used for the somatic embryogenesis of carrot (Feirer et al. 1985) and during this phase of *Alfalfa* explant (Cvikrová et al. 1999). Monteiro et al. (2002) also confirmed the role of polyamine, especially spermidine in *Panax ginseng* somatic embryogenesis. Studies on the embryogenesis potential of *Panax ginseng* on callus when using spermidine showed relatively good efficacy (Kevers et al. 2000). The use of polyamine in combination with auxin or cytokinin helps to increase the frequency of embryogenesis. In this study, we have documented the role of spermidine in improving the frequency of somatic embryogenesis in Ngoc Linh ginseng when used at a concentration of 0.1 mM in combination with 1.0 mg/L 2,4- D, 0.5 mg/L NAA and 0.2 mg/L KIN.

8.3.5 Effects of Carbohydrate Sources on Increasing of Somatic Embryogenesis Frequency

Plant cells, tissues, or organelles normally require a supply of carbohydrates to satisfy the energy requirements. In micropropagation, carbohydrates act as osmotic agents and assist in cell growth. The main source of carbon added in the culture medium is sugar; plant cells will use this sugar source to create a carbon framework for growth and organogenesis. Sugar is also a determinant of the plant's ability to initiate organs.

Carbohydrate composition in culture media has been shown to influence somatic embryogenesis in many plant species (Lou and Kako 1995). The results in this experiment showed that three types of sugar (sucrose, glucose, and fructose) were used to have the ability to induce embryogenesis. However, the rate of embryogenesis using sucrose was the highest, while glucose and fructose gave the lower result (Fig. 8.7). According to the number of embryos formed on the explant, sucrose also showed the highest result (167 embryos), glucose reached only 75 embryos, the lowest was fructose (35 embryos) (Fig. 8.8). The rate of embryogenesis and the number of embryos formed increased gradually with the increase in sugar concentration and the highest results were at a concentration of 50 g/L, then decreased gradually with a continued increase in concentration.

In this study, an increase in sugar concentrations promoted an increase in pre-embryonic and embryo clusters, which is similar to studies that have been reported in many other species (Luo et al. 1996). Callus or organs of plant require the incorporation of the carbon source in the culture medium and sucrose is used as

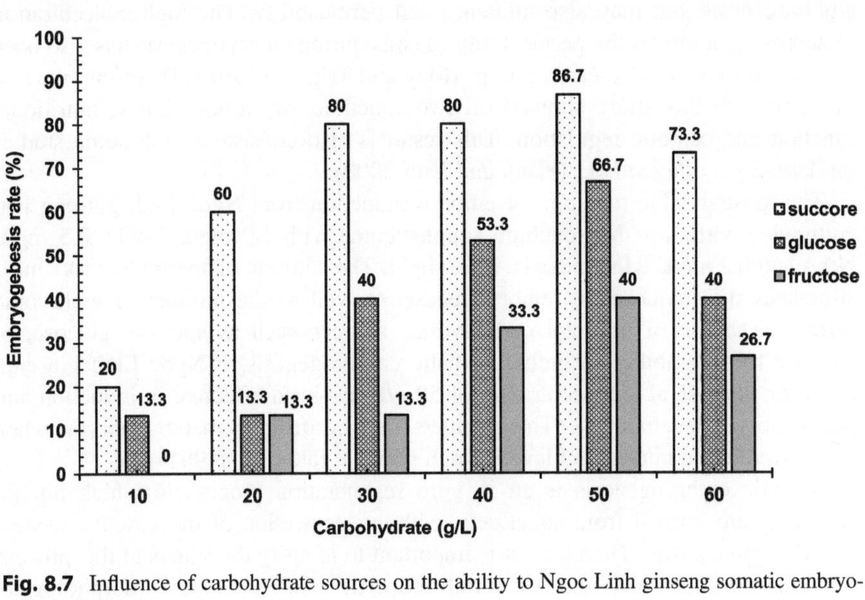

Fig. 8.7 Influence of carbohydrate sources on the ability to Ngoc Linh ginseng somatic embryogenesis frequency

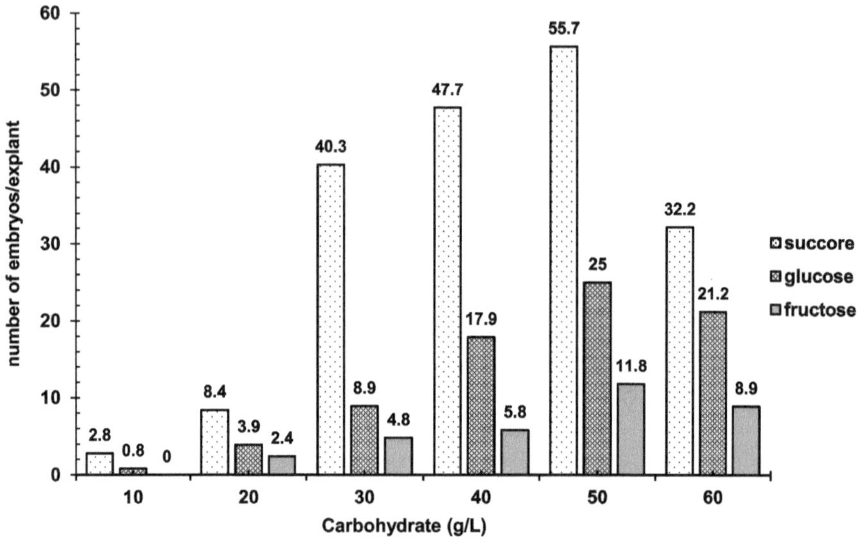

Fig. 8.8 Influence of carbohydrate sources on the number of embryos Ngoc Linh ginseng formed

the primary carbon source in tissue culture (Fuentes et al. 2000). Although sucrose is used as the primary carbon source in somatic embryogenesis, there are a few studies that have demonstrated the need for a high carbohydrate concentration, which serves as not only a source of nutrients but also osmotic conditioning properties.

Thus, high sugar concentrations not only play an important role in somatic embryogenesis, but may also influence cell permeability. The high concentration of sucrose that affects the permeability of cells during embryogenesis has also been mentioned by several research groups (May and Trigiano 1991). Therefore, the role of sucrose in this study showed on two functions of carbohydrates: nutritional function and osmotic regulation. This result is also consistent with some studies on *Asparagus officinalis* L. (Kanji and Yuji 2000).

The most suitable medium for embryo induction from Ngoc Linh ginseng leaf cultured in vitro was MS medium supplemented with 1.0 mg/L 2,4-D, 0.5 mg/L NAA and 0.2 mg/L KIN under 16-h/day light. The addition of spermidine (0.1 mM) stimulates the frequency of embryogenesis as well as the number of embryonic formed. The use of different carbohydrate sources such as sucrose, glucose, or fructose has the ability to induce somatic embryogenesis in Ngoc Linh ginseng; however, sucrose at a concentration of 50 g/L has been effective in induction and best embryo development. The plantlets derived from somatic embryos when transferred to the nursery achieved a high survival rate (Fig. 8.9h1–h8).

Somatic embryogenesis is an in vitro regeneration process in which bipolar structures are formed from soma cells without connection of the vascular system to the original tissue. Therefore, it is important to identify the stages in this process through morphological and structural properties. In this study, morphological

Fig. 8.9 The process somatic embryogenesis of Ngoc Linh ginseng (*Panax vietnamensis* Ha *et* Grushv.). (**a1**, **a2**, **a3**) Callus was formed from leaf, petiole, and rhizome of Ngoc Linh ginseng in vitro after 8 weeks of culture; (**b1**, **b2**, **f1**, **g1**) Somatic embryos formed from callus at 0, 4, 8, and 12 weeks of culture on MS medium supplemented with 1.0 mg/L 2,4-D, 0.5 mg/L NAA, 0.2 mg/L KIN, and 0.1 mM Spd; (**c, e, g2**) Spherical embryo structure observed under fluorescence microscope; (**d**) Spherical embryo observed under scanning electron microscope; (**h1, h2, h3, h4, h5, h6, h7, h8**) Ngoc Linh ginseng plantlets are derived from somatic embryos in vitro

changes were noted by observing histological slices of embryos at all specific stages under fluorescence microscopy and scanning electron microscopy. The first morphological changes of the explant on MS medium supplemented with 1.0 mg/L 2,4-D, 0.5 mg/L NAA, and 0.2 mg/L KIN were noted after 7 days of incubation with changes in shape and size of some callus cells in the culture (Fig. 8.9b1). Callus cells are large in size, have many different shapes, arranged loosely, and divided in many different directions. On the surface of callus mass there appear spherical structures (Fig. 8.9a1–a3). These structures have a darker yellow color than the surrounding callus mass. A cluster of calli containing these embryonic spherical structures is called pro-embryogenic mass (PEM). These PEM is usually located at the periphery of the callus mass and scattered with large vacuole callus cells (Fig. 8.9d).

It is difficult to differentiate embryogenic and non-embryogenic cells in the PEM. A double staining method has been developed by Gupta and Holmstrom (2005) to differentiate these two cell types. This method is called the double staining method because two dyes Acetocarmine and Evan's blue are used to dye cells. Embryogenic-capable cells have a large nucleation and concentrated cytoplasm-stained bright red with acetocarmine. The cytoplasmic constituents also have an affinity for acetocarmine and exhibit a bright red color. Acetocarmine is commonly used to detect glycoproteins, chromatin, and DNA in cell chemistry studies (Sharma et al. 1980). Suspension cells, derived from embryonic cells with a smaller nucleus, react with Evan's blue dye to differentiate with PEM. In contrast, cells that are incapable of embryogenesis have large vacuoles with small nuclei that allow Evan's blue dye to penetrate inside. The cells in the non-embryonic callus have a very small nucleation, so the material inside the cell-dyed red by acetocarmine is difficult to detect while the entire cell captures the green color of Evan's blue dye.

Cells in the PEM are small with large nucleus and dense cytoplasm. Over the next 7–14 days, these cells undergo an early embryogenesis process. The spherical embryo consists of cytoplasmic-rich small cells surrounded by a layer of epidermal cells. The somatic embryo grows gradually by the cell division, leading to the formation of an early heart-shaped embryo. The elongation of the embryo axis in the late heart-shaped stage shows that the embryo has a bipolar structure. The growth in the height of the cotyledons surrounding the embryo axis produces a mine-shaped embryo and the embryo stage of the cotyledons.

Starch particles present during the formation of the embryonic region and the subsequent development of the somatic embryos were determined by staining with IKI. Starch is considered the main source of energy for cell proliferation and growth. Starch is rapidly utilized during the formation of the embryonic regions and then gradually decreased during the spherical and heart-shaped embryo stages. However, starch accumulation was noted again in embryos at later stages, especially in the cotyledon region. This model of starch accumulation and use has been previously reported by Arnold (1987), Hartweck et al. (1988), and Barciela and Vieitez (1993) in systems of organogenesis and embryogenesis. The micropropagation of *Panax vietnamensis* Ha *et* Grushv. through somatic embryogenesis is shown in Fig. 8.10.

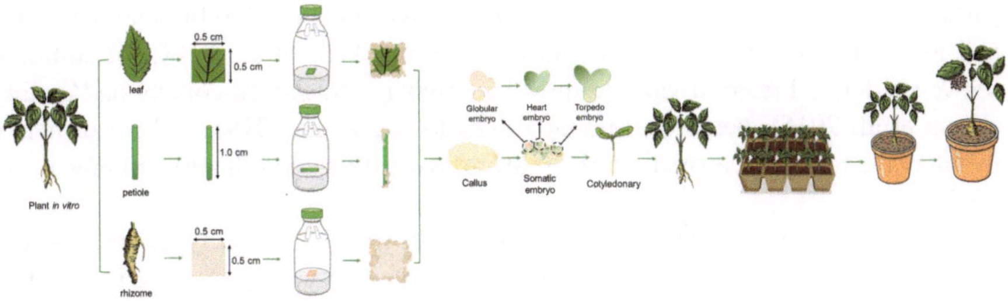

Fig. 8.10 The micropropagation of *Panax vietnamensis* Ha *et* Grushv. through somatic embryogenesis

8.4 Somatic Embryogenesis in *Kappaphycus striatus*

8.4.1 General Introduction

Kappaphycus striatus belongs to the phylum Rhodophyta, distributed mainly in tropical seas, in coastal open seas and bays, where there are water exchange, stable high salinity, clear water, high light intensity, etc. (Trono 1992). *K. striatus* not only is rich in crude fiber, iron, mega-3 fatty acids (Adharini et al. 2018), and antioxidants, but also contains biological compounds that serve the pharmaceutical and biological industries (Hayashi and Reis 2012). Therefore, *K. striatus* has economic value and is used as a raw material for Kappa-carrageenan extraction, agricultural fertilizer (Shah et al. 2013; Pramanick et al. 2013), and bioethanol (Meinita et al. 2012; Khambhaty et al. 2012; Hargreaves et al. 2013).

Propagation by vegetative reproduction is the only method used in wild culture. Seaweed is propagated by vegetative reproduction for many years without selection, which has led to a decline in the vitality of the seaweed; moreover, under the influence of ecological conditions, the seaweed has changed in biological characteristics such as growth rate, content, and quality of carrageenan. Especially, in recent years, the effects of climate change (change in temperature, sea level rise, salinity, high light intensity, nutrient source, etc.) make seaweed very susceptible to disease outbreaks, leading to a significant reduction in yield and quality of seaweed (Vairappan et al. 2014). Therefore, seaweed is degraded after a long time propagated by vegetative reproduction.

In vitro culture is a propagation method that is less dependent on weather, meets a large number of seaweed varieties, and is a way to create a source of disease-free, high-yielding, and well-developed seaweed in nutrient-poor conditions. It has been considered as an alternative method to produce disease-free seaweed for sustainable cultivation and provides raw materials for various industries including food production. Seaweed derived from in vitro cultures also had lower concentrations of Arsenic, Cadmium, and Lead compared with seaweeds of natural origin (Yong et al. 2017). In addition, the growth rate of seaweeds derived from in vitro culture is 1.5–1.8 times faster than that of traditional varieties (Reddy et al. 2003). In particular, seaweed has the ability to grow well when exposed to high temperatures and is resistant to pathogens that cause white rot (Azizi et al. 2018). In addition, seaweeds derived from in vitro cultures had higher carrageenan content and quality (Yong et al. 2015), better nutritional values (Yong et al. 2014), and lower heavy metal content than seaweeds derived from vegetative reproduction (Yong et al. 2017).

This chapter focuses on studying the effect of media culture, type, and concentration of PGRs on somatic embryogenesis of *K. striatus*, a phylogenetic pattern that has not been described yet previously in this species.

8.4.2 Callus-Derived Somatic Embryogenesis in K. striatus

8.4.2.1 Culture Media

In in vitro culture, agar is commonly used as a coagulant. This is a type of polysaccharide extracted from seaweed, added to the culture medium to solidify the medium to avoid the phenomenon of tissue sinking in the medium or dying from lack of oxygen in liquid medium.

The 16-week-old callus cluster (2 mm × 2 mm) and 10 mg fresh weight of *K. striatus* were placed in a 100 mL glass flask containing 40 mL of solid PES medium (15 g/L agar), or semi-solid PES medium (4 g/L agar), or liquid PES medium to study the effect of culture medium on the ability to callus-derived somatic embryogenesis of *K.* striatus. The experiment was repeated three times with ten glass flasks.

The ability of somatic embryogenesis on PES media showed significant differences. After 1 week of culture, callus clusters grown in liquid PES medium broke off, then necrosis. In solid and semi-solid PES media, callus clusters continued to grow, increasing in size. After 8 weeks of culture, the results are shown in Table 8.1 and Fig. 8.11.

The solidity of the medium has an effect on the ability to somatic embryogenesis. In solid PES medium, callus cells continued to divide (Fig. 8.11a) and callus clusters grew larger (82.33 mg fresh weight and 7.90 mg dry weight) (Table 8.1). In the semi-solid PES medium, the callus mass grew rapidly and the callus fibers had numerous pigmented callus (Fig. 8.11b, c). In addition, the number of somatic embryos (15.67 somatic embryos), fresh weight (135.67 mg) and dry weight (14.50 mg) on semi-solid PES medium were higher than those on solid PES medium after 8 weeks of culture. In the liquid PES medium, the callus was completely submerged in water, after a period of cell clusters disjointed and necrosis (Fig. 8.11d).The multicellular form called the somatic embryo was crescent-shaped (Fig. 8.15h), spherical (Fig. 8.15i) or shoe-shaped (Fig. 8.15j) and it was large in size (150–500 μm).

The ability to somatic embryogenesis depends on the solidity of the medium and depends on the species (Yokoya et al. 2004). The results of this study are similar to the study of Reddy et al. (2003) in cartilage algae; callus clusters transferred to the medium for 2 months continued to grow into callus fibers and were able to maintain growth in 2 years if they continue to be transferred to a new medium. Callus cultured

Table 8.1 Effect of culture medium on callus-derived somatic embryogenesis of *K. striatus* after 8 weeks of culture

Culture media	No. of somatic embryos per explant	Fresh weight (mg)	Dry weight (mg)	Morphology
Solid PES	$0.00 \pm 0.00^{b*}$	82.33 ± 6.43^{b}	$7.90^{b} \pm 0.36$	Callus cells continued to divide
Semi-solid PES	15.67^{a}	$135.67^{a} \pm 5.03$	$14.50^{a} \pm 0.50$	Callus fibers develop densely, many cells contain pigment in each callus fiber, the surface of callus clusters is brown after 2 weeks of culture, somatic embryogenesis
Liquid PES	–	–	–	Callus clusters broken off, white, necrosis after 1 week of culture

*Note:*Different letters (a, b, c, etc.) in the same column represent statistical differences with Duncan's test (with $P < 0.05$)*

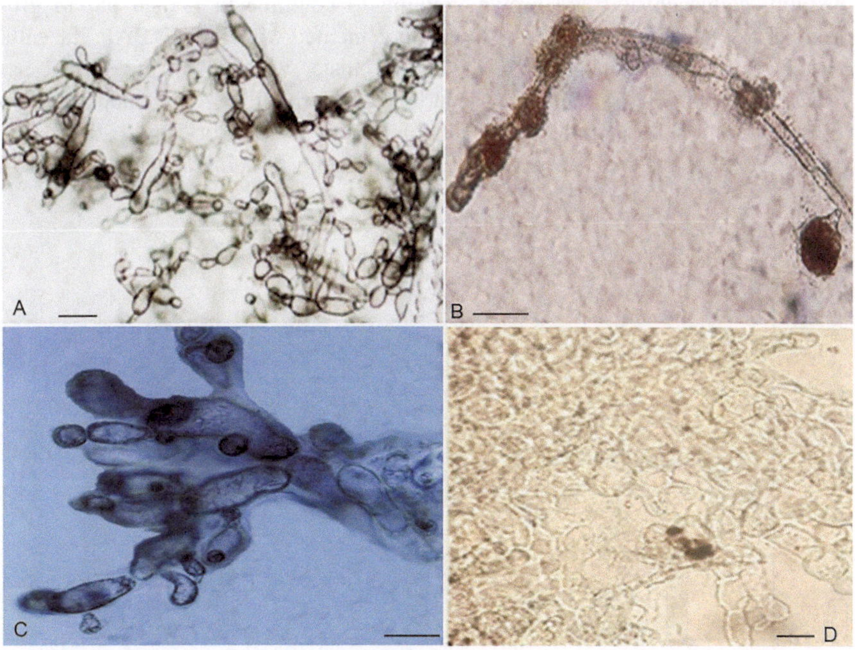

Fig. 8.11 Anatomical morphology of callus-derived somatic embryogenesis of *K. striatus* in different culture media (*Bars: 100 µm*). (**a**) Solid PES medium; (**b, c**) Semi-solid PES medium; (**d**) Liquid PES medium

in semi-solid PES medium enhanced cell division and stimulated callus to differentiate into somatic embryos in *K. alvarezii* (Reddy et al. 2003) and seaweeds (Polne-Fuller and Gibor 1987). However, some studies have reported that callus enhanced differentiation into somatic embryos in the solid medium (Yeong et al. 2014) or in the liquid medium on *Solieria filiformis* (Yokoya and Handro 2002). In this study, callus-derived somatic embryogenesis of *K. striatus* in a semi-solid medium (4 g/L agar) was optimal after 8 weeks of culture.

8.4.2.2 Combination of NAA and BAP

The 16-week-old callus cluster (2 mm × 2 mm) around 10 mg fresh weight of *K. striatus* was placed in a 100 mL glass flask containing 40 mL of semi-solid PES medium (4 g/L agar supplemented with NAA (1; 2; 3 mg/L) or BAP (1; 2; 3 mg/L)) alone or in a combination to determine the appropriate type and concentration of PGRs on the ability to somatic embryogenesis of *K. striatus*. The experiment was repeated three times with ten glass flasks.

On semi-solid PES medium without NAA and BAP, the biomass (fresh weight and dry weight) and the number of somatic embryos per explant (20 somatic embryos) of *K. striatus* were low after 8 weeks of culture (Fig. 8.12B). On this medium, callus clusters are brown, tending to continue to branch out to grow (Fig. 8.14). On semi-solid PES medium supplemented with 1 mg/L BAP, the callus cluster proliferated quite quickly, the tissue cluster was brown, the surface saw growing callus fibers, and the anatomical morphology contained embryos (Figs. 8.12 and 8.14c). In the treatments of 1–2 mg/L NAA or 2 mg/L BAP, there were some explants with necrosis, slow proliferation of callus clusters (Fig. 8.14)

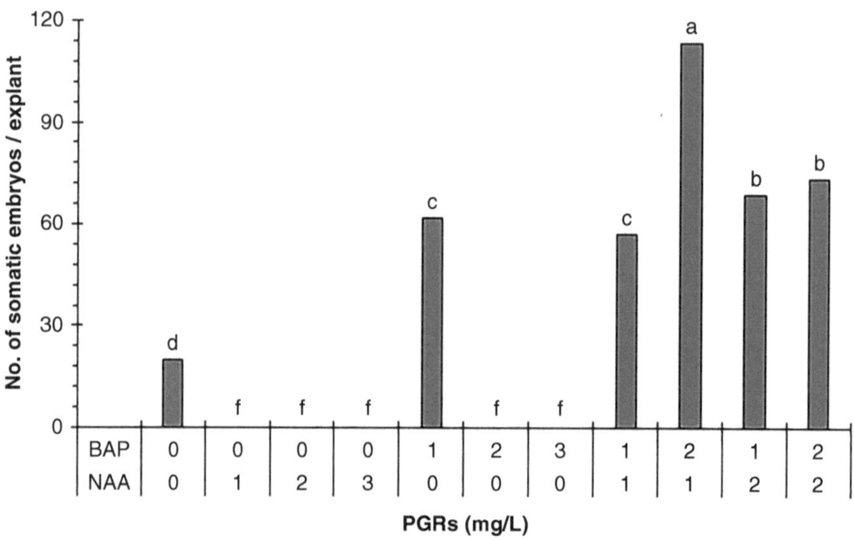

Fig. 8.12 Effect of NAA and BAP only or in combination on the number of somatic embryos per explant of *K. striatus* after 8 weeks of culture

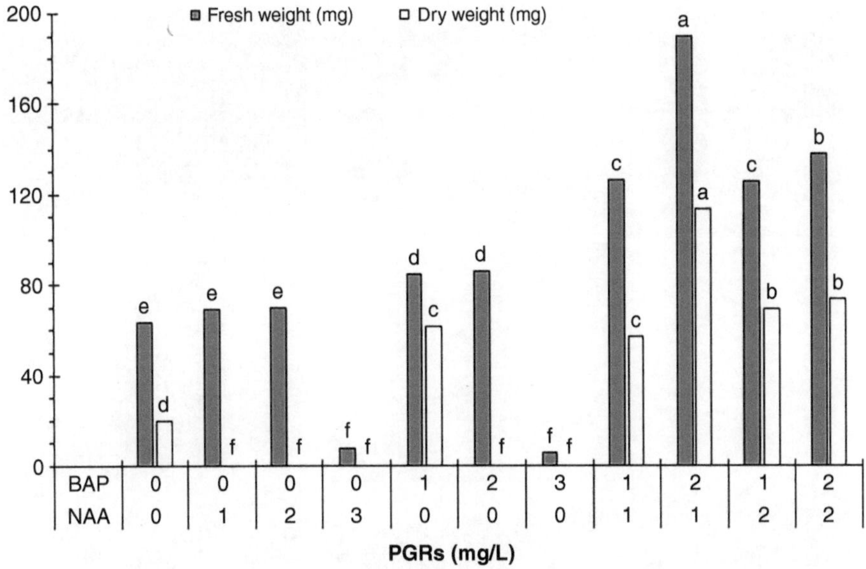

Fig. 8.13 Effect of NAA and BAP only or in combination on mass of *K. striatus* after 8 weeks of culture

and milky color, somatic embryogenesis not recognized. On the 3 mg/L NAA and 3 mg/L BAP treatments, callus underdevelopment (Fig. 8.12), necrosis, and somatic embryogenesis were not observed (Fig. 8.14d).

The combination of NAA (1–2 mg/L) and BAP (1–2 mg/L) increased somatic embryogenesis more than either NAA or BA treatment alone (Figs. 8.12 and 8.14e–i). The treatment of 1 mg/L NAA combined with 2 mg/L BAP recorded a rapidly proliferating callus cluster, the number of somatic embryos per explant (113.33 somatic embryos/explants), fresh weight (189.67 mg), and the dry weight (18.10 mg) was highest compared with other treatments after 8 weeks of culture (Figs. 8.13 and 8.14f), whereas the combined NAA and BAP treatments showed the possibility of somatic embryogenesis; however, the somatic embryo size was small (100–200 μm) (Figs. 8.12 and 8.14g–i).

Factors affecting somatic embryogenesis were diverse, but mainly PGRs and other factors. The concentration and type of PGRs have an effect on the morphological response of somatic embryos. The presence of NAA and BAP in the culture medium showed a marked effect on cell growth and differentiation (Reddy et al. 2003).

In higher plants, preembryonic cells are produced from callus cells with large cell spaces when stimulated by 2,4-D, cytokinins, or some other stress. In seaweed, NAA and BAP supplementation stimulated cell division and somatic embryogenesis (Reddy et al. 2003). This result is similar to some studies on *K. alvarezii* (Reddy et al. 2003) and *Agardhiella subulata* (Cheney et al. 1987).

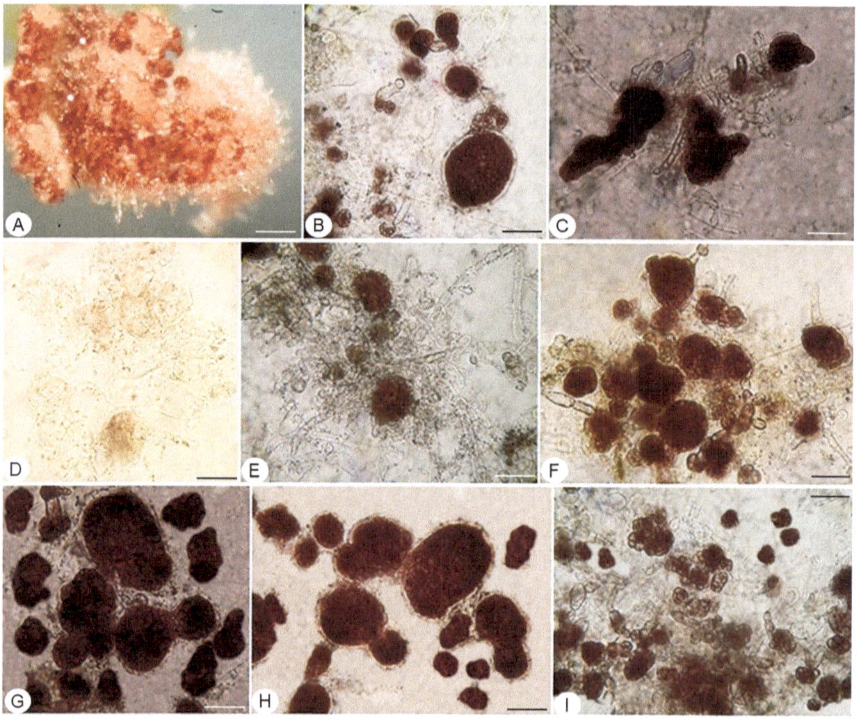

Fig. 8.14 Anatomical morphology of callus-derived somatic embryogenesis of *K. striatus* cultured on semi-solid PES medium supplemented with NAA and BAP only or in combination after 8 weeks of culture (*Bars: 100 μm*). (**a**) Original explant; (**b**) Without PGRs; (**c**) 1 mg/L BAP; (**d**) 3 mg/L BAP; (**e**) 1 mg/L NAA + 1 mg/L BAP; (**f**) 1 mg/L NAA + 2 mg/L BAP; (**g, h**) 2 mg/L NAA + 1 mg/L BAP; (**i**) 2 mg/L NAA + 2 mg/L BAP

8.4.2.3 The Process of Callus-Derived Somatic Embryogenesis of K. striatus

The results of morphological anatomy and observation under electron microscopy showed that the developmental shapes of somatic embryos obtained from the treatments had no morphological differences. After 2–3 weeks of culture, callus samples inoculated into semi-solid PES medium (4 g/L agar) supplemented with NAA and BAP alone or in combination developed into embryogenic cell clusters and dark brown (Fig. 8.15). After 4 weeks of culture, anatomical morphology showed dense pigmented callus (Fig. 8.15b).

Initially, pre-embryonic cells were formed at the wall of two callus cells. These cells were initially very small, brownish in color; each septum usually forms two symmetrical cells on either side, then grow larger, reaching a size of about 50–100 μm. These cells had nuclei, brown color (Fig. 8.15c), and then appeared a cytoplasmic division septum, dividing the old cell into two daughter cells after 5 weeks of culture (Fig. 8.15d). These cells continued to grow and then divide into four cells after 6 weeks of culture, then into eight cells after 7 weeks of culture

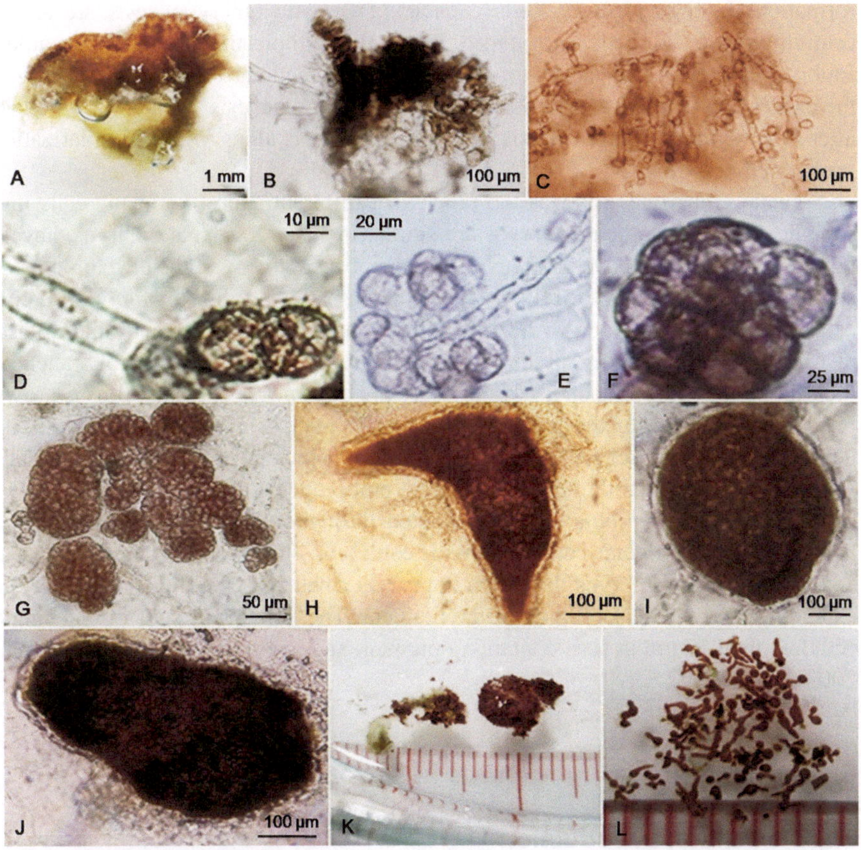

Fig. 8.15 Callus-derived somatic embryogenesis of *K. striatus*. (**a**) Embryogenic callus; (**b**) Callus contains pigment; (**c**) Callus fibers carry pigment-containing callus cells; (**d**) Cells duplicate; (**e**) Callus fibers carrying clusters of cells that are tripling; (**f**, **g**) Multicellular cluster; (**h**) Rescent-shaped somatic embryos form shoot poles; (**i**) Spherical somatic embryo; (**j**) Shoe sole-shaped somatic embryo; (**k**) Cluster of somatic embryos; (**l**) Discrete somatic embryos

(Figs. 8.15e, f) and multicellular after 8 weeks of culture (Fig. 8.15g). The multicellular form called the somatic embryo was crescent-shaped (Fig. 8.15h), spherical (Fig. 8.15i) or shoe-shaped (Fig. 8.15j) and it was large in size (150–500 µm).

Observation under the microscope shows that the cluster of somatic embryos is brown, the small embryos are stacked but essentially separate (Fig. 8.15k). These clusters of somatic embryos cultured in liquid shaking condition separated from each other (size 0.5–0.6 mm) and continued to develop. The shoot pole of the somatic embryo will germinate, and the first sprout will appear (Fig. 8.15l). At this location the sprout can appear from one-to-many branches.

Somatic embryos contain nutrients similar to sexual embryos. Somatic tissues and cells directly give rise to somatic embryos through intermediate callus formation. In higher plants, cells capable of somatic embryogenesis are small, centrally nucleated cells surrounded by dense cytoplasm and small vacuoles. Embryogenic cells that meet intracellular and extracellular conditions that permit the expression of embryogenesis will develop embryogenesis, which occurs either after or concurrently with meiosis. Subsequent mitosis along with cell polarization leads to embryogenesis.

Somatic embryogenesis was the process of differentiation of somatic cells into somatic embryos. Somatic embryos are similar in shape to sexual embryos, have a bipolar structure, and have all the parts of an embryo. However, the development of somatic and sexual embryos follows two different pathways. In *Kappaphycus*, callus cells are 10–15 μm wide 20–200 μm long, containing vacuoles and chloroplasts distributed throughout the cell. Initially, the callus cells are fibrous but not branched; the callus at the fibrous apex is usually short and oval in shape. If the callus is grown in a semi-solid medium, the callus goes through two stages. First, the callus fibers are densely packed with cells, pigment-containing cells located at the apex of the callus fibers. The cells containing this pigment are embryogenic cells arranged in clusters of dark red-brown globular cells distributed throughout the surface of the callus mass. These embryogenic cells then divide continuously to produce spherical or oval embryos, somatic embryos that often appear as small buds on the surface of cells extending near the apical region of the filament. These embryos start from a single cell, less than 10 μm in size, containing pigment and dense cytoplasm (Reddy et al. 2003).

8.5 Somatic Embryogenesis in *Jatropha curcas* L.

8.5.1 General Introduction

Jatropha curcas L., a shrub plant, belongs to the family Euphorbiaceae including 8000 species. Oilseeds (30–40%) and crude oil from seeds can be processed into biodiesel and many other valuable products (organic fertilizers, bio-pesticide, pharmaceuticals, etc.). In recent years, many countries around the world have paid much attention to the research and development of *J. curcas*, especially the seed quality, and the planting of leading varieties to get raw materials for biodiesel production (Rajore and Batra 2005).

The direct seeding method is a simple method that has been applied on this plant; however, the survival and germination rate of seed are poor, and the yield and oil content is unstable. Meanwhile, plants propagated by cuttings have a shorter lifespan and are less resistant to drought and disease than those propagated by seeds. Besides, micropropagation has also been performed to rapidly propagate *J. curcas*; plantlets were induced from in vitro culture of various plant parts such as axillary buds, shoot tips, nodes and leaves (Sujatha and Mukta 1996; Sardana et al. 1998; Rajore and Batra 2005; Sujatha et al. 2005).

There are a few studies on somatic embryogenesis that have also been successfully applied to *J. curcas* (Jha et al. 2007; Siang et al. 2012). However, to date there is no study on the increase of somatic embryo formation frequency reported yet.

8.5.2 Callus Induction from Leaf Transverse Thin Cell Layers

The leaves of 6-month-old *J. curcas* plants were collected and washed under running tap water for 30 min, then leaf surfaces were sterilized in 70% (v/v) alcohol for 30 s followed by three rinses with sterilized distilled water. Then, leaves were sterilized in NaOCl solution (50%, v/v) with two drops of Tween-20 for each 100 mL solution for 10 min and rinsed four to five times in sterilized distilled water in a laminar air flow chamber. The disinfected leaf was cut into tTCLs (0.5 × 10 mm) and cultured on MS medium supplemented with kinetin (KIN) (0.5, 1.0, 1.5, and 2.0 mg/L) combined with indole-3-butyric acid (IBA) (0.1, 0.5, and 1.0 mg/L) or 2,4-dichlorophenoxyacetic acid (2,4-D) (1.0, 1.5, and 2.0 mg/L). The callus induction was selected after 4 weeks of culture.

In this study, the leaf tTCLs cultured on MS medium supplemented with KIN showed swelling after 5–7 days of culture. On PGR-free medium, leaf tTCLs also swelled and then slowly turned to brownish and necrotized. The combination between KIN and IBA obtained callus induction, which was compact and bright in color (yellow or green) (Fig. 8.16). The callus formation rates on MS medium supplemented with KIN and IBA were lower than those on MS medium supplemented with KIN and 2,4-D, in which the soft, friable, and yellow calli were observed.

The embryogenic callus formation (many small cream nodules) in *J. curcas* from leaf tTCLs on medium MS supplemented with 2 mg/L KIN was obtained after 4 weeks of culture (Jha et al. 2007). In addition, cotyledon-derived embryogenic calli of *J. curcas* on medium containing dicamba were obtained (Siang et al. 2012). These results showed that callus induction rate (89.3%) was highest on medium

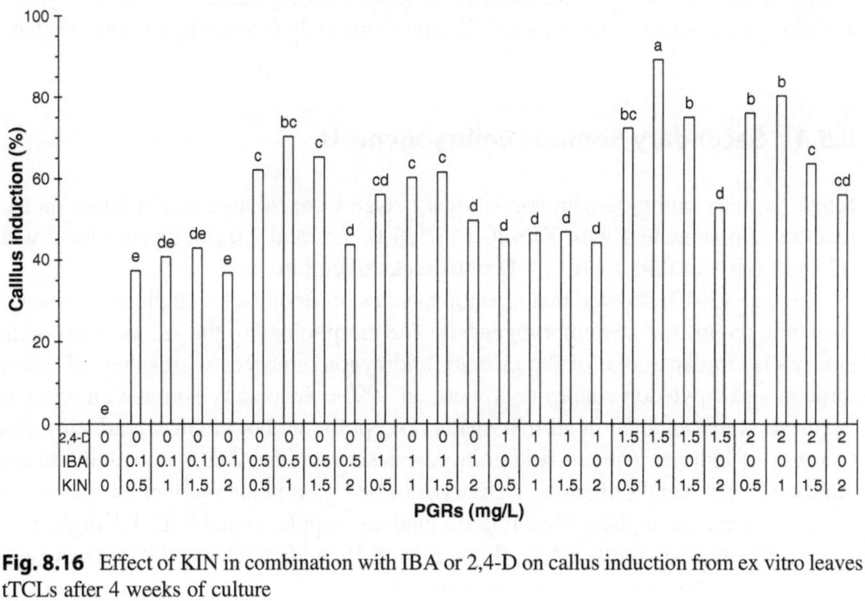

Fig. 8.16 Effect of KIN in combination with IBA or 2,4-D on callus induction from ex vitro leaves tTCLs after 4 weeks of culture

supplemented with 1.0 mg/L KIN and 1.5 mg/L 2,4-D (Fig. 8.16). Callus clusters were soft and friable with light cream color and showed the somatic embryogenesis. During the somatic embryogenesis process, the tissue integrity was disrupted (Smith and Krikorian 1989). Merkle et al. (1995) showed that the separation of cells made soft and friable calli leading to the cutin layer was formed and transformed into somatic embryos.

8.5.3 Primary Somatic Embryogenesis via Embryogenic Callus

Embryogenic callus clumps were cut into square (1.0 × 1.0 cm) and transferred to MS medium containing 2,4-D (0.01, 0.03, 0.05, and 0.07 mg/L) or KIN (0.5, 1.0, 1.5, and 2.0 mg/L) in order to obtain somatic embryogenesis after 4 weeks of culture.

In this study, the highest of somatic embryogenesis rate (76.67%), number of embryos (39.70 embryos), and fresh weight of embryos clusters (0.058 g) were achieved on medium supplemented with only 1.0 mg/L KIN (Fig. 8.17). The results of this study differ from those reported by Jha et al. (2007); the highest somatic embryogenesis rate and number of embryos were obtained on medium supplemented with 0.5 mg/L KIN in combination with 0.25 mg/L IBA.

Cytokinins have been shown to be essential for somatic embryogenesis; however, each plant responds differently. KIN is effective in somatic embryogenesis in some woody species (Dunstan et al. 1995). Single cytokinins used to generate somatic embryos from immature zygote embryos have also been successful in some species of both angiosperms and gymnosperms. In addition, cytokinins also stimulate somatic embryogenesis from non-embryogenic calli in other species such as peanut (Gill and Saxena 1992). Kalimuthu et al. (2007) showed that somatic embryos of *J. curcas* were also recorded on media containing only 6-benzylaminopurine (BA).

8.5.4 Secondary Somatic Embryogenesis

Single primary embryos with cotyledonary stage were isolated and cultured on MS medium supplemented with KIN (0.1, 0.5, 1.0, 1.5, and 2.0 mg/L) combined with 0.2 mg/L IBA or 0.05 mg/L 2,4-D in 6 weeks of culture.

Shi et al. (2010) showed that repeating cycles of secondary somatic embryogenesis would record somatic embryogenesis. The morphology of the secondary somatic embryo is similar to that of the primary and zygote embryos with spherical, heart, torpedo and cotyledon shapes (Zimmerman 1993). Secondary somatic embryos of *J. curcas* were obtained in all treatments (except the without PGRs treatment) after 6 weeks of culture. The somatic embryogenesis rate, the total number of somatic embryos, fresh weight of somatic embryos (0.087 g), and the number of embryos at all stages were the highest obtaining on medium supplemented with 1.5 mg/L KIN combined with 0.05 mg/L 2,4-D. (Figs. 8.18, 8.19, 8.20, 8.21 and 8.22). The results of this study provide an efficient approach for rapid multiplication of somatic

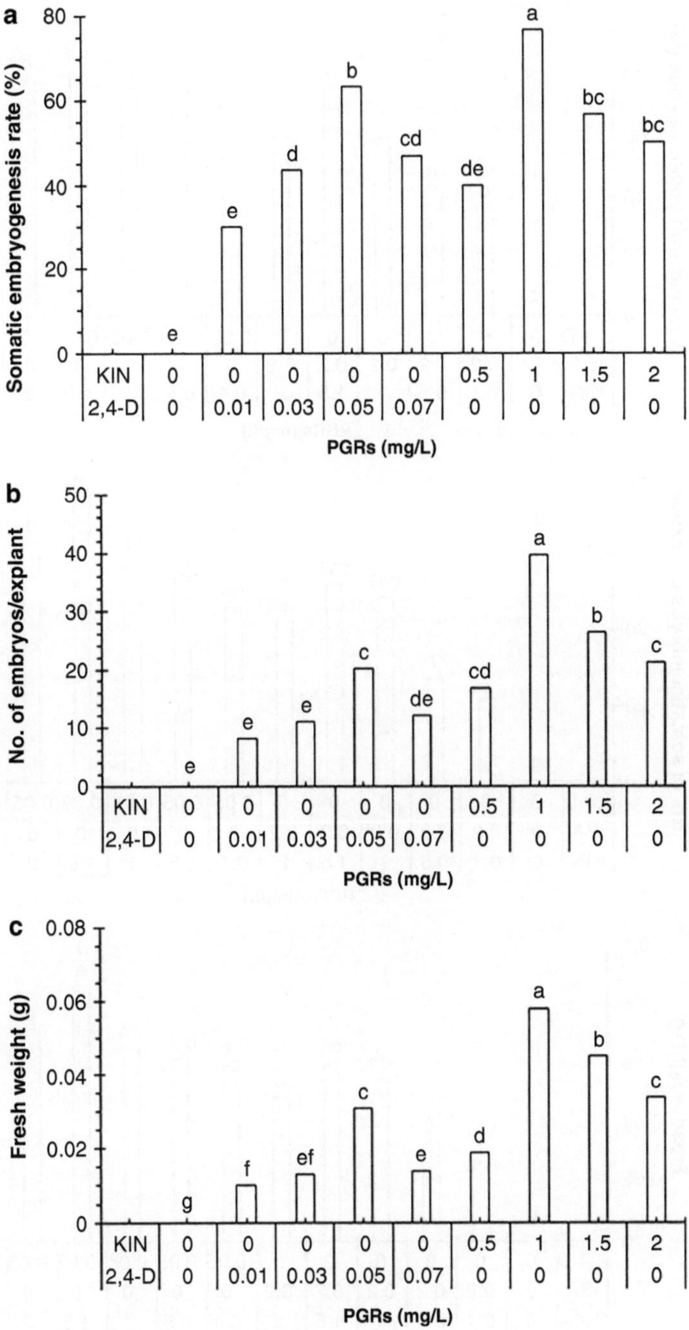

Fig. 8.17 Effect of 2,4-D and KIN on callus-derived somatic embryogenesis of *J. curcas* after 4 weeks of culture

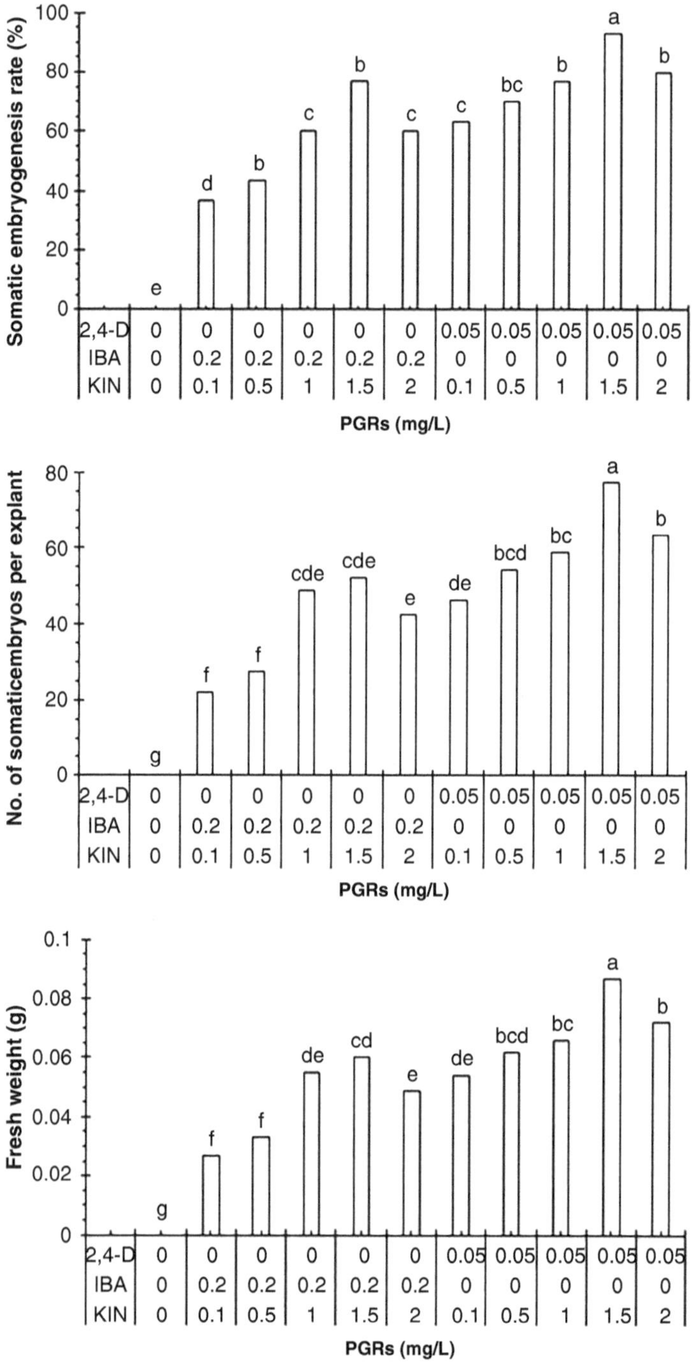

Fig. 8.18 Effect of KIN in combination with IBA or 2,4-D on secondary somatic embryogenesis of *J. curcas* after 6 weeks of culture

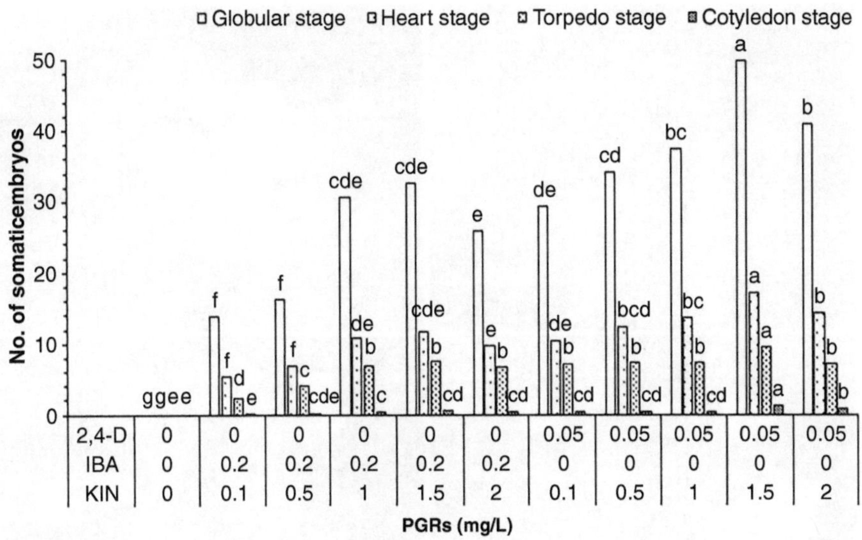

Fig. 8.19 Effect of KIN in combination with IBA or 2,4-D on numbers of secondary somatic embryos in all stages of *J. curcas* after 6 weeks of culture

embryos from primary somatic embryos in *J. curcas*. This secondary somatic embryogenesis was able to regenerate in vitro plants after 4 weeks of culture.

Secondary somatic embryogenesis from primary somatic embryos on medium supplemented with PGRs has been studied on a number of plants such as *Carthamus tinctorius* (Kumar and Kumari 2010), *Cinnamomum camphora* (Shi et al. 2010), and no secondary somatic embryos were observed when cultured on PGR-free medium (Singh and Chaturvedi 2009).

Until now, primary and secondary somatic embryogenesis has been recorded on *Cyclamen persicum* plant (Fig. 8.23). *C. persicum* plantlets regenerated from somatic embryos (Fig. 8.24).

8.6 Conclusion

This chapter showed the somatic embryogenesis in some plants such as *P. vietnamensis*, *K. striatus*, and *J. curcas*. Several factors affecting somatic embryogenesis have been investigated. The results of this study will open a new direction in micropropagation of these three plants and potential applications on other crops.

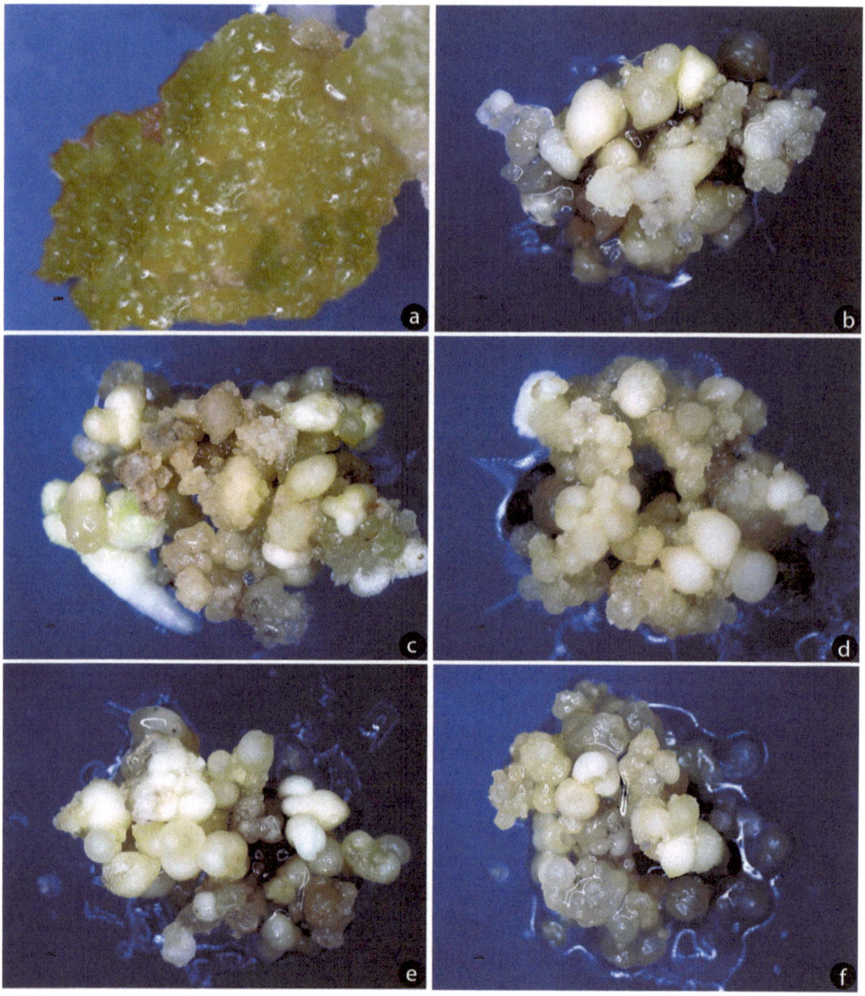

Fig. 8.20 Effect of KIN in combination with 0.2 mg/L IBA on secondary somatic embryogenesis of after *Jatropha curcas* L. after 6 weeks of culture. (**a**) Without PGRs; (**b**) 0.1 mg/L KIN combined with 0.2 mg/L IBA; (**c**) 0.5 mg/L KIN combined with 0.2 mg/L IBA; (**d**) 1.0 mg/L KIN combined with 0.2 mg/L IBA; (**e**) 1.5 mg/L KIN combined with 0.2 mg/L IBA; (**f**) 2.0 mg/L KIN combined with 0.2 mg/L IBA

Fig. 8.21 Effect of KIN in combination with 0.05 mg/L 2,4-D on secondary somatic embryogenesis of *J. curcas* after 6 weeks of culture. (**a**) Without PGRs; (**b**) 0.1 mg/L KIN combined with 0.05 mg/L 2,4-D; (**c**) 0.5 mg/L KIN combined with 0.05 mg/L 2,4-D; (**d**) 1.0 mg/L KIN combined with 0.05 mg/L 2,4-D; (**e**) 1.5 mg/L KIN combined with 0.05 mg/L 2,4-D; (**f**) 2.0 mg/L KIN combined with 0.05 mg/L 2,4-D

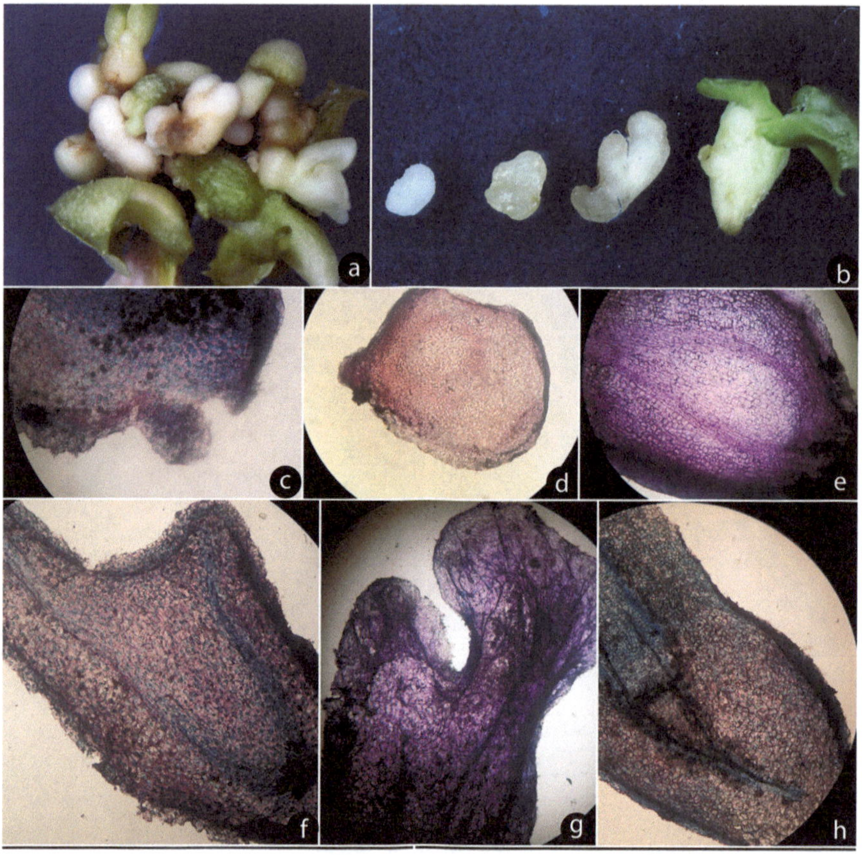

Fig. 8.22 Secondary somatic embryogenesis and in vitro plant regeneration of *J. curcas*. (**a**) Secondary somatic embryogenesis; (**b**) Spherical, heart, torpedo, and cotyledon shapes; (**c**) Primary somatic embryogenesis formed secondary somatic embryogenesis; (**d–h**) Anatomical morphology of the secondary somatic embryo

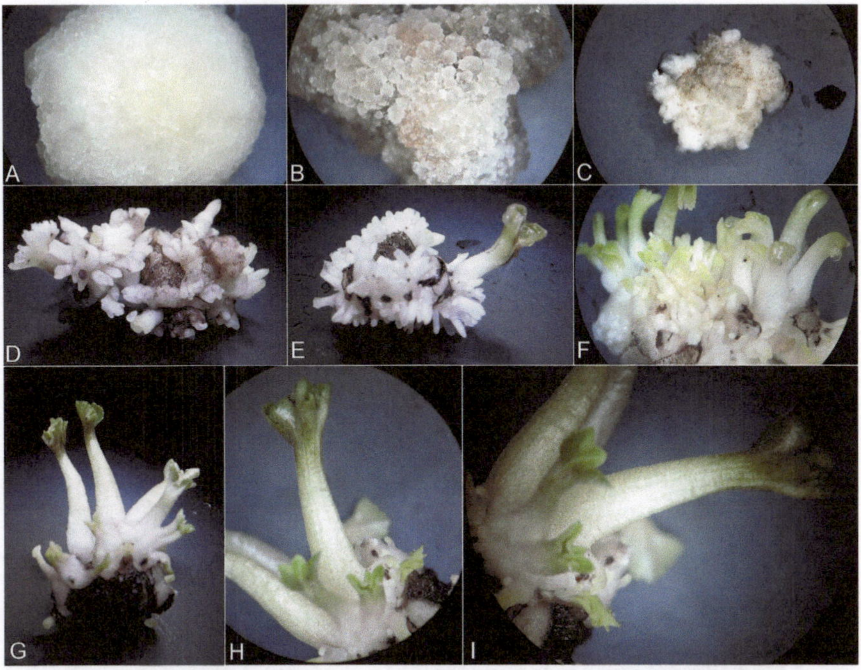

Fig. 8.23 Primary and secondary somatic embryogenesis of *C. persicum*. (**a**, **b**) Embryogenic callus; (**c**, **d**) Primary somatic embryogenesis; (**c**) Primary somatic embryogenesis formed secondary somatic embryogenesis; (**d–i**) Secondary somatic embryogenesis

Fig. 8.24 *C. persicum* plantlets regenerated from somatic embryos

References

Adharini R, Suyono E, Suadi S, Jayanti A, Setyawan A (2018) A comparison of nutritional values of *Kappaphycus alvarezii*, *Kappaphycus striatum*, and *Kappaphycus spinosum* from the farming sites in Gorontalo Province, Sulawesi, Indonesia. J Appl Phycol 31(1):725–730

Ananya P, Kalyan M, Sarmistha SR (2009) Effect of polyamines on *in vitro* somatic embryogenesis in *Momordica charantia* L. Plant Cell Tissue Organ Cult 97:303–311

Arnold SV (1987) Effect of sucrose on starch accumulation in an adventitious bud formation on embryos of *Picea abies*. Ann Bot 59:15–22

Arumugam N, Bhojwani SS (1990) Somatic embryogenesis in tissue cultures of Po (*Dophyllum hexandrum*). Can J Bot 68:487–491

Aurelia SJ, Aleksandra P, Zygmunt K (2005) Influence of cultivar, explant source and plant growth regulator on callus induction and plant regeneration of *Cannabis sativa* L. Acta Biol Crac Ser Bot 47(2):145–151

Azizi A, Hanafi NM, Basiran MN, Teo CH (2018) Evaluation of disease resistance and tolerance to elevated temperature stress of the selected tissue cultured *Kappaphycus alvarezii* Doty 1985 under optimized laboratory conditions. 3 Biotech 8(8):321–330

Barciela J, Vieitez AM (1993) Anatomicai sequence and morphometric analysis during somatic embryogenesis on cultured cotyledon explants of *Camellia japonica* L. Ann Bot 71:395–404

Bhaskaran S, Smith RH (1990) Regeneration in cereal tissue cultures. Crop Sci 30:1328–1337

Borkird C, Choi JH, Sung ZR (1986) Effect of 2,4- dichlorophenoxyacetic acid on the expression of embryogenic program of carrot. Plant Physiol 81:1143–1145

Carlos M, Martinez FX (1998) The potential uses of somatic embryogenesis in agroforestry are not limited to synthetic seed technology. Rev Bras Fis Veg 10:1–12

Cheney DP, Luistro AH, Bradley PM (1987) Carrageenan analysis of tissue cultures and whole plants of *Agardhiella subulata*. In: Twelfth international seaweed symposium. Springer, Dordrecht, pp 161–166

Cvikrová M, Binarová P, Cenklová V, Eder J, Machácková I (1999) Reinitiation of cell division and polyamine and monoamine levels in alfalfa explants during somatic embryogenesis. Physiol Plant 105:330–337

Dong NT, Luan TC, Huong NTT (2007) Vietnamese ginseng and some medicinal plants of the Ginseng family. Science and Technology Publisher, Vietnam, pp 109–110

Dougall DK, Verma D (1978) Growth and embryo formation in wild carrot suspension cultures with ammonium ion as a sole nitrogen source. In Vitro 14:180–188

Dudits D, Bögre L, Gyrögyey J (1991) Molecular and cellular approaches to the analysis of plant embryo development from somatic cells *in vitro*. J Cell Sci 99:475–484

Dunstan DI, Tautorus TE, Thrope TA (1995) Somatic embryogenesis in woody plants. In: Thorpe TA (ed) In vitro embryogenesis in plants. Kluwer Academic Publishers, Dordercht, pp 471–538

Feher A, Pasternak TP, Dudits D (2003) Transition of somatic plant cells to an embryogenic state. Plant Cell Tissue Organ Cult 74:201–228

Feirer RP, Wann SR, Einspahr DW (1985) The effects of spermidine synthesis inhibitors on *in vitro* plant development. Plant Growth Regul 3:319–327

Fleming AJ (2002) The mechanism of leaf morphogenesis. Planta 216(1):17–22

Fuentes SRL, Calheiros MBP, Manetti-Filho J, Vieira LGE (2000) The effects of silver nitrate and different carbohydrate sources on somatic embryogenesis in *Coffea canephora*. Plant Cell Tissue Organ Cult 60:5–13

Gill R, Saxena PK (1992) Direct somatic embryogenesis and regeneration of plants from seedling explants of peanut (*Arachis hypogaea*): promotive role of thidiazuron. Can J Bot 70:1186–1192

Gupta PK, Holmstrom D (2005) Double staining technology for distinguishing embryogenic cultures. In: Jain SM, Gupta PK (eds) Protocol for somatic embryogenesis in woody plants. Springer Publishers, pp 573–575

Haliloglu K (2006) Efficient regeneration system from wheat leaf base segments. Biol Planta 50(3):326–330

Hamidou SF, Peggy O, Lloyd M, Peng WC (2005) Putrescine enhances somatic embryogenesis and plant regeneration in upland cotton. Plant Cell Tissue Organ Cult 81:91–95

Hargreaves PI, Barcelos CA, da Costa ACA, Pereira NJ (2013) Production of ethanol 3G from *Kappaphycus alvarezii*: evaluation of different process strategies. Bioresour Technol 134:257–263

Hartweck LM, Lazzeri PA, Cui D, Collins GB, Williams EG (1988) Auxin-orientation effects on somatic embryogenesis from immature soybean cotyledons. In Vitro Cell Dev Biol Plant 24: 821–828

Hayashi L, Reis RP (2012) Cultivation of the red algae *Kappaphycus alvarezii* in Brazil and its pharmacological potential. Brazilian J Pharmacogn 22(4):748–752

Jha TB, Mukherjee P, Datta MM (2007) Somatic embryogenesis in *Jatropha curcas* Linn., an important biofuel plant. Plant Biotech Rep 1:135–140

Joshi R, Shukla A, Kumar P (2010) Interactive effect of GA_3 and polyamines on *in vitro* somatic embryogenesis from immature embryos in Maize (*Zea mays* L.). Maydica 55:111–119

Kalimuthu K, Paulsamy S, Senthilkumar R, Sathya M (2007) *In vitro* propagation of the biodiesel plant *Jatropha curcas* L. Plant Tissue Cult Biotech 17(2): 137–147

Kanji M, Yuji S (2000) Effects of sugar concentration and strength of basal medium on conversion of somatic embryos in *Asparagus officinalis* L. Sci Hortic 84(1–2):15–26

Kevers C, Le Gal N, Monteiro M, Dommes J, Gaspar T (2000) Somatic embryogenesis of *Panax ginseng* in liquid cultures: a role for polyamines and their metabolic pathways. Plant Growth Regul 31:209–214

Khambhaty Y, Mody K, Gandhi M, Thampy S, Maiti P, Brahmbhatt H, Eswaran K, Ghosh PK (2012) *Kappaphycus alvarezii* as a source of bioethanol. Bioresour Technol 103(1): 180–185

Kochba J, Ben-Hayyim G, Speigel-Roy P, Saad S, Neumann H (1982) Selection of table salt-telerant callus cell lines and embryos in *Citrus sinensis* and *C. auranticum*. Zeitschrift Fur Pflanzenphysiologie 106:111–118

Krishna RS, Vasil IK (1995) Somatic embryogenesis in herbaceous monocots. In: Thorpe TA (ed) *In vitro* embryogenesis in plants. Kluwer Academic Publishers, Dordrecht, pp 471–540

Kumar PS, Kumari RBD (2010) Effect of primary and secondary somatic embryogenesis in safflower (*Carthamus tinctorius* L.) at morphological and biochemical levels. Am Eur J Agric Environ Sci 8(6):784–792

Lou H, Kako S (1995) Role of high sugar concentrations in inducing somatic embryogenesis from cucumber cotyledons. Sci Hortic 64:11–20

Luan TC (2003) Research on Vietnamese ginseng chemistry. Workshop on conservation and development of Ngoc Linh ginseng in Quang Nam Province, Vietnam, pp 62–75

Luo H, Obara-Okeyo P, Tamaki M, Kako S (1996) Influence of sucrose concentration on *in vitro* morphogenesis in cultured cucumber cotyledon explant. J Am Soc Hortic Sci 71:497–502

Malik KA, Saxena PK (1992) Regeneration in *Phaseolus vulgaris* L.: high-frequency induction of direct shoot formation in intact seedlings by N 6-benzylaminopurine and thidiazuron. Planta 186(3):384–389

May RA, Trigiano RN (1991) Somatic embryogenesis and plant regulation from leaves of Dendrarthema grandiflora. J Am Soc Hortic Sci 116:366–371

Meinita MD, Hong YK, Jeong GT (2012) Detoxification of acidic catalyzed hydrolysate of *Kappaphycus alvarezii* (cottonii). Bioprocess Biosyst Eng 35(1–2):93–98

Méndez-Hernández HA, Ledezma-Rodríguez M, Avilez- Montalvo RN, Juárez-Gómez YL, Skeete A, Avilez-Montalvo J, De-la-Peña C, Loyola-Vargas VM (2019) Signaling overview of plant somatic embryogenesis. Front Plant Sci 10:77

Merkle SA, Parrott WA, Flinn BS (1995) Morphogenic aspects of somatic embryogenesis, in *In vitro* embryogenesis in plants, vol V, T A Thorpe. Kluwer Academic Publishers, Dordercht, pp. 155–203

Moltrasio R, Robredo CG, Gomez MC, Paleo AHD, Diaz DG, Rios RD, Franzone PM (2004) Alfalfa (*Medicago sativa*) somatic embryogenesis: genetic control and introduction of favourable alleles into elite Argentinean germplasm. Plant Cell Tissue Organ Cult 77:119–124

Monteiro M, Kevers C, Dommes J, Gaspar T (2002) A specific role for spermidine in the initiation phase of somatic embryogenesis in *Panax ginseng* CA Meyer. Plant Cell Tissue Organ Cult 68: 225–232

Mullin RH, Schlegel DE (1976) Cold storage maintenance of strawberry meristem plantlets. Hortic Sci 11:100–101

Murashige T, Skoog F (1962) A revised medium for rapid growth and bio assays with tobacco tissue cultures. Physiol Plant 15(3):473–497

Neumann KH (2000) Some studies on somatic embryogenesis: a tool in plant biotechnology. Based on a lecture at the 87th Indian Science congress Jan in Pune, India, pp 1–27

Nhut DT, Chien HX, Truc NB, Nam NB, Tinh TX, Luan VQ, Binh NV, Hien VT, Huong TT, Nhan NCT, Thuy LNM, Nga LTM, Hien TT, Hai NT (2010) Micropropagation of *Panax vietnamensis* Ha *et* Grushv. Vietnam J Biotech 8(3B):1211–1219

Nhut DT, Vinh BVT, Hien TT, Huy NP, Nam NB, Chien HX (2012) Effects of spermidine, proline and carbohydrate sources on somatic embryogenesis from main root transverse thin cell layers of Vietnamese ginseng (*Panax vietnamensis* Ha *et* Grushv.). Africa. J Biotech 11(5):1084–1091

Pasternak TP, Prinsen E, Ayaydin F, Miskolczi P, Potters G, Asard H, Van Onkelen HA, Dudits D, Feher A (2002) The role of auxin, pH and stress in the activation of embryogenic cell division in leaf protoplast-derived cells of *Alfalfa*. Plant Physiol 129:1807–1819

Polne-Fuller M, Gibor A (1987) Calluses and callus-like growth in seaweeds: induction and culture. Hydrobiologia 151/152, 131–138

Pramanick B, Brahmachari K, Ghosh A (2013) Effect of seaweed saps on growth and yield improvement of green gram. Afr J Agric Res 8(13):1180–1186

Rajam MV (1997) Polyamines. In: Prasad MNV (ed) Plant ecophysiology. Wiley, New York, pp 343–374

Rajore S, Batra A (2005) Efficient plant regeneration via shoot tip explant in *Jatropha curcas* L. J Plant Biochem Biotech 14:73–75

Reddy CRK, Kumar GRK, Siddhanta AK, Tewari A, Eswaran K (2003) *In vitro* somatic embryogenesis and regeneration of somatic embryos from pigmented callus of *Kappaphycus alvarezii* (Doty) Doty (Rhodophyta, Gigartinales). J Phycol 39(1): 610–616

Šamaj J, Baluska F, Pretová A, Volkmann D (2003) Auxin surge following fertilization in carrots: a mechanism for regulating plant totipotency. Planta 214:505–509

Sardana J, Batra A, Ali DJ (1998) *In vitro* plantlet formation and micropropagation of *Jatropha curcas* L. Adv Plant Sci 11(2): 167–169

Savita V, Virk GS, Nagpal A (2010) Effect of explant type and different plant growth regulators on callus induction and plantlet regeneration in *Citrus jambhiri* Lush. Int J Sci Tech 5:97–106

Shah MT, Zodape ST, Chaudhary DR, Eswaran K, Chikara J (2013) Seaweed sap as an alternative liquid fertilizer for yield and quality improvement of wheat. J Plant Nutr 36(2):192–200

Sharma HR, Ingalls JR, Delvin TJ (1980) Apparent digestibility of tower and candle rapeseed meals by Holstein bull calves. Can J Anim Sci 60:915–918

Shasthree BT, Coroline VJE, Savitha R (2012) Micropropagation of *Bryonopsis laciniosa* L. Naud an endangered and medicinally important cucurbit. Adv. Plant Sci 25(II):373–376

Shi X, Dai X, Liu G, Zhang J, Ning G, Bao M (2010) Cyclic secondary somatic embryogenesis and efficient plant regeneration in camphor tree (*Cinnamomum camphora* L.). In Vitro Cell Dev Biol Plant 46:117–125

Shohana P, Yeon-Yu K, Rama KP, Yu-Jin K, Neha GW, Deok-Yung K (2009) Identification and characterization of spermidine synthase gene from *Panax ginseng*. Mol Biol Rep 37:923–932

Siang TC, Soong ST, Yien ATS (2012) Plant regeneration studies of *Jatropha curcas* using induced embryogenic callus from cotyledon explants. Afr J Biotechnol 11:8022–8031

Singh M, Chaturvedi R (2009) An efficient protocol for cyclic somatic embryogenesis in neem (*Azadirachta indica* A. Juss.). Int J Civil Environ Eng 1:49–51

Smith DL, Krikorian A (1989) Release of somatic embryogenic potential from excised zygotic embryos of carrot and maintenance of proembryonic cultures in hormone-free medium. Am J Bot 76:1832–1843

Sujatha M, Mukta N (1996) Morphogenesis and plant regeneration from tissue cultures of *Jatropha curcas*. Plant Cell Tissue Organ Cult 44:135–141

Sujatha M, Makkar HPS, Becker K (2005) Shoot bud proliferation from axillary nodes and leaf sections of non-toxic *Jatropha curcas* L. Plant Growth Regul 47:83–90

Trono GCJ (1992) *Eucheuma* and *Kappaphycus*: taxonomy and cultivation. Bull Mar Sci Fish Kochi Univ 12: 51–65

Umehara M, Kamada H (2005) Development of the embryo proper and the suspensor during plant embryogenesis. Plant Biotechnol 22:253–260

Vairappan CS, Razalie R, Elias UM, Ramachandram T (2014) Effects of improved post harvest handling on the chemical constituents and quality of carrageenan in red alga, *Kappaphycus alvarezii* Doty. J Appl Phycol 26(2):909–916

Vanildo S, Eny SF, Walter H, Migue PG (2004) Effect of plant growth regulators on the cellular growth and levels of intracellular protein, starch and polyamines in embryogenic suspension of *Pinus taeda*. Plant Cell Tissue Organ Cult 76:53–60

Xie DY, Hong Y (2001) Regeneration of *Acacia mangium* through somatic embryogenesis. Plant Cell Rep 20:34–40

Yeong HY, Phang SM, Reddy CRK, Khalid N (2014) Production of clonal planting materials from Gracilaria changii and Kappaphycus alvarezii through tissue culture and culture of G. changii explants in airlift photobioreactors. J Appl Phycol 26(2):729–746

Yokoya NS, Handro W (2002) Effects of plant growth regulators and culture medium on morphogenesis of *Solieria filiformis* (Rhodophyta) cultured *in vitro*. J Appl Phycol 14:97–102

Yokoya NS, West JA, Luchi AE (2004) Effects of plant growth regulators on callus formation, growth and regeneration in axenic tissue cultures of *Gracilaria tenuistipitata* and *Gracilaria perplexa* (Gracilariales, Rhodophyta). Phycol Res 4:244–254

Yong WTL, Chin JYY, Thien VY, Yasir S (2014) Evaluation of growth rate and semi-refined carrageenan properties of tissue cultured *Kappaphycus alvarezii* (Rhodophyta, Gigartinales). Phycol Res 62(4):316–321

Yong YS, Yong WTL, Thien VY, Ng SE, Anton A, Yassir S (2015) Acclimatization of micropropagated *Kappaphycus alvarezii* (Doty) Doty ex Silva (Rhodophyta, Solieriaceae) in outdoor nursery system. J Appl Phycol 27(1):413–419

Yong WTL, Chin JYY, Thien VY, Yasir S (2017) Heavy metal accumulation in field cultured and tissue cultured *Kappaphycus alvarezii* and *Gracilaria changii*. Int Food Res J 24(3):970–975

Zimmerman JL (1993) Somatic embryogenesis: a model for early development in higher plants. Plant Cell 5:1411–1423

Light-Emitting Diodes (LEDs) in Plant Regeneration, Growth, and Secondary Metabolite Accumulation

9

Nguyen Ba Nam, Hoang Thanh Tung, Michio Tanaka, and Duong Tan Nhut

Abstract

Light is the key factor influencing plant growth. Plants react to light mainly via photosynthetic, photomorphogenic, and phototropic responses. In addition, light also affects the synthesis of bioactive compounds of plants. In the following case studies, the positive effects of light-emitting diodes (LEDs) on the shoot regeneration and growth of *Chrysanthemum morifolium* Ramat. cv. "Jimba," and the secondary metabolism of *Panax vietnamensis* shoot culture are demonstrated. With the newly developed capability and the trends of brighter intensity and lower price, the LED-based light source has made it a promising light source for studies on biological processes. With current studies and further investigations to be made, LEDs have been increasingly applied for promoting plant production. From the results shown above and by other researchers, and the many attractive features of LEDs, it is clear that LEDs are an effective alternative to fluorescent lamps in plant micropropagation.

Keywords

Chrysanthemum morifolium · *Panax vietnamensis* · Light-emitting diodes · Plant regeneration · Growth · Secondary metabolites accumulation

---

N. B. Nam
University of Dalat, Dalat City, Vietnam

H. T. Tung · D. T. Nhut (✉)
Tay Nguyen Institute for Scientific Research, VAST, Dalat City, Vietnam

M. Tanaka
Kagawa University, Takamatsu, Kagawa, Japan

D. T. Nhut et al. (eds.), *Plant Tissue Culture: New Techniques and Application in Horticultural Species of Tropical Region*,
https://doi.org/10.1007/978-981-16-6498-4_9

9.1 Introduction

Light directly influences plant growth and development. Plants use photons of light as one of the major energy sources for photosynthesis. The spectral quality and photoperiod of light serve as the signal for photoperiodism and photomorphogenesis, especially in plant micropropagation laboratories. Therefore, artificial lights can control plant growth and development in greenhouses and in in vitro cultures. In general, fluorescent lamps are the main source of light for micropropagation. This light source emits a wide-spectrum light ranging from 350 to 750 nm, many of which are not conducive to plant growth. Light-emitting diodes (LEDs) are an effective light source for photobiological systems. The LEDs can eliminate low-quality wavelengths that are inactive for photosynthesis and promoting growth. Numerous studies have investigated the effects of LED lighting on plant growth, morphogenesis, and secondary metabolites accumulation. Tanaka et al. (1998) reported enhanced growth of *Cymbidium* plantlets cultured in vitro under super bright red and blue LEDs. Nhut (2002) also showed that both in vitro and subsequent growth of banana plantlets were improved under blue and red LEDs. Anzelika et al. (2008) analyzed the morphogenesis of Chrysanthemum cultured in vitro under illumination with various spectra and photon flux densities using LEDs.

Furthermore, several plant species have been grown successfully under LEDs, such as grape (Poudel et al. 2008), potato (Miyashita et al. 1995; Jao and Fang 2004), *Lilium* (Lian et al. 2002), Chrysanthemum (Kim et al. 2004), *Eucalyptus* (Nhut 2002), *Rehmannia glutinose* (Hahn et al. 2000), *Zantedeschia* (Jao et al. 2005), *Euphorbia milii* (Dewir et al. 2006), *Spathiphyllum* (Nhut et al. 2005), *Withania somnifera* (Lee et al. 2007), and *Phalaenopsis* orchids (Wongnok et al. 2008). Many studies show that LEDs are more suitable for plant growth than fluorescent lamps. This chapter provides a brief development history of LEDs and a broad base review on LEDs in plant propagation.

9.2 Historical Development of LEDs

The first light-emitting diode which could light up when an electric current flowed through was discovered in 1907 by an Englishman named Henry Joseph Round (Round 1907). However, Round's research was neglected for many decades until it was demonstrated and improved by Holonyak and Bevacqua (1962) at the General Electric's Solid-State Device Research Laboratory in Syracuse, New York (USA) in 1962, albeit the light emitted could only be seen in a darkened room. As an important invention, the blue LED was announced on November 12, 1993, by Shuji Nakamura at a press conference in Japan. Nakamura et al. (1997) combined a blue gallium nitride (GaN) LED with a yellow-emitting phosphor to create white LEDs. The blue, red, and green lights are the three primary colors; the correct combination of ratios could generate all the other color lights. This process requires the design of very complex electrical circuits, both hardware and software controls. With the continuous improvement in production and cost reduction, LEDs are already widely used and commercialized in many fields, such as traffic signals, large outdoor displays, lighting in cars and buses, and backlighting for cell phones.

An LED is a small electronic device that emits a specific wavelength without producing heat (infra-red). The waveband of the emitted light depends on the composition and the condition of the material used and can be infrared, visible, or near-ultraviolet. Some important advantages of this light are:

- LEDs produce more light per watt than incandescent and fluorescent lamps.
- LEDs can emit light of an intended wavelength which stimulate plant growth.
- LEDs have an extremely long lifespan (up to 100,000 h).
- LEDs emit little heat except for devices that convert alternating current into direct current.
- LEDs are a convincing replacement for incandescent and fluorescent lamps for commercial micropropagation laboratories, greenhouse, and many other applications.

The inventors of the efficient blue LED: Isamu Akasaki, Hiroshi Amano, and Shuji Nakamura were awarded the 2014 Nobel Prize in Physics, verifying the beneficial impact of LEDs in economics, environment, and quality of life.

9.3 Structure of LEDs

An LED's simple structure consists of two components: *n*-type semiconductors and *p*-type semiconductors placed directly connected and form a diode. Light is emitted with the *p–n* junction when a direct current flows through it. A block (n-type) contains negatively charged electrons, and the other (p-type) has many positive charge pores. A current flow to the junction, electrons, and pores tends to move diffusely to the junction. At the border on either side of the junction, the electrons meet the hole to form neutral electrons. This process can release energy in the form of neutral electrons. This process can release energy in the form of light and is called electroluminescence. Semiconductors are an essential element of LEDs. But the light emitted from LEDs also depends on the conductive frame placed between two semiconductors and the shell to protect the semiconductor block. The development of LEDs began with a semiconductor material, gallium arsenide, which emits infrared and red light. The standard LED is based on GaAsP, which is made up of elements from three substances: gallium, arsenic, and phosphorus, emitting red light with a wavelength of 655 nm and brightness levels ranging from 1 to 10 mcd at 20 mA.

Earlier, low-intensity LEDs are mainly used in indicator lights. Following the use of GaAsP material, the red LEDs were developed. The advantage of this material is that it can produce more intense light with low voltage. LED technology exploded in the 1970s. Many of the resulting semiconductor materials can emit light of different colors or wavelengths. The most common materials are GaP, which emits red and green, and GaAsP, which emits orange, red, and yellow. In the 1980s, a new material

GaAlAs (gallium aluminum arsenide) was developed. LEDs based on GaAlAs could emit high-efficiency light than ever before. LEDs can be used for signs or messages because it is easy to design circuits and assemble any shapes out of them. In addition, LEDs have been used in the design of bar codes, scanners, fiber-optic cables, and medical equipment. The improvement of the crystal material allows the LED to emit a yellow, green, or orange light. Depending on the semiconductor materials, LED's wavelength is determined. Examples: Aluminum gallium arsenide (AlGaAs) emits red and infrared waveband; aluminum gallium phosphide (AlGaP) green light; aluminum gallium indium phosphide (AlGaInP) orange, yellow, and green light; gallium arsenide phosphide (GaAsP) red, orange-red, orange, and yellow light; indium gallium nitride (InGaN) near UV, blue light; silicon carbide (SiC), sapphire (Al_2O_3), and zinc selenide (ZnSe) blue light; diamond (C) UV (Steigerwald et al. 2002).

LEDs can be controlled to emit different colors and are brighter and more energy-efficient than tube lights and incandescent bulbs. Hence, LEDs are a suitable replacement for standard light sources in commercial plant cultivation. It is possible to select the output wavelength with LEDs that match plant photoreceptors' absorption peak with a narrow spectral region. Blue light (440 nm) is the absorption peak of cryptochrome and carotenoid; the red-light area (640 nm) is the absorption peak of phytochrome and chlorophyll. Furthermore, the light intensity of LEDs can be adjusted readily, along with the adjustment of CO_2 concentration, relative humidity, and temperature; plant growth can be optimized under in vitro and ex vitro conditions (Pinho et al. 2004).

9.4 Application of LEDs in Plant Regeneration, Growth, and Secondary Metabolites Accumulation on Some Economically Important Plants

Plant cell, tissue, and organ cultures are important tools in basic research and commercial applications. Research subjects comprise flowers, fruit, vegetables, and forestry plants (Vestberg et al. 2002). The quality of micropropagated plants depends on the culture conditions. Environmental factors have a decisive influence on plant quality, such as temperature, carbon dioxide (CO_2) concentration, nutrients, and light. These factors affect CO_2 assimilation, water absorption, evapotranspiration rate, plant growth, and morphology. Light especially plays an essential role as the energy source for photosynthesis and as the source of information for photoperiodism (night/day length), phototropism (light direction), and photomorphogenesis (light quantity and quality). These responses depend on the photon flux density (PFD), light quality, duration, and photoperiod (Taiz and Zeiger 1991).

In controlled environments as in vitro culture, light quality plays a determining role in the plant's morphologic characteristics such as elongation, shoot regeneration, leaf shape, and root formation (Hoenecke et al. 1992; Brown et al. 1995). Several recent advances in LED technology have paved the way for its utilization in applications that require a high photon flux, such as for plant lighting in controlled environments. In vitro morphogenesis, plant growth and development, and secondary metabolite accumulation of various plant species under white, blue, red, green, yellow, and orange LEDs have been reported.

9.4.1 LEDs in Plant Regeneration

Micropropagation is presently used as an advanced biotechnological system to produce young plants for agriculture and forestry. However, the high cost of plant production is one reason to limit the commercial application of micropropagation for all plant species. In addition, the survival of the plant in the final ex vitro stage is usually poor. For this reason, quality improvement and cost reduction of tissue-cultured plants are the pursued objectives of plant producers.

Fluorescent lamps are the primary light source commonly used for in vitro cultivation of plants. Fluorescent lamps have fixed emission spectra composing many bands in the wavelength range from 320 to 800 nm. However, this light source contains unnecessary wavelengths for promoting plant growth. Due to the development of a new generation of high-power LEDs with a wide diversity of emission wavelengths, LED-based illuminators have become versatile and commercially attractive light sources for different applications in plant cultivation. LED-based illuminators have improved features, including smaller mass and volume, longer lifetime, and tailored spectrum than conventional fluorescent lamps. Thus, LED-based illuminators provide an alternative to fluorescent lamps as a light source with a controllable spectrum that can be used for plant cultivation.

Shoot regeneration is the first stage in the process of micropropagation. Quality shoots will determine the ability of the continuous growth stage of plantlets. Numerous studies in different plant species have demonstrated that LED treatments are superior to conventional fluorescent lamps in in vitro shoot regeneration (Gupta and Pradhan 2017). However, due to the difference of light-harvesting photoreceptors in tune with the genetic make-up of the plant species, there are differential responses to different spectral regions (Kim et al. 2004).

In shoot organogenesis, red LEDs promote regeneration and elongation of various species such as *Rehmannia glutinosa* (Hahn et al. 2000), grapes (Heo et al. 2006; Poudel et al. 2008) and *Oncidium* (Chung et al. 2010). Under red LEDs, a larger number of shoots can be regenerated from nodal segments of *Stevia rebaudiana* (Ramírez-Mosqueda et al. 2017). However, leaves are yellowish-green when grown under red LEDs (Nhut et al. 2003). On the other hand, the blue LEDs inhibit in vitro shoot length in Chrysanthemum (Anzelika et al. 2008) and cotton (Li et al. 2010). Moreover, the effects of red or blue LEDs in inhibition or promotion depend on the plant species (Lotfi et al. 2019). Blue LEDs are a more effective light in the shoot formation of *Anthurium* from callus than red LEDs (Budiarto 2010). Similar findings on the action of blue LEDs in shoot regeneration on other plants, like *Dendrobium* (Lin et al. 2011) and *Curculigo orchioides* (Gupta and Sahoo 2015), have been noted.

The results indicate that the use of red LEDs in combination with blue LEDs improved shoot regeneration in various species (Dewir et al. 2006; Chung et al. 2010; Lee et al. 2014; Ramírez-Mosqueda et al. 2017; Hung et al. 2016a, b; Kim et al. 2004; Anzelika et al. 2008; Park and Kim 2010). Higher shoot regeneration was observed in *Abeliophyllum distichum* under the B/R LED combination (Lee et al. 2014). Conversely, the blue and red mix LED treatment was not suitable for improving the shoot organogenesis in Bananas (Wilken et al. 2014), Bacopa (Karatas et al. 2016), and *Vanilla* (Bello-Bello et al. 2016).

The combination of red and far-red LEDs stimulated somatic embryo formation in the *Doritaenopsis* with a low level of endoreduplication (Park and Kim 2010). LED lighting also affects the tuberous regeneration of the *Lilium* (Lian et al. 2002). The combination of blue and red LEDs stimulated an increase in fresh and dry weights, volume, and size of lily tubers (Lian et al. 2002). Red LEDs combined with blue LEDs also increased the protocorm-like body (PLB) formation of *Phalaenopsis* (Nhut et al. 2005). In *Cymbidium*, many PLBs formed at 25% red LEDs combined with 75% blue LEDs (Huan and Tanaka 2004). Blue LED light is believed to be the best lighting condition for shoot formation from PLBs of in vitro cultured *Dendrobium officinale* (Lin et al. 2010). Many studies tried to determine the role of light quality in plant morphogenesis, but the results were unclear. The influences depend on the emitted wavelengths, the selected plants, the stages of plant growth, and media conditions such as light intensity, environmental composition, or ventilation conditions (Hahn et al. 2000).

9.4.2 LEDs in In Vitro and Subsequent Growth of Plants

LEDs are a suitable replacement for their predecessors, which include fluorescent lamps (TFL), high-pressure sodium (HPS), halogen (MHL), and incandescent lamps because of their emitted spectral flexibility. The various responses from different explants and plant species cultured under the LED light spectral regions have been shown in different studies (Gupta and Jatothu 2013). The lighting system requires LEDs to emit high intensity particularly in the red and blue spectral regions. In the past, blue LEDs only emitted low-intensity light and had a high cost (Bula et al. 1991). Since 1993, Nichia Chemical Company (Japan) has produced blue LEDs with increased intensity. This event paved the way for the development of the LED lighting system.

Red light is essential to plant photosynthesis and enhances growth. However, plants will have a stretched and elongated appearance when grown under monochromatic red LEDs. It has early been reported that red LEDs significantly affect stem elongation, leaf expansion, chlorophyll synthesis (Tripathy and Brown 1995), and photosynthesis (Tennessen et al. 1994). Red light can significantly enhance photosynthesis and the in vitro growth and development of plantlets. However, plants grown under only red light have an elongated appearance. Strawberry, *Eucalyptus*, *Cymbidium*, *Phalaenopsis*, and banana stems elongate under red LEDs (Nhut 2002). Their leaves are long, thin, and turned yellowish-green, leading to an increasing growth imbalance under in vitro culture conditions (Nhut 2002). The effects of red light on the stem elongation in Chrysanthemum (Kim et al. 2004) and *Azorina vidalii* (Moreira da Silva and Debergh 1997) have also been reported. However, in a few studies, red light did not affect stem elongation in other plant species. Hahn et al. (2000) reported the opposite effect of red light on in vitro rhubarb growth; although the stem length of plantlets was not inhibited, the fresh biomass and the photosynthesis rate were lower when cultured under other lights. Shoot growth inhibition was observed under red light in herbaceous plants such as Marigold and *Salvia* (Heo et al. 2002).

Unlike the influence of red LEDs, stem elongation was inhibited by blue LEDs in Chrysanthemum (Kim et al. 2004) and *Zantedeschia* (Jao et al. 2005). The effects of blue light on inhibiting growth have been reported in various plant species (Kim et al. 2004). Moreira da Silva and Debergh (1997) showed that the plant height of *Azorina vidalii* was shortened when grown under blue light. Mortensen and Stromme (1987) also observed blue-light growth inhibition in many plant species in a greenhouse. Blue light plays a role in plant chlorophyll synthesis (Akoyunoglou and Anni 1984) and increases growth parameters such as fresh and dry weights, when in vitro plantlets were cultivated under the combined blue and red LEDs than that of the monochromatic LEDs (Kim et al. 2004; Li et al. 2010; Lian et al. 2002; Nhut 2002; Poudel et al. 2008; Shin et al. 2008). Of all the light treatments, the highest shoot fresh and dry weights of upland cotton plantlets were obtained under B:R = 1:1 LED light, whereas fluorescent lamps resulted in the lowest weight gains of upland cotton plantlets (Li et al. 2010). Similar results have been observed on Chrysanthemum (Kim et al. 2004). Hahn et al. (2000) reported that the photosynthesis rate of in vitro cultured rhubarb was very high under mixed LED systems (50% red LEDs and 50% blue LEDs). In contrast, plants cultivated under blue or red LEDs have photosynthesis at a very low rate.

In some cases, plant growth is best with 10% blue LEDs added, while in others, the proportion of blue LEDs must be up to 30% when combined with red LEDs (Nhut and Nam 2010). Different blue/red ratio affects morphogenesis on *Anthurium* plantlet during in vitro culture. The highest percentage of shoot regeneration was obtained when the ratio of red LEDs is higher than blue LEDs, but the highest number of shoots was obtained under blue LEDs (Budiarto 2010).

The effects of different wavelengths of LEDs on somatic embryo growth in three southern pine species were also studied. The results showed that red LEDs stimulated somatic embryo germination with a higher frequency in all the three species (*Pinus taeda* L.; *Pinus elliottii* Engelm.; *Pinus palustris* Mill.) than blue LEDs and fluorescent lamps (Merkle et al. 2005).

The effects of light quality on the sucrose, starch, and soluble sugar contents and the photosynthesis rate of in vitro explants were also studied. Increases in sucrose, starch, and soluble sugar content in cotton were observed when cultured under red LEDs (Li et al. 2010). In contrast, a combination of blue and red LEDs was effective in grape (Heo et al. 2006) and *Doritaenopsis* (Shin et al. 2008). The red light was able to inhibit the transport of photosynthetic products and therefore increased the starch content of leaves (Soebo et al. 1995). In contrast, blue LED light affected the increase in the anthocyanin and sugar content of the grape (Kondo et al. 2014) and the phenolic acid and flavonoid content in rose and Chrysanthemum plants (Ouzounis et al. 2014). Some other studies had similar conclusions. Blue light increased quercetin content of *Kalanchoe pinnata* plant (Nascimento et al. 2013) and increased carotenoid, chlorophyll, and glucosinolate levels in broccoli (Kopsell and Sams 2013).

In vitro rooting is also influenced by LEDs. The highest number of roots of *Anthurium plantlets* was obtained during in vitro culture under red plus blue LEDs (Budiarto 2010). Blue LEDs increased root formation of in vitro cultured *Doritaenopsis* (Shin et al. 2008). The light quality promotes root formation and growth of in vitro *Cunninghamia lanceolata*, which shortens the culturing time. That is also why LEDs had been considered at the rooting stage in the micropropagation process (Xu et al. 2020). Moreover, in addition to the influence of light quality, in vitro rooting may also depend on the genotype used and environmental conditions.

The acclimatization of cultured plantlets is the last step in micropropagation. The viability of plantlets at the acclimatization stage with low cost will determine the success of the whole propagation process (Chandra et al. 2010). During this process, the plantlets must adapt to overcome new environmental conditions in the greenhouse or field. The in vitro rooting of micropropagated plantlets is critical to their adaptability to ex vitro conditions. The growth of in vitro cultures under 80% red LEDs combined with 20% blue LED improves the growth and survival rate at ex vitro conditions (Nhut 2002). Nhut (2002) showed that *Eucalyptus citriodora*, *Phalaenopsis*, *Musa*, and *Spathiphyllum* grew well under 80% red LEDs combined with 20% blue LEDs. Strawberry grew well when cultured in vitro under 70% red LEDs combined with 30% blue LEDs (Nhut et al. 2003). Tanaka et al. (1998) demonstrated that the growth of *Cymbidium* could be improved when plantlets were cultured under the B:R = 1:1 blue and red LEDs. Furthermore, the red plus blue LEDs in various combinations promoted in vitro plant growth and development (Lian et al. 2002; Kim et al. 2004; Moon et al. 2006; Shin et al. 2008; Hung et al. 2015; Al-Mayahi 2016).

9.4.3 Influence of LEDs on Secondary Metabolites Accumulation

The supply of quality light could control photomorphogenesis, plant growth, biomass, and secondary metabolism. Although the effects of light on growth and development are known, the role of light in the synthesis of bioactive compounds is unclear (Ouzounis et al. 2014). Kreuzaler and Hahlbrock (1973) suggested that light plays a role in activating flavonoid glycoside synthesis in the cell suspension cultures of *Petroselinum hortense*. Some other studies have also investigated the effects of light on the accumulation of bioactive compounds such as anthocyanins in *Perilla fruttiescnens* (Zhong et al. 1991) and artemisinin from the hairy root culture of *Artimisia annua* (Liu et al. 2002). Also, Wang et al. (2001) demonstrated that red light improved *Artemisia annua* hairy root growth and artemisinin production. Yu et al. (2005) studied the effects of light on the synthesis of ginsenoside in ginseng hairy roots (*Panax ginseng* C. A. Mayer). The results showed that the fluorescent lamps stimulated the synthesis of ginsenoside, whereas red LEDs and dark conditions improved the biomass of hairy roots (Yu et al. 2005). Red lights appear to be an important factor influencing secondary metabolism. Younas et al. (2018) reported that red light enhanced silymarin accumulation, whereas yellow light significantly reduced silymarin contents in the callus culture of *Silybum marianum*. Similar to the study of Younas et al. (2018), red light also increased phytochemical contents in in vitro cultures of *Myrtus communis* (Cioć et al. 2018). Khurshid et al. (2020) demonstrated that red light significantly enhanced phenolics and flavonoids accumulation in the callus culture of Eclipta alba compared to other LED lights and Cool-White fluorescent tubes. Previous studies by Gupta and Jatothu (2013) and Fazal et al. (2016) reported that blue LEDs enhanced the production of phytochemical contents in the shoot culture of *Swertia chirata* and the callus culture of *Prunella vulgaris*, respectively. From preceding examples, although the exact mechanism is unclear, light quality and quantity can certainly influence *metabolite accumulation* in plants.

9.5 Case Studies

9.5.1 Chrysanthemum morifolium

Chrysanthemum morifolium, of the family Asteraceae, consists mainly of herbaceous angiosperms. Chrysanthemum is an industrial flower with the second most important economic value after roses (Teixeira da Silva 2004). The commercial value of Chrysanthemums was about $145 million in the United States in 2009 (United States Department of Agriculture 2010).

In the production of Chrysanthemums to bloom at the desired time and with high quality, people use light control measures for them, in which the light from the segmental luminaire during the night is used. Yulian et al. (1995) indicated that Chrysanthemum are short-day plants that require a lot of light and low-temperature nights. In the early stages of new plant roots, the plant needs little light; during growth, too much light will slow down the plant growth and reduce flower quality. Photoperiod affects flowering: the photoperiod is equal to or shorter than the critical illumination length, flower buds are formed; when the photoperiod is longer than the critical illumination length, it cannot form flower buds (Narumon 1998).

In this study, the leaves of in vitro Chrysanthemum plant were cut into round discs of 0.8 cm diameter and used as explants. In addition, longitudinal thin cell layer explants (10 mm × 0.5 mm) were excised from the stem. These explants were cultured on MS medium supplemented with 30 g/L sucrose, 8 g/L agar, 0.5 mg/L NAA, and 2 mg/L BA.

Nodes (2 cm in height) of in vitro Chrysanthemum were transplanted on MS basal media supplement with 30 g/L sucrose and 8 g/L agar. All explants were cultured under 16 h photoperiod with different lighting conditions, that is, 100% red LED

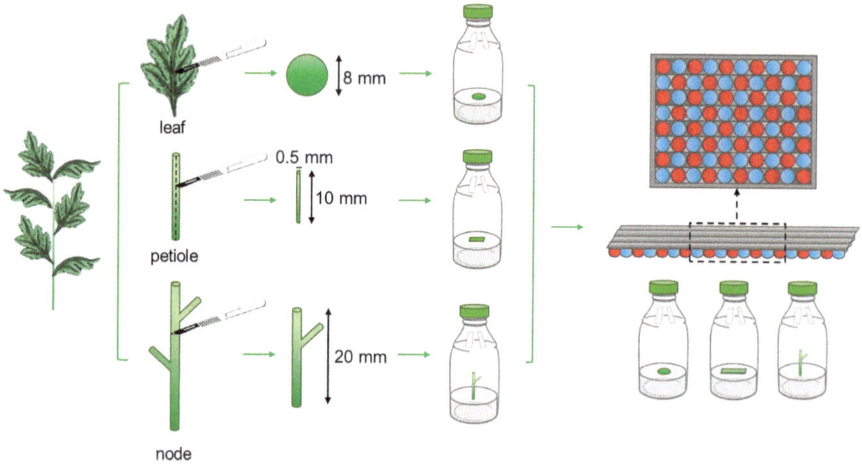

Fig. 9.1 Diagram depicting shoot regeneration of Chrysanthemum under LEDs condition

(100R), 100% blue LED (100B), 50% red LED + 50% blue LED (50R:50B), 60% red LED + 40% blue LED (60R:40B), 70% red LED + 30% blue LED (70R:30B), 80% red LED + 20% blue LED (80R:20B), 90% red LED + 10% blue LED (90R:10B), and fluorescent lamps (FL). For the experimental research on the growth and development of Chrysanthemum in response to supplemental LED lighting during the nighttime in the greenhouse, *Chrysanthemum morifolium* cultivars "Yellow, Sapphire, and Diamond" seedling-cuttings with an approximate height of 5 cm with four to five leaves were planted and supplemented at night for 6 h (from 8 P.M. to 2 A.M. the next morning) for 4 weeks with the following lighting conditions: Compact 3U (3U), 100R, 90R:10B, 80R:20B, 70R:30B, 60R:40B, 50R:50B. Compact 3U-FL were used as a control treatment. Diagram depicting shoot regeneration of Chrysanthmum under LEDs condition was shown in Fig. 9.1.

For shoot regeneration of Chrysanthemum, the significant effects of LEDs on the leaf discs and longitudinal thin cell layer explants of Chrysanthemums are shown in Fig. 9.2a–d. The highest percentage of regeneration and the greatest number of shoots were obtained when leaf explants were cultured under 100R. The greatest height (1.51 cm) and percentage of shoots taller than 1 cm (81.53%) were obtained when leaf explants were cultured under 70R:30B. The effects of different lighting conditions on the shoot regeneration from longitudinal thin cell layers are shown in Fig. 9.3a–d. Results of the shoot regeneration of Chrysanthemum under different lighting sources are shown in Table 3.6. Shoot regeneration rate (50.60%), number of shoots per explant (5.24 shoots), and percentage of shoots taller than 1 cm (44.20%) of explants under 70R:30B were better than those under the other light treatments. The 70R:30B is the most suitable lighting condition for shoot regeneration directly from leaf discs and indirectly from stem longitudinal thin cell layers.

For Chrysanthemum explants grown in vitro and subsequent ex vitro, fresh weight (1.14 g/plantlet), dry weight (94.83 mg/plantlet), and leaf area (3.52 cm^2)

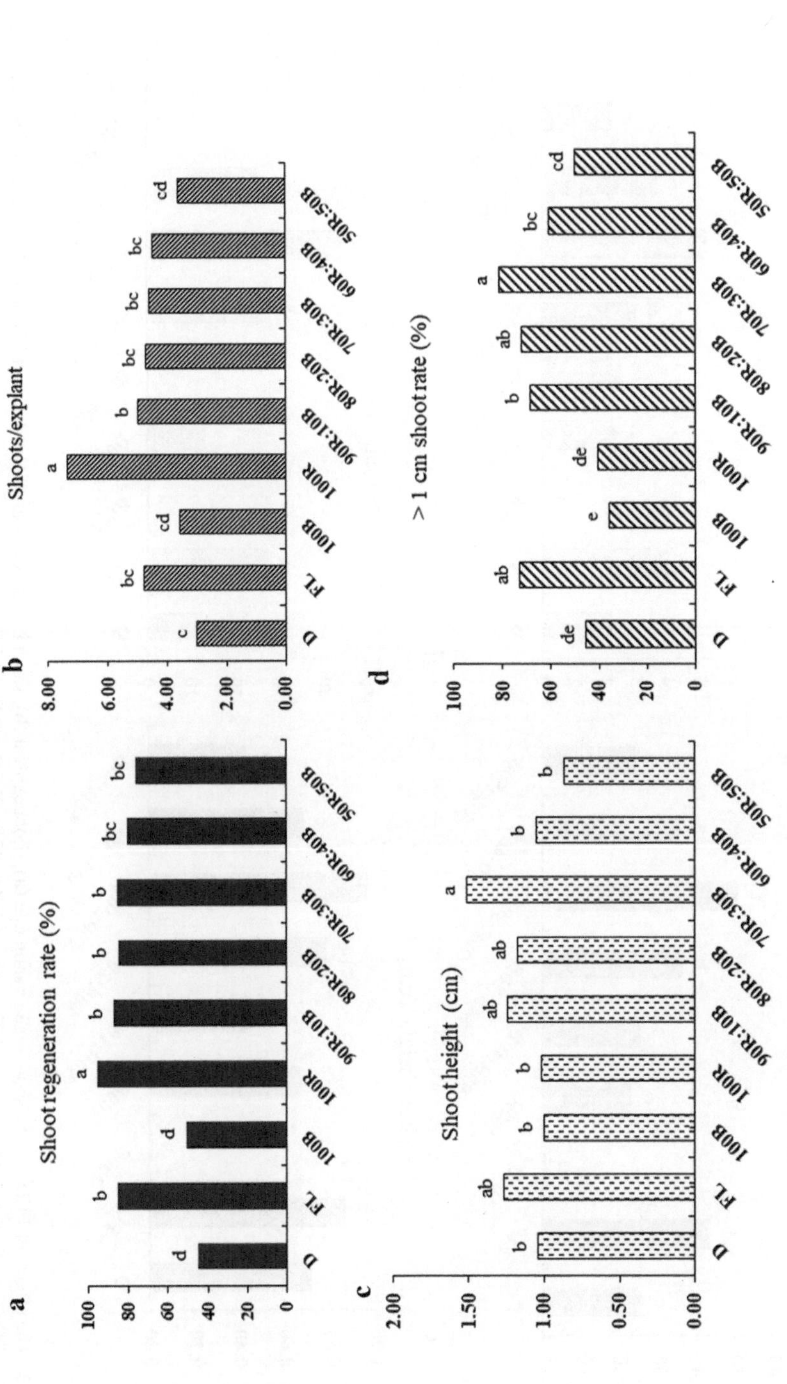

Fig. 9.2 The effects of LEDs on the shoot regeneration rate (**a**), shoot/explant (**b**), shoot height (**c**), and >1 cm shoot rate from the leaf discs of Chrysanthemum. *D* Darkness, *FL* fluorescent light, *B* Blue LEDs, *R* Red LEDs

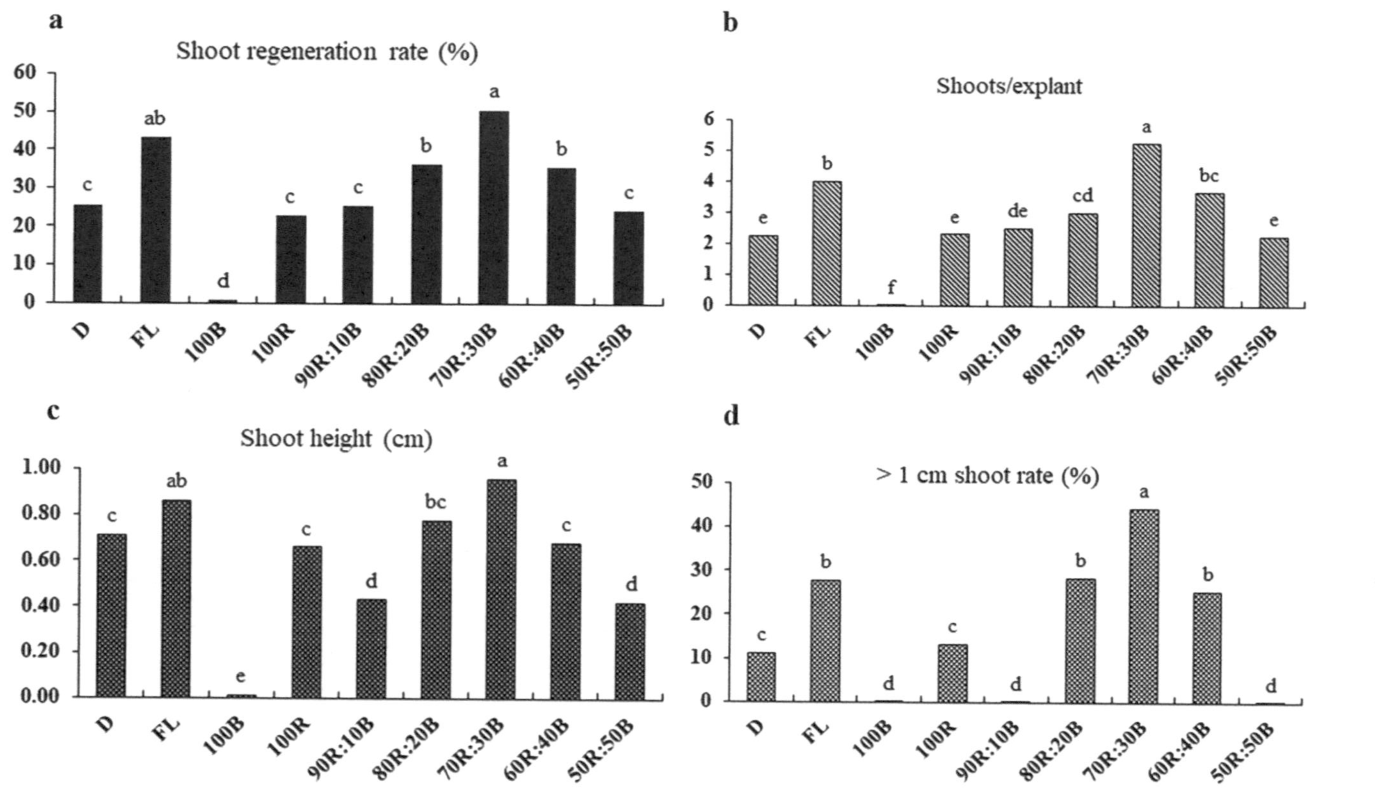

Fig. 9.3 The effects of LEDs on the shoot regeneration rate (**a**), shoot/explant (**b**), shoot height (**c**), and >1 cm shoot rate from the longitudinal thin cell layer explants of Chrysanthemum. *D* Darkness, *FL* Fluorescent light, *B* Blue LEDs, *R* Red LEDs

of plantlet under 70R:30B were higher than those under the other light treatments (Table 9.1 and Fig. 9.4). The results indicated that the growth of Chrysanthemum plantlets under LED lighting systems was significantly higher than that of the FL treatment, especially since plantlets cultured under 70R:30B were vigorous after 4 weeks of being transferred to soil. In vitro and subsequent ex vitro Chrysanthemum growth were best at 70R:30B; however, the highest shoot formation from leaf explants confirmed the effect of 100R.

In the greenhouse, LEDs can replace compact 3U fluorescent tubes to supplemental light during the nighttime for Chrysanthemum. Results showed that 70R:30B is suitable for the growth and development of *Chrysanthemum morifolium* cultivars "Sapphire and Diamond." In comparison, 60:40B is suitable for the growth and development of *C. morifolium* cultivar "Yellow" (Fig. 9.5).

9.5.2 Panax vietnamensis

Panax vietnamensis or *Vietnamese ginseng* is medicinally valuable and is now endangered of being over-collected. This species of ginseng is among the 250 endangered and highly endangered species in Vietnam's Red Data Book (Nhut et al. 2011). *P. vietnamensis* is one of the ginseng species with saponin content about 12–15% higher than other species of the genus *Panax* in the world. Pharmacological and clinical studies show that *P. vietnamensis* has effects very similar to Korean ginseng. Recently, pharmacological trials have shown that majonoside-R2, the major saponin (Rg1; Rb1; MR2) of Ngoc Linh ginseng, has anti-stress effects and is an important anti-cancer promoter (Duc et al. 1997).

For *P. vietnamensis*, in vitro shoots (2 cm of height) were inoculated on SH medium supplemented with 0.5 mg/L BA, 0.5 mg/L NAA, 30 g/L sucrose, 9 g/L agar, and 1 g/L activated carbon. Culture bottles were placed under different lighting conditions including 12 treatments: 100R, 100B, 90R:10B, 80R:20B, 70R:30B, 60R:40B, 50R:50B, 40R:60B, 30R:70B, 20R:80B, 10R:90B, and FL. Twelve-week-old plantlets were planted on the greenhouse to assess acclimatization and subsequent growth. The relative growth rate (mg/mg/week) and net assimilation rate (NAE, mg/cm^2/day) were recorded (Hunt et al. 2002).

The effects of LEDs on the growth and saponin biosynthesis of *P. vietnamensis* are shown in Fig. 9.6. Plantlets' growth of *P. vietnamensis* varies in response to the different lighting conditions. Fresh weight (540 mg), dry weight (82 mg), plant height (5.4 cm), SPAD value (27.7), and leaf area (9.38 cm^2) were greater in plantlets cultured under the 70R:30B than those under other treatments. Different lighting conditions affected plant growth and development and the saponin biosynthesis of *P. vietnamensis*. The highest MR_2 content was recorded when plants were maintained under 20R:80B. However, the highest Rg_1 and Rb_1 contents were found under FL light.

Judging from the results, the 80R:20B light appears suitable for the growth of *P. vietnamensis* in the greenhouse, and the 60R:40B can be used for the in vitro

Table 9.1 The effects of different lighting conditions on the in vitro growth of Chrysanthemum explants

Treatment	FW (g)	DW (mg)	RN[1]	RL (cm)[1]	PH (cm)	LN[1]	LA (cm^2)	IL (cm)[1]	SPAD	RGR	NAR[1]
FL	0.88^{c*}	75.27^{c}	$12.99a^{a,b}$	5.51^{a}	$5.53^{b,c}$	$13.66^{a,b}$	2.53^{b}	0.40^{b}	$34.60^{b,c}$	0.459^{c*}	$0.141^{a,b}$
100R	0.65^{d}	56.28^{d}	$11.99^{b,c}$	4.00^{b}	6.23^{a}	10.58^{c}	1.86^{c}	0.57^{a}	31.62^{c}	0.386^{d}	$0.140^{a,b}$
90R:10B	0.86^{c}	71.83^{c}	12.30^{b}	4.02^{b}	6.00^{a}	14.33^{a}	2.68^{b}	0.42^{b}	35.73^{b}	0.447^{c}	$0.123^{b,c}$
80R:20B	0.98^{b}	85.30^{b}	12.66^{b}	4.10^{b}	$5.86^{a,b}$	13.66^{a}	3.17^{a}	0.42^{b}	$37.56^{a,b}$	0.490^{b}	$0.139^{a,b,c}$
70R:30B	1.14^{a}	94.83^{a}	12.31^{b}	3.43^{b}	5.24^{c}	$12.65^{a,b}$	3.52^{a}	0.41^{b}	39.40^{a}	0.517^{a}	0.155^{a}
60R:40B	1.04^{b}	86.67^{b}	$13.66^{a,b}$	3.39^{b}	4.55^{d}	$11.99^{a,b}$	3.34^{a}	0.38^{b}	$38.10^{a,b}$	0.494^{b}	$0.150^{a,b}$
50R:50B	0.99^{b}	84.07^{b}	14.99^{a}	3.80^{b}	4.27^{d}	10.99^{b}	3.11^{a}	0.38^{b}	$37.48^{a,b}$	0.487^{b}	0.160^{a}
100B	0.60^{d}	51.80^{d}	10.30^{c}	3.55^{b}	4.70^{d}	$11.99^{a,b}$	2.01^{c}	0.39^{b}	$37.73^{a,b}$	0.361^{e}	0.111^{b}
F-test	**	**	**	**	**	**	**	**	**	**	**
CV%	4.78	3.49	4.27	4.47	4.78	4.47	8.81	2.66	5.00	2.22	1.24

Note: ***: Different letters in the same column indicate significant differences in Duncan's test (p-value ≤ 0.05)*
FW Fresh weight, ***DW*** Dry weight, ***RN*** Root number, ***RL*** Root length, ***PH*** Plant height, ***LN*** leaf number, ***LA*** Leaf area, ***IL*** Internode length, ***RGR*** Relative growth rate (mg/mg/week), ***NAR*** Net assimilation rate ($mg/cm^2/day$). *(1) Data were converted by $(x + 0.5)^{0.5}$ before being analyzed statistically; the values in the table are original average values*

Fig. 9.4 Growth of Chrysanthemum plants in vitro under LEDs after 4 weeks of culture. *FL* Fluorescent light, *B* Blue LEDs, *R* Red LEDs

Fig. 9.5 The greenhouse where we carried out the LED experiments with Chrysanthemum. (**a**, **b**) Greenhouses were divided into various combinations of red and blue LEDs, and the Compact 3U-FL was used as a control treatment; (**c**) Chrysanthemum growing under different lighting conditions

culture system (Table 9.2 and Fig. 9.7). The saponin biosynthesis is better in plantlets grown under FL than those under LED lights.

9.6 Conclusion

This chapter has shown that 70R:30B affects plant regeneration and growth of *Chrysanthemum morifolium* Ramat. cv. "Jimba"; meanwhile, 60R:40B can be used for the plant regeneration and 80R:20B for the acclimatization and growth in the greenhouse of *P. vietnamensis*. The saponin biosynthesis is better in plantlets grown under FL than those under LED lights.

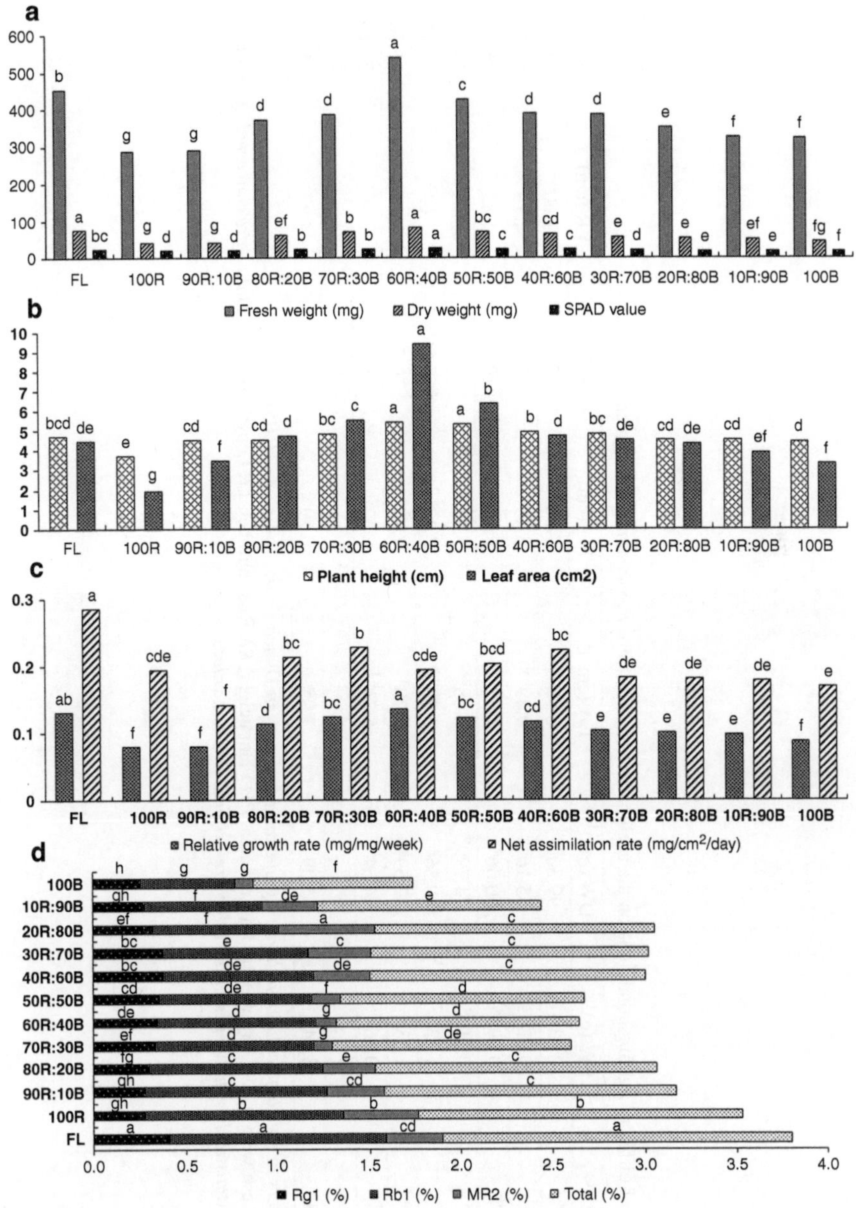

Fig. 9.6 The effects of LEDs on the growth and saponin biosynthesis of *Panax vietnamensis* are shown by the growth indices and saponin content

Table 9.2 The effect of different lighting conditions on the acclimatization of in vitro *P. vietnamensis* in greenhouse

Treatment	SR (%)	FW (g)[1]	DW (mg)[1]	LA (cm^2)[1]	PH (cm)[1]	RN[1]	LR (cm)[1]	SPAD[1]
Compact	$70^{d,c}$	0.93^{c}	99.30^{b}	$8.62^{c,d}$	7.66^{a}	7.88^{a}	$1.60^{a,b}$	21.16^{c}
100R	$70^{d,c}$	0.62^{e}	55.16^{d}	$9.49^{c,d}$	5.49^{c}	2.31^{b}	1.13^{b}	$23.73^{b,c}$
50R:50B	65^{d}	$0.69^{d,e}$	60.06^{d}	4.73^{d}	4.66^{c}	10.57^{a}	$1.82^{a,b}$	$25.03^{b,c}$
60R:40B	65^{d}	$0.71^{d,e}$	$74.79^{c,d}$	$10.28^{b,c}$	$5.75^{b,c}$	11.13^{a}	2.13^{a}	28.02^{b}
70R:30B	$90^{a,b}$	1.23^{b}	130.56^{a}	$13.70^{b,c}$	$7.33^{a,b}$	9.36^{a}	$1.62^{a,b}$	27.43^{b}
80R:20B	95^{a}	1.53^{a}	152.59^{a}	25.70^{a}	7.98^{a}	10.73^{a}	2.38^{a}	36.70^{a}
90R:10B	80^{bc}	$0.86^{c,d}$	$94.64^{b,c}$	$18.21^{a,b}$	5.32^{c}	8.97^{a}	2.16^{a}	27.10^{b}
F-test	**	**	**	**	**	**	**	**
CV%	7.82	3.94	1.09	7.95	6.79	7.29	4.87	5.73

Note: ***: Different letters in the same column indicate significant differences in Duncan's test (p-value $\leq$ 0.05)*
SR Survival rate, *FW* Fresh weight, *DW* Dry weight, *LA* Leaf area, *PH* Plant height, *RN* Root number, *LR* Length root. *(1): Data were converted by $(x + 0.5)^{0.5}$ before being analyzed statistically; the values in the table are original average values*

Fig. 9.7 Effects of LED light on the acclimatization of Ngoc Linh ginseng after 12 weeks in ex vitro conditions under LEDs. *Compact* 3U compact lamp light, *B* green LED light, *R* red LED light. *A* Fresh weight (mg); Dry weight (mg); SPAD value, *B* Plant height (cm); Leaf area (cm^2); *C* Relative growth rate (mg/mg/week); Net assimilation rate ($mg/cm^2/day$); *D* Saponin content Rg1 (%), Rb1 (%), MR2 (%), Total (%)

References

Akoyunoglou G, Anni H (1984) Blue light effect on chloroplast development in higher plants. In: Senger H (ed) Blue light effects in biological systems. Springer, Berlin, pp 397–406

Al-Mayahi AMW (2016) Effect of red and blue light emitting diodes "CRB-LED" on *in vitro* organogenesis of date palm (*Phoenix dactylifera* L.) cv. Alshakr. World J Microbiol Biotechnol 32:160

Anzelika K, Renata MC, Stase D, Silva Z, Genadij K, Gintautas T, Pavelas D, Arturas Z (2008) *In vitro* culture of chrysanthemum plantlets using light-emitting diodes. Cent Eur J Biol 3: 161–167

Bello-Bello JJ, Martínez-Estrada E, Caamal-Velázquez JH, Morales-Ramos V (2016) Effect of LED light quality on in vitro shoot proliferation and growth of vanilla (Vanilla planifolia Andrews). Afr J Biotech 15(8):272–277

Brown CS, Schuerger AC, Sager JC (1995) Growth and photomorphogenesis of pepper plants under red light-emitting diodes with supplemental blue or far-red lighting. J Am Soc Hortic Sci 120:808–813

Budiarto K (2010) Spectral quality affects morphogenesis on *Anthurium* plantlet during *in vitro* culture. Agrivita 32(3):234–240

Bula RJ, Morrow RC, Tibbitts TW, Barta DJ (1991) Light-emitting diodes as a radiation source for plants. HortSci 26:203–205

Chandra S, Bandopadhyay R, Kumar V, Chandra R (2010) Acclimatization of tissue cultured plantlets: from laboratory to land. Biotechnol Lett 32:1199–1205

Chung JP, Huang CY, Dai TE (2010) Spectral effects on embryogenesis and plantlet growth of *Oncidium* 'Gower Ramsey'. Sci Hortic 124:511–516

Cioć M, Szewczyk A, Żupnik M, Kalisz A, Pawłowska B (2018) LED lighting affects plant growth, morphogenesis and phytochemical contents of *Myrtus communis* L. *in vitro*. Plant Cell Tissue Organ Cult 132(3):433–447

Dewir YH, Chakrabarty D, Hahn EJ, Paek KY (2006) Flowering of *Euphorbia millii* plantlets *in vitro* as affected by paclobutrazol, light emitting diodes (LEDs). Acta Hortic 764:169–173

Duc NM, Kasai R, Nham NT, Yamasaki K, Tanaka O (1997) New dammarane saponins from Vietnamese ginseng. Proceed Inter Symp Plant Glycos, Kunming, China

Fazal H, Abbasi BH, Ahmad N, Ali SS, Akbar F, Kanwal F (2016) Correlation of different spectral lights with biomass accumulation and production of antioxidant secondary metabolites in callus cultures of medicinally important *Prunella vulgaris* L. J Photochem Photobiol B Biol 159:1–7

Gupta SD, Jatothu B (2013) Fundamentals and applications of light-emitting diodes (LEDs) in *in vitro* plant growth and morphogenesis. Plant Biotechnol Reports 7:211–220

Gupta SD, Pradhan S (2017) Influence of LED lighting on *in vitro* plant regeneration and associated cellular redox balance. In: Gupta SD (ed) Light emitting diodes for agriculture. Springer, Singapore, pp 273–303

Gupta SD, Sahoo TK (2015) Light emitting diode (LED)-induced alteration of oxidative events during *in vitro* shoot organogenesis of *Curculigo orchioides* Gaertn. Acta Physiol Plant 37:233

Hahn EJ, Kozai T, Paek KY (2000) Blue and red light-emitting diodes with or without sucrose and ventilation affect *in vitro* growth of *Rehmannia glutinosa* plantlets. J Plant Biol 43:247–250

Heo JW, Lee CW, Chakrabarty D, Paek KY (2002) Growth responses of marigold and salvia bedding plants as affected by monochromic or mixture radiation provided by a Light-Emitting Diode (LED). Plant Growth Regul 38:225–230

Heo JW, Shin KS, Kim SK, Paek KY (2006) Light quality affects *in vitro* growth of grape 'Teleki 5BB'. J Plant Biol 49(4):276–280

Hoenecke ME, Bula RJ, Tibbitts TW (1992) Importance of 'blue' photon levels for lettuce seedlings grow under red light-emitting diodes. HortSci 27:427–430

Holonyak N, Bevacqua SF (1962) Coherent (visible) light emission from Ga(As1-xPx) junctions. Appl Phys Lett 1(4):82–83

Huan LVT, Tanaka M (2004) Effects of red and blue light-emitting diodes on callus induction, callus proliferation and protocorm like body formation from callus in *Cymbidium* orchid. Environ Control Biol 42(1):57–64

Hung CD, Hong CH, Jung HB, Kim SK, Van Ket N, Nam MW, Choi DH, Lee HI (2015) Growth and morphogenesis of encapsulated strawberry shoot tips under mixed LEDs. Sci Hortic 194: 194–200

Hung CD, Hong CH, Kim SK, Lee KH, Park JY, Dung CD, Nam MW, Choi DH, Lee HI (2016a) *In vitro* proliferation and ex vitro rooting of microshoots of commercially important rabbiteye blueberry (*Vaccinium ashei* Reade) using spectral lights. Sci Hortic 211: 248–254

Hung CD, Hong CH, Kim SK, Lee KH, Park JY, Nam MW, Choi DH, Lee HI (2016b) LED light for *in vitro* and *ex vitro* efficient growth of economically important highbush blueberry (*Vaccinium corymbosum* L.). Acta Physiol Plant 38:152

Hunt R, Causton DR, Shipley B, Askew AP (2002) A modern tool for classical plant growth analysis. Ann Bot 90(4):485–488

Jao RC, Fang W (2004) Growth of potato plantlets *in vitro* is different when provided concurrent versus alternating blue and red light photoperiods. HortSci 39:380–382

Jao RC, Lai CC, Fang W, Chang SF (2005) Effect of red light on the growth of *Zantedeschia* plantlets *in vitro* and tuber formation using light emitting diodes. HortSci 40(2):436–438

Karatas M, Aasim M, Dazkirli M (2016) Influence of light-emitting diodes and benzylaminopurine on adventitious shoot regeneration of water hyssop (*Bacopa monnieri* (L.) Pennell) *in vitro*. Arch Biol Sci 68(3):501–508

Khurshid R, Ullah MA, Tungmunnithum D, Drouet S, Shah M, Zaeem A, Hameed S, Hano C, Abbasi BH (2020) Lights triggered differential accumulation of antioxidant and antidiabetic secondary metabolites in callus culture of *Eclipta alba* L. PLoS One 15:e0233963

Kim SJ, Hahn EJ, Heo JW, Paek KY (2004) Effects of LEDs on net photosynthetic rate, growth and leaf stomata of Chrysanthemum plantlets *in vitro*. Sci Hortic 101:143–151

Kondo S, Tomiyama H, Rodyoung A, Okawa K, Ohara H, Sugaya S, Terahara N, Hirai N (2014) Abscisic acid metabolism and anthocyanin sysnthesis in grape skin are affected by light emitting diode (LED) irradiation at night. J Plant Physiol 171:823–829

Kopsell DA, Sams CE (2013) Increases in shoot tissue pigments, glucosinolates, and mineral elements in sprouting broccoli after exposure to short-duration blue light from light emitting diodes. J Am Soc Hortic Sci 138:31–37

Kreuzaler F, Hahlbrock K (1973) Flavonoid glycosides from illuminated cell suspension cultures of *Petroselinum hortense*. Phytochemistry 12:1149–1152

Lee J, Nam J, Park HC, Na G, Miura K, Jin JB, Yoo CY, Baek D, Kim DH, Jeong JC, Kim D, Lee SY, Salt DE, Mengiste T, Gong Q, Ma S, Bohnert HJ, Kwak SS, Bressan RA, Hasegawa PM, Yun DJ (2007) Salicylic acidmediated innate immunity in Arabidopsis is regulated by SIZ1 SUMO E3 ligase. Plant J 49(1):79–90

Lee NN, Choi YE, Moon HK (2014) Effect of LEDs on shoot multiplication and rooting of rare plant *Abeliophyllum distichum* Nakai. J Plant Biotechnol 41:94–99

Li H, Xu Z, Tang C (2010) Effect of light-emitting diodes on growth and morphogenesis of upland cotton (*Gossypium hirsutum* L.) plantlets *in vitro*. Plant Cell Tissue Organ Cult 3:155–163

Lian ML, Murthy HN, Paek KY (2002) Effect of light emitting diodes (LEDs) on the *in vitro* induction and growth of bulblets of *Lilium* oriental hybrid 'Pesaro'. Sci Hort 94:365–370

Lin X, Sze SCW, Tong Y, Zhang Z, Feng Y, Chen JP, Ng TB, Lin X, Shaw PC, Zhang KY (2010) Protective effect of Dendrobium officinale polysaccharides on experimental Sjogren's syndrome. J Complement Integr Med 7(1):1

Lin X, Shaw PC, Sze SCW, Tong Y, Zhang Y (2011) Dendrobium officinale polysaccharides ameliorate the abnormality of aquaporin 5, pro-inflammatory cytokines and inhibit apoptosis in the experimental Sjögren's syndrome mice. Int Immunopharmacol 11(12):2025–2032

Liu CZ, Guo C, Wang YC, Ouyang F (2002) Effect of light irradiation on hairy root growth and artemisinin biosynthesis of *Artemisia annua* L. Biochemistry 38:581–585

Lotfi M, Zamanzadeh V, Valizadeh L, Khajehgoodari M (2019) Assessment of nurse–patient communication and patient satisfaction from nursing care. Nurs Open 6(97):1–8

Merkle SA, Montello PM, Xia X, Upchurch BL, Smith DR (2005) Light quality treatments enhance somatic seedling production in three southern pine species. Tree Physiol 26:187–194

Miyashita Y, Kitaya Y, Kozai T (1995) Effects of red and farred light on the growth and morphology of potato plantlets *in vitro* using light emitting diode as a light source for micropropagation. Acta Hortic 393: 189–194

Moon HK, Park SK, Kim YW, Kim CS (2006) Growth of Tsuru-rindo (*Tripterospermum japonicum*) cultured *in vitro* under various sources of light-emitting diode (LED) irradiation. J Plant Biol 49:174–179

Moreira da Silva MH, Debergh PC (1997) The effect of light quality on the morphogenesis of *in vitro* cultures of *Azorina vidalii* (Wats.) Feer. Plant Cell Tissue Organ Cult 51:187–193

Mortensen LM, Stromme E (1987) Effects of light quality on some greenhouse crops. Sci Hortic 33: 27–36

Nakamura S, Pearton S, Fasol G (1997) The blue laser diode. Springer, Berlin, 368p

Narumon C (1998) The effect of growth regulation on quality and vase-life of Chrysanthemum. Thai Lan, Bangkok, pp 143–146

Nascimento LB, Leal-Costa MV, Coutinho MA, Moreira Ndos S, Lage CL, Barbi Ndos S, Costa SS, Tavares ES (2013) Increased antioxidant activity and changes in phenolic profile of *Kalanchoe pinnata* (Lamarck) persoon (Crassulaceae) specimens grown under supplemental blue light. Photochem Photobiol 89:391–399

Nhut DT (2002) *In vitro* growth and physiological aspects of some horticultural plantlets cultured under red and blue light-emitting diodes (LEDs), Doctoral thesis, Kagawa University, Japan

Nhut DT, Nam NB (2010) Light emitting diodes (LEDs): an artificial lighting source for biological studies. In: Proceed 3rd inter conf devel BME in Vietnam, 11–14 January 2010

Nhut DT, Takamura T, Watanabe H, Okamoto K, Tanaka M (2003) Responses of strawberry plantlets cultured *in vitro* under super bright red and blue light-emitting diodes (LEDs). Plant Cell Tissue Organ Cult 73:43–52

Nhut DT, Takamura T, Watanabe H, Okamoto K, Tanaka M (2005) Artificial light source using light-emitting diodes (LEDs) in the efficient micropropagation of *Spathiphyllum* plantlets. Acta Hortic 692:137–142

Nhut DT, Huy NP, Binh NV, Nam NB, Thuy LN, Ha DT, Chien HX, Huong TT, Cuong LK, Hien VT (2011) Shoot regeneration and micropropagation of *Panax vietnamensis* Ha *et* Grushv. from *ex vitro* leaf-derived callus. Afr J Biotechnol 10:19499–19504

Ouzounis T, Fretté X, Rosenqvist E, Ottosen CO (2014) Spectral effects of supplementary lighting on the secondary metabolites in roses, Chrysanthemums, and campanulas. J Plant Physiol 171: 1491–1499

Park SY, Kim MJ (2010) Development of zygotic embryos and seedlings is affected by radiation spectral compositions from light emitting diode (LED) system in Chestnut (*Castanea crenata* S. et Z.). J Korean Soc Forest Sci 99(5):750–754

Pinho P, Moisio O, Tetri E, Halonen L (2004) Photobiological aspects of crop plants grown under light emitting diodes. In: Proceed CIE symp 04, LED light sources: physical measurements and visual and photobiological assessment, Tokyo, Japan, CIE (026): 71–74

Poudel RP, Kataoka I, Mochioka R (2008) Effect of red-and bluelightemitting diodes on growth and morphogenesis of grapes. Plant Cell Tissue Organ Cult 92:147–153

Ramírez-Mosqueda MA, Iglesias-Andreu LG, Bautista-Aguilar JR (2017) The effect of light quality on growth and development of *in vitro* plantlet of *Stevia rebaudiana* Bertoni. Sugar Tech 19:331–336

Shin KS, Murthy HN, Heo JW, Hahn EJ, Paek KY (2008) The e ffect of light quality on the growth and development of *in vitro* cultured *Doritaenopsis* plants. Acta Physiol Plant 30:339–343

Soebo A, Krekling T, Appelgren M (1995) Light quality effects photosynthesis and leaf anatomy of birch plantlets *in vitro*. Plant Cell Tissue Organ Cult 41:177–185

Steigerwald DA, Bhat JC, Collins D, Fletcher RM, Holcomb MO, Ludowise MJ, Martin PS, Rudaz SL (2002) Illumination with solid state lighting technology. IEEE J Sel Top Quantum Electron 8:310–320
Taiz L, Zeiger E (1991) Plant physiology. Benjamin/Cummings, California, US
Tanaka M, Takamura T, Watanabe H, Endo M, Yanagi T, Okamoto K (1998) *In vitro* growth of Cymbidium plantlets cultured under superbright red and blue light-emitting diodes (LEDs). J Hortic Sci Biotechnol 73(1): 39–44
Teixeira da Silva JA (2004) Ornamental Chrysanthemums: improvement by biotechnology—review of plant biotechnology and applied genetics. Plant Cell Tissue Organ Cult 79:1–18
Tennessen DJ, Singsaasl EL, Sharkeyl TD (1994) Light-emitting diodes as a light source for photosynthesis research. Photosynth Res 39:85–92
Tripathy BC, Brown CS (1995) Root-shoot interaction in the greening of wheat seedlings grown under red light. Plant Physiol 107:407–411
United States Department of Agriculture (2010) Floriculture crops 2009 summary. Nat Agr Stat Serv, 76p
Vestberg M, Cassells AC, Schubert A, Cordier C, Gianinazzi S (2002) AMF and micropropagation of high value crops. In: Gianinazzi S, Schüepp H, Barea JM, Haselwandter K (eds) Mycorrhizal technology in agriculture. Birkhäuser Verlag, Basel, pp 223–233
Wang YC, Zhang HX, Zhao B, Yuan XF (2001) Improved growth of *Artemisia annua* L. hairy roots and artemisinin production under red light conditions. Biotechnol Lett 23:1971–1973
Wilken D, Gonzalez EJ, Gerth A, Gómez-Kosky R, Schumann A, Claus D (2014) Effect of immersion systems, lighting, and TIS designs on biomass increase in micropropagating banana (*Musa* spp. cv. 'Grande naine' AAA). In Vitro Cell Dev Biol Plant 50:582–589
Wongnok A, Piluek C, Techasilpitak T, Tantivivat S (2008) Effects of light emitting diodes on micropropagation of *Phalaenopsis* orchids. Acta Hortic 788:149–156
Xu Y, Yang M, Cheng F, Liu S, Liang Y (2020) Effects of LED photoperiods and light qualities on *in vitro* growth and chlorophyll fluorescence of *Cunninghamia lanceolata*. BMC Plant Biol 20(1):269
Younas M, Drouet S, Nadeem M, Giglioli-Guivarc'h N, Hano C, Abbasi BH (2018) Differential accumulation of silymarin induced by exposure of *Silybum marianum* L. callus cultures to several spectres of monochromatic lights. J Photochem Photobiol B 184:61–70
Yu KW, Murthy HN, Hahn EJ, Paek KY (2005) Ginsenoside production of hairy root cultures of Panax ginseng: influence of temperature and light. Biochem Eng J 23:53–56
Yulian FY, Okuda N, Fukada N (1995) Effects of plant growth regulators on budding of garland Chrysanthemum (*Chrysanthemum coronarium* L.). Tech Bullf Facult Agric 47(2):107–113
Zhong JJ, Seki T, Kinoshita SI, Yoshida T (1991) Effect of light irradiance on anthocyanin production by suspended culture of *Perilla frutescens*. Biotechnol Bioeng 38:653–658

In Vitro Hydroponic Culture System in Plant Micropropagation 10

Duong Tan Nhut, Ha Thi My Ngan, Nguyen Thi Nhu Mai, and Hoang Thanh Tung

Abstract

A combined approach of in vitro culture and hydroponics in plant propagation is detailed in this chapter. The background technical information and the positive effects of a novel in vitro hydroponic system on the rooting of Chrysanthemum (*Chrysanthemum morifolium*) explants and microtuber development in Potato (*Solanum tuberosum*) are documented. Simple creative designs and common laboratory supplies enable the construction of apparatuses for the successful propagation of explants.

Keywords

Chrysanthemum · In vitro hydroponic · Microtubers · Rooting · *Solanum tuberosum*

10.1 Introduction

Micropropagation proves to be an effective method for multiplying plants on a largescale within a short time. Moreover, problems remain. The capped-vessel with a gelled-medium results in high humidity and high levels of carbon dioxide and ethylene within the culture environment (Buddendorf-Joosten and Woltering 1994). These conditions can lead to browning of tissues, vitrification of explants, and altered structural development (Ziv 1991). In micropropagated explants, stomata may not function properly, the root system is poorly developed, and the epidermal layer is thin, resulting in desiccation intolerance (Mathur et al. 2008; Kumar and Rao

D. T. Nhut (✉) · H. T. M. Ngan · N. T. N. Mai · H. T. Tung
Tay Nguyen Institute for Scientific Research, VAST, Dalat City, Vietnam

D. T. Nhut et al. (eds.), *Plant Tissue Culture: New Techniques and Application in Horticultural Species of Tropical Region*,
https://doi.org/10.1007/978-981-16-6498-4_10

2012). All these issues can lead to difficulties in plantlet acclimatization at the nursery stage.

Under in vitro culture conditions, the plantlets are accustomed to controlled conditions during growth. In the acclimatization stage, plants are exposed to adverse external factors such as microbial pathogens (mostly fungi and bacteria), temperature fluctuations, low humidity, and poor nutrition. Together with sudden changes in growth conditions, these factors reduce the survival percentage of plants significantly (Mathur et al. 2008; Kumar and Rao 2012). Therefore, improving the culture conditions and procedures at the micropropagation and acclimatization stages can improve plant growth vigor, adaptability, and survival of plantlets at the nursery stage.

Hydroponics is a plant culture technique, which enables plant growth in a nutrient solution with mechanical support from an inert substrate. A vast amount of information is available in the literature, for example, Schwartz (1995), Asao (2012), and Maucieri et al. (2019). The growth conditions, that is, nutrient levels, light intensity, photoperiod, and relative humidity for plant growth, are readily controlled and optimized. Many adverse situations, such as diseases, can be minimized with proper management practices.

In recent years, the techniques of hydroponics have been applied to plant micropropagation procedures. A combined approach of micropropagation followed by hydroponic culture of explants for subsequent acclimatization is becoming popular. A microponic technique in which in vitro Chrysanthemum explants are subsequently maintained in a hydroponic system to produce disease-free plantlets has been reported by Hahn et al. (1996) and, more recently, by Nhut et al. (2005) and Tung et al. (2018). For the microponic system, it does not require completely sterile conditions. However, plants or cuttings should be kept under clean conditions to minimize contamination (Nhut et al. 2005). Other modifications to the combined approach are noted in the literature, for example, for the growth of *Thymus* (Sargsyan et al. 2011) and cassava (Castañeda-Méndez et al. 2017) explants. Most recently, Purohit et al. (2020) published a hydroponic-based hardening protocol for in vitro raised kiwifruit. The combined approach inherits many advantages of micropropagation and hydroponic methods. It can overcome some inherent issues of plant micropropagation, such as vitrification, rooting, and poor development of the newly generated plantlets. As a result, the fragile in vitro grown plants are readily hardened and better adapted to the acclimatization process.

To further improve the system, depending on the explants and the objective of the experiment, the microponic technique can be conducted under a more stringent sterile condition. This approach is termed "*in vitro* hydroponics" (IH)—a propagation system that combines in vitro micropropagation and sterile hydroponic techniques. For IH, one can design a simple apparatus to suit one's purpose. This approach has been successfully employed to produce microtubers in Potatoes (Nhut et al. 2006) and wasabi nursery plants (Hoang et al. 2019). An improved, low-cost, aseptic hydroponic system has been established to grow Arabidopsis for various biological studies (Alatorre-Cobos et al. 2014).

This chapter's primary objective is to provide technical information and demonstrate the IH system's usefulness in *Chrysanthemum morifolium* and *Solanum tuberosum* plantlets' growth and development. We also advocate that simple home-made designs are useful and adequate for use in IH systems.

10.2 In Vitro Hydroponic

10.2.1 In Vitro Hydroponic and Plant Growth and Development

The term hydroponics was first introduced and described by Gericke (1937) for growing plants in water or soilless conditions for commercial purposes. The hydroponic system has been practiced for centuries (Jones 1982). In the 1990s, specialized hydroponic kits became popular for the production of commercial vegetables. Various porous materials were introduced as substrates suitable as inorganic and organic growing media. The nutrient solution in commercial hydroponic systems and equipment and automation protocols continue to improve over the past decades. It is recognized that hydroponics can be a useful tool for producing clean and safe vegetables (Asao 2012).

The IH approach to plant micropropagation is a bridge between in vitro culture and large-scale hydroponic plant growth. In vitro hydroponics offers many advantages similar to the conventional hydroponics but in a smaller scale. In IH, plants grow in a nutrient solution with inert substrates such as vermiculite, perlite, rockwool, cotton fibers, and nylon film, replacing traditional gelling agents, that is, agar and Gelrite (Oh et al. 2012; Tung et al. 2018). The use of these supporting media improves the culture system's aeration, and the liquid medium enables a more efficient uptake of nutrients by the explants. Since the explants are maintained under sterile conditions, contamination is kept at a minimum. This culture method prepares the explants for subsequent acclimatization events, resulting in improved vigor and success in plant propagation.

10.2.2 General Information on How to Establish In Vitro Hydroponic Systems

With clearly defined objectives, one can design suitable apparatuses for culturing explants using the IH approach. The apparatus need not to be complicated. The key requirement is that sterile conditions can be maintained during the apparatus's assembly and treatment period. The following systems serve as examples of such devices in promoting rooting in Chrysanthemum (Tung et al. 2018) and microtuber formation in Potato (Nhut et al. 2006).

For promoting rooting in Chrysanthemum, small plastic tubes, for example, nylon film tubes (1.5 cm × 1.5 cm) (Flexoffice, Thien Long Stationery Group, Vietnam), serve as support to shoots derived from cultured plantlets. The tubes are housed in glass bottles (250 mL) containing 40 mL of liquid culture medium, sealed with a

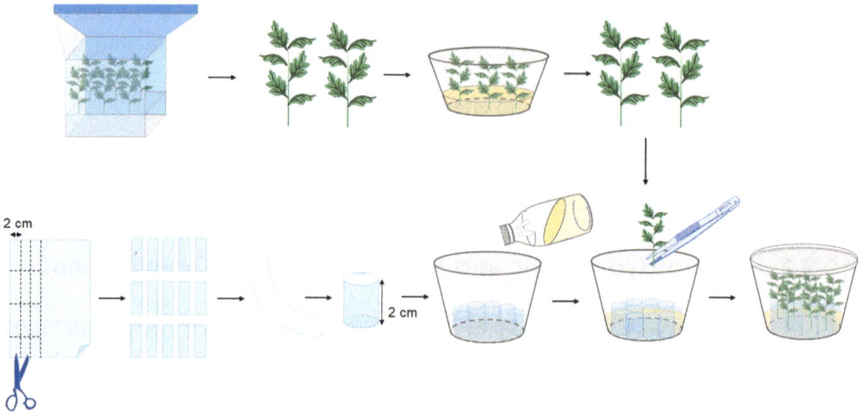

Fig. 10.1 Establishment of the in vitro hydroponic system for rooting stage of *Chrysanthemum morifolium*

plastic cap. Each bottle includes ten nylon film tubes (Fig. 10.1). The IH system using small plastic tubes as support helps to simplify the culturing procedure. This system resulted in faster and healthier plantlets and rooting responses, ensuring its quality (Tung et al. 2018). This effectively saves time and costs of the operation.

Micropropagation of Potato through microtuber formation in vitro greatly improves its seed production capability. This method generates virus-free seed Potatoes. The small size of the microtubers allows ease of storage and transport. An IH system can further enhance microtuber formation and at a low cost (Nhut et al. 2006). The IH system consists of the following components: (1) Cylindered polypropylene capped boxes serve as culture vessels (11.5 cm in diameter, 8.5 cm in height, Dai Dong Tien Plastics Company, Vietnam, manufacture code: L621). (2) Polypropylene rings are cut from nylon bags, which help to hold the carrier sheets and keep them above the liquid medium level (36 cm in length, 2 cm in large, 0.05 cm in thickness). (3) Thin layers of carrier substrate made of a pad of cotton fibers or filter papers serve to support and retain nutrients for the explants (11.4 cm in diameter, 0.2 cm in thickness). (4) Carrier plates (11.4 cm in diameter, 0.2 cm in thickness), made from the lid of plastic boxes, serve as a support to the carrier substrate. A notch is cut at the edge of the carrier plate, allowing for a cotton string (4 cm in length) for nutrient transfer by capillary action. These parts are sterilized separately at 121 °C, 15 psi, for 30 min and assembled under aseptic conditions (Fig. 10.2).

To set up a three-level IH system, place 50 mL of a nutrient medium into the plastic box. Add a polypropylene ring as a spacer to the bottom of the box and place the plastic carrier plate with a cotton fiber pad or filter papers onto the spacer plastic ring. Apply 2 mL of the liquid medium to the cotton or filter paper substrate and gently place 15 stem nodes on top of the substrate. Repeat the process to build the second and third levels, if necessary. The capillary cotton string was finally placed in the system from top to bottom through the notch at the edge of each level. To finish,

Fig. 10.2 Establishment of the in vitro hydroponic culture system for microtuberization in Potato

cap the cylindrical box and cover it with sterilized polypropylene wrap to keep the system under aseptic conditions (Fig. 10.2).

Recently Hoang et al. (2017) reported the successful growth of wasabi plantlets (*Wasabia japonica*) using an IH culture system. Each culture unit consists of a culture flask containing a liquid nutrient medium with vermiculite as the supporting material. The culture flasks are covered using a vented cover and maintained in a sterile environment. After 28 days of culture, most growth parameters such as fresh and dry weights, shoot/root dry weight ratio, and leaf area ratio were highest in 100% nutrient solution with vermiculite as substrate. The use of vermiculite instead of rockwool improves oxygen concentration allowing wasabi plantlets to multiply rapidly (Hoang et al. 2019).

The following sections illustrate the application of IH systems in the rooting of Chrysanthemum explants and the formation of microtubers in Potatoes.

10.3 Enhanced Growth and Rooting of *Chrysanthemum morifolium* Using In Vitro Hydroponic System

Chrysanthemum is one of the most important cut flower crops in Vietnam, occupying about 25% of the cut flower production area (Linh 2002). As production expanded, demand for seedlings also increased. Chrysanthemum is propagated mainly by cuttings because it is simple, economical, and can be carried out under ex vitro conditions. However, this method has some limitations, such as a low multiplication coefficient and poor seedling quality. Consequently, there is a need for a more efficient system for propagation. Micropropagation proves to be an effective method for multiplying plants on a large scale in a short time. The success of micropropagation methods depends heavily on the success of acclimatization in nursery conditions. During micropropagation, the growth of plants is controlled, while in the acclimatization stage, plants are exposed to numerous adverse external factors such as microbial pathogens (mostly fungi and bacteria), temperature fluctuation, low humidity, and poor nutrition. These factors significantly reduce plants' survival rate (Valero-Aracama et al. 2006; Mathur et al. 2008). When transferred to ex vitro conditions, physiological changes within plant organs also lead to morphological and anatomical abnormalities. Moreover, by combining micropropagation and hydroponic techniques, plantlets' quality is improved (Tung et al. 2018).

10.3.1 Materials and Methods

Wash axillary buds of 3-month-old *Chrysanthemum morifolium* plants under running tap water for 30 min and soak in 70% ethanol for 30 min, followed by sterilization with 0.1% w/v $HgCl_2$ for 7 min. Then rinse the buds three times with sterilized distilled water and cultured in 100 mL glass vessels containing 20 mL MS (Murashige and Skoog 1962) medium supplemented with 0.5 mg/L 6-benzyladenine (BA), 30 g/L sucrose, and 8 g/L agar to induce adventitious shoots.

Place the newly generated adventitious shoots (3 cm) with two pairs of leaves inside nylon tubes with a liquid medium (MS medium supplemented with 0.5 mg/L IBA, 30 g/L sucrose) of an IH system to evaluate the effects of the culture system on the growth and rooting efficiency of Chrysanthemum plantlets. For the control, culture Chrysanthemum shoots on a solid medium. Adjust the pH of all media to 5.8 before autoclaving at 121 °C for 30 min.

10.3.2 Results and Discussion

Rooting is the final stage of micropropagation and determines the quality of plantlets. Thus, improving rooting efficiency is one of the essential goals of the micropropagation process. After 4 weeks of culture, the effects of culture systems on the in vitro growth and rooting of Chrysanthemum plantlets are shown in Fig. 10.3. The results indicated that the IH system was most suitable for the Chrysanthemum rooting stage (Fig. 10.3).

The results show remarkable differences in shoot height parameters, the number of roots, root length, fresh weight, dry weight, and dry mass ratio. The IH system promoted optimum growth of in vitro Chrysanthemum plantlets compared to the control treatment (in vitro micropropagation) in all growth and development parameters (Fig. 10.3). The use of the liquid medium enables a more efficient uptake of nutrients resulting in better plantlets' growth. Also, the IH system gave the best rooting efficiency with healthy-looking roots and longer roots. The rooting time was even faster compared to the control (Fig. 10.4).

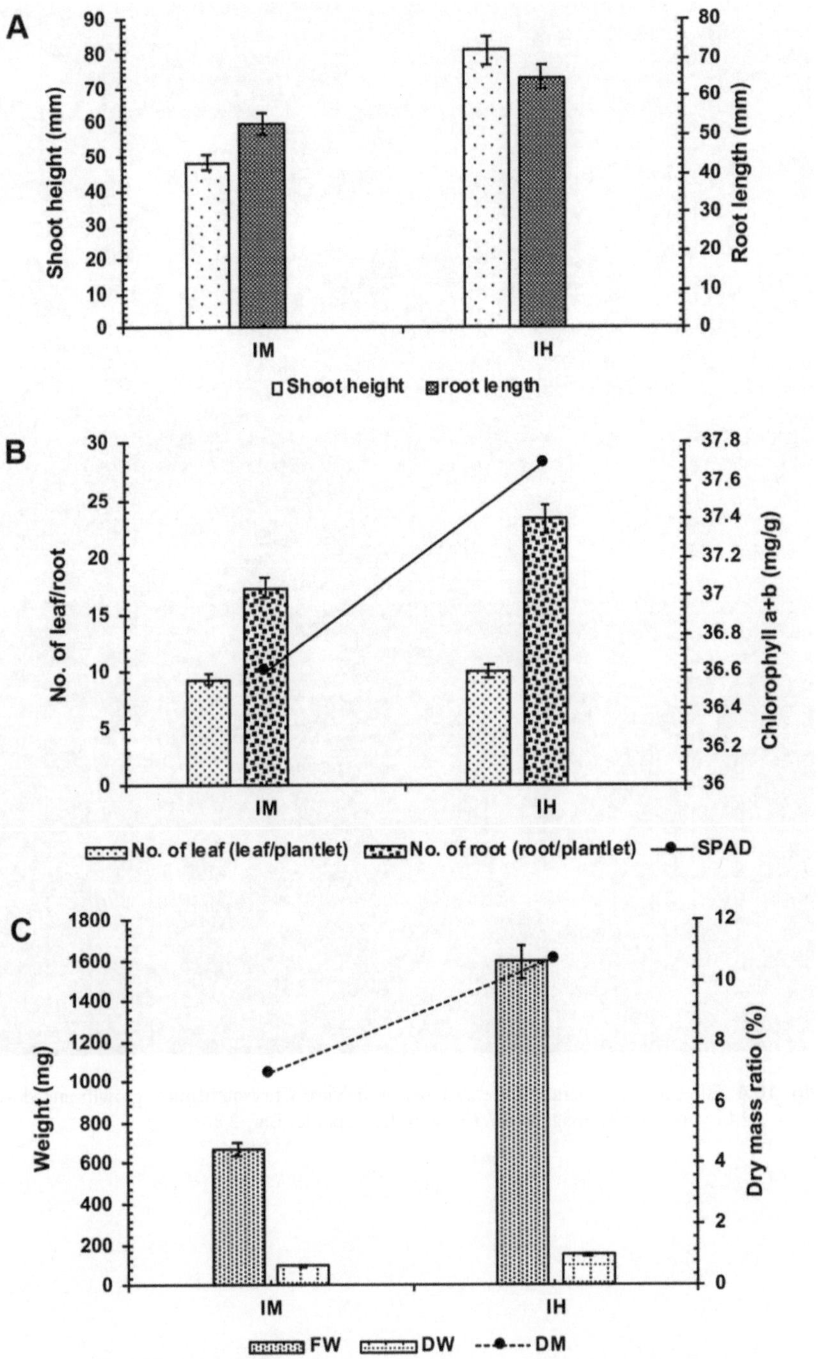

Fig. 10.3 Effects of different culture systems on in vitro Chrysanthemum growth and development. *IM* In vitro micropropagation, *IH* In vitro hydroponic, *SPAD* Chlorophyll a + b content, *FW* Fresh weight, *DW* Dry weight, *DM* Dry mass ratio

Fig. 10.4 Effects of different culture systems on in vitro Chrysanthemum growth and development. *IM* In vitro micropropagation, *IH* In vitro hydroponic. Bar: 2 cm

10.4 Enhanced Potato Microtubers' Formation and Germination Capacity Using a Novel in Vitro Hydroponic System

Potato (*Solanum tuberosum* L.) is one of the most important agricultural plants of the world. Its annual worldwide output is just after Rice (*Oryza sativa*), Wheat (*Triticum aestivum*), and Barley (*Hordeum vulgare*). In some countries, Potato constitutes the main daily food due to their low price and highly nutritive value (Vinterhalter et al. 2008). In vitro propagated Potato plantlets are commonly used in Potato seed production programs for the production of tubers.

Microtubers (in vitro tubers) can now be produced and stored in the laboratory year-round to be directly transported to the market without transferring to fresh media (Nhut et al. 2004). Besides, the production of microtubers also ensures a disease-free seed source, which can be stored in the laboratory all year round. Also, these microtubers can be transplanted directly to the field without an acclimatization stage (Jimerez et al. 1999). However, there are limitations in microtuber production originating from the components of the culture environment and to the low photosynthetic ability of the explants or plantlets. Most systems currently used for microtuber production suffer from inefficiencies regarding the number and size of microtubers produced per cycle. A model version of a low-cost IH system was constructed using a simple cylindrical plastic box consisting of one or three Stories (Nhut et al. 2006). The ability to produce Potato microtubers using this novel system and the influence of the number of stories on microtuber formation rate and quality were investigated, and two different porous materials used as carrier substrates were also tested (Nhut et al. 2006). The system is described below.

10.4.1 Materials and Methods

Dissect, aseptically Potato meristems from 2-month-old Potato plants under a stereoscopic microscope and culture on ½MS (MS half-strength) medium supplemented with 0.5 mM naphthaleneacetic acid (NAA), 0.3 mM gibberellic acid (GA_3), 3.7 mM adenine sulfate, 10% coconut water, 0.5 g/L activated charcoal, 20 g/L sucrose, and 8 g/L agar for 30 days. Use the in vitro Potato shoots as the primary explant source. Place stem-node cuttings (1.5 cm in length) on solid (in vitro micropropagation system) or semi-solid (in vitro hydroponic system) media depending on the purpose of the experiments.

Adjust the pH of the media to 5.8. In the control medium, adjust the pH before the addition of Gelrite. Dispense the media in the IH system under aseptic conditions after autoclaving at 121 °C, 15 psi for 30 min. Maintain all in vitro cultures at 25 ± 2 °C, 70–80% relative humidity, using fluorescent light with an intensity of 40–45 μmol m^{-2} s^{-1}, and a photoperiod 16 h/day (Rang Dong, Vietnam).

For the germination capacity experiment, cultivate microtubers in the greenhouse under natural light/dark circle of sunlight at 27 ± 3 °C and 75–85% relative humidity. Water the plantlets gently every 2 days with tap water.

10.4.2 Results and Discussion

The effects of the numbers of stories in the IH culture system on microtuber formation rate and growth were recorded and shown in Fig. 10.5. The formation rate (Fig. 10.5a), diameter, and fresh weight (Fig. 10.5b) of Potato microtuber varied in different stories of the system. Accordingly, in the first layer, which was closest to

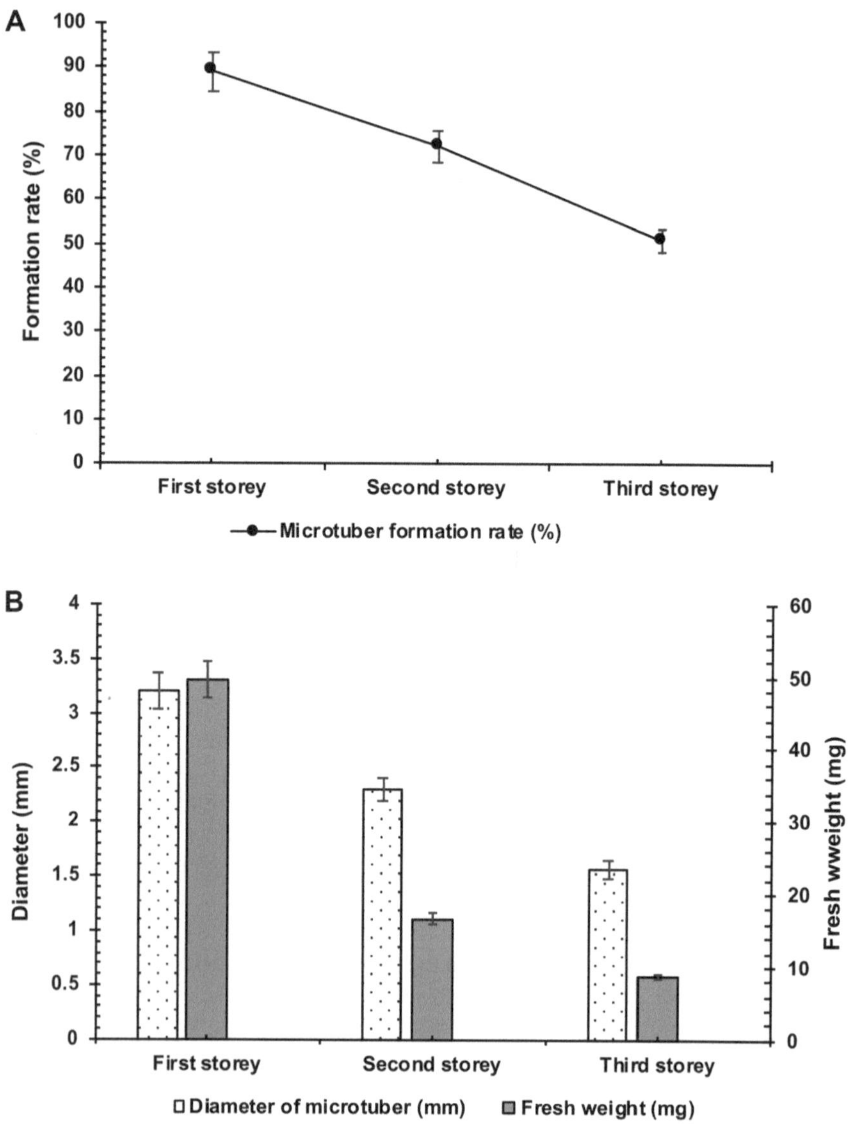

Fig. 10.5 Effects of story position in in vitro hydroponic culture system on Potato microtuberization. (**a**) Microtuber formation rate (%); (**b**) Growth of Potato microtuber

the nutrient medium, more microtubers (8–10 microtubers from one in vitro Potato plantlet) were obtained with bigger size and heavier weights than those from the second and third stories. In this IH system, since the lower levels are closer to the source of nutrients, explants are likely to obtain nutrients more efficiently than those in the upper levels. This results in more microtuber formation and better quality at the same time (Fig. 10.5).

After 30 days of cultivation, the formation of microtuber on solid media (agar substrate) and semi-solid media (in vitro hydroponic with cotton layer as substrate) was recorded (Figs. 10.6 and 10.8). The culture system, media/substrate used, significantly influences microtuber formation. The results showed the highest percentage (95%) of microtuber formation when the stem nodes were cultured on a solid medium (Fig. 10.6a). The diameter and fresh weight of microtubers formed from stem nodes cultured in IH systems (first floor) were higher than those on solid media (Fig. 10.6b). In IH system, bright yellow microtubers were obtained with uniform size. This result suggests that the cotton layer can be used as a substrate in a hydroponic system for Potato microtuberization. The cotton layer substrate can store a large amount of water and nutrient solution without restricting gas exchange in the root zone, thereby enhancing the growth and nutrient accumulation of the microtubers.

Shoot formation was also observed when the explants were cultured on solid media (control treatment) and the IH system's second or third story. These growing conditions severely affect the size and quality of the microtubers, as some of the nutrients have been channeled into new shoots development. The stem node segments on the control treatment and the higher stories of the IH system tend to develop more shoots than microtubers. This may be because they are closer to the light source. As a result, microtuber formation is inhibited. This observation is similar to Gopal et al. (1998); they indicated an increase in microtuber numbers with decreasing light intensity. Besides, high light intensity also caused shoot formation and growth on newly formed microtubers. In order to improve the quality of microtubers, further experiments on the effects of light intensity on microtuber growth are required.

The germination capacity and survival percentage of Potato microtuber derived from IH and control treatment were recorded after 30 days in greenhouse conditions (Figs. 10.7 and 10.8). In this experiment, the high germination and survival percentages of microtubers from the IH culture system demonstrated this culture system's effectiveness in Potato micropropagation. The germination rate (96%) and survival rate (97%) of microtubers formed in the first story from the IH system were higher than those in the second and third levels and control treatments. The larger the microtubers, the better acclimatization, germination, and the development of stronger plantlets that are more resistant to disease. The IH culture system allows microtuberization in Potato to be completed in 30 days, significantly reducing the cost and labor compared to the traditional method (4–5 months) (Gopal et al. 1998). Therefore, the IH system with the cotton layer as substrate effectively produces many high-quality microtubers in a shorter time with a low cost. This method can be expanded, reaching a commercial-scale production. In commercial production, the

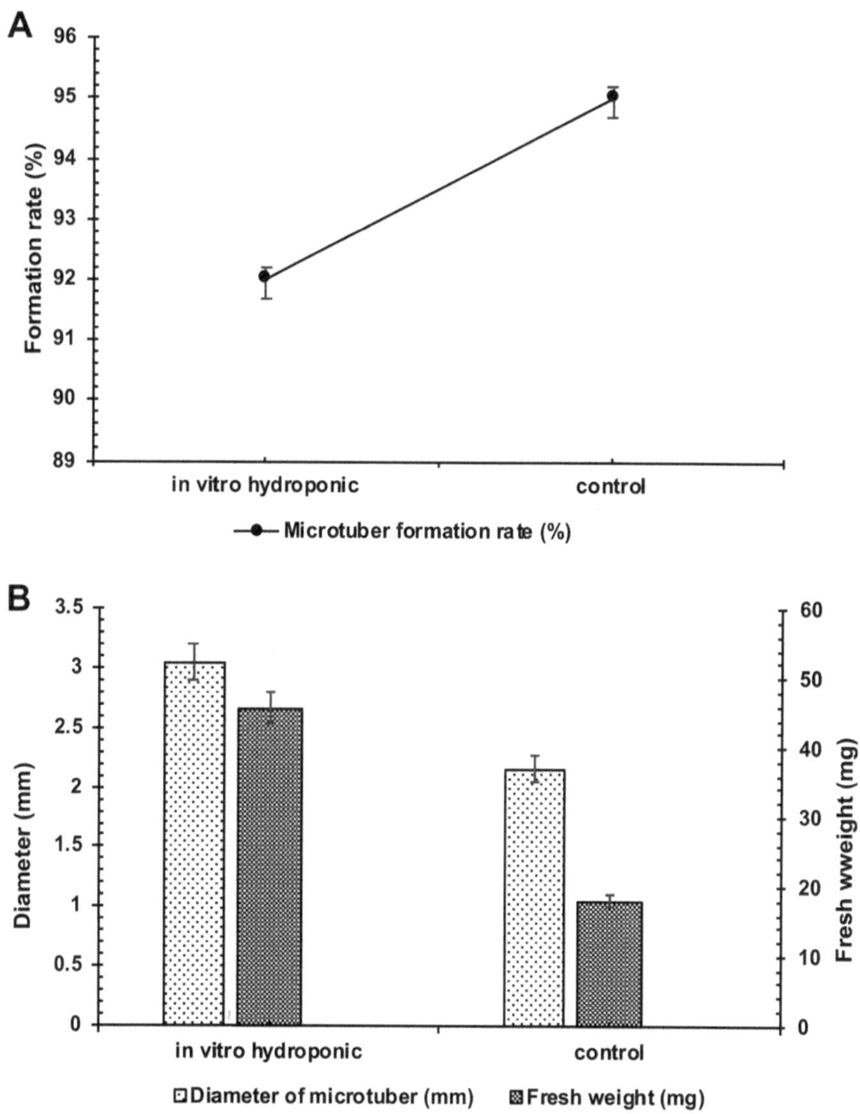

Fig. 10.6 Effects of cultural systems on Potato microtuberization. (**a**) Microtuber formation rate (%); (**b**) Growth of Potato microtuber; *Control* Agar solid media

quality and size of the microtubers are the decisive factors for growing efficiency when transplanting to the field conditions (Jimerez et al. 1999). Subsequent growth *of* in vitro hydroponic derived Potato plantlets in the greenhouse (Fig. 10.9).

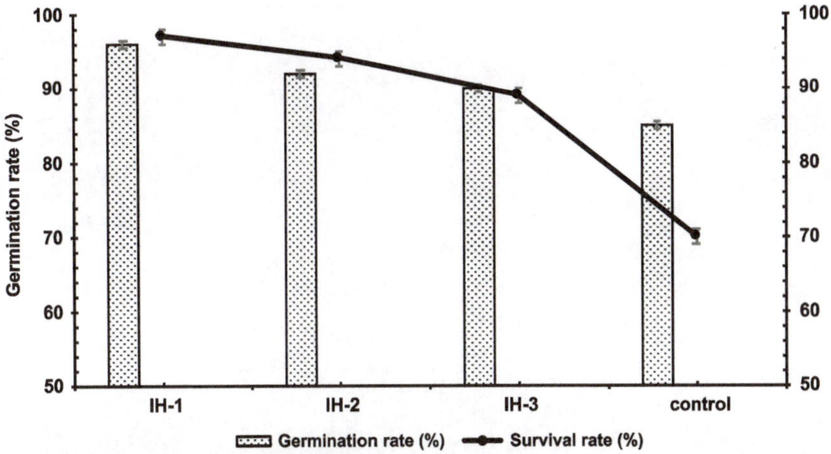

Fig. 10.7 Germination and survival rate of microtubers derived from different culture systems in greenhouse conditions. *IH-1* In vitro hydroponic-first story, *IH-2* In vitro hydroponic-second story, *IH-3* In vitro hydroponic-third story; *Control* Agar solid media

Fig. 10.8 Effects of different culture systems on in vitro Potato growth and development. (**a**) Stem node formation process; (**b**) In vitro micropropagation; (**c**) In vitro hydroponic

Fig. 10.9 Subsequent growth *of* in vitro hydroponic derived Potato plantlets in the greenhouse

10.5 Conclusion

The most significant advantage of an in vitro hydroponic system is that it is simple and easy to set up. The equipment used for the system is easy to find and can be purchased at a local store or over the Internet. The cost of the equipment can be kept to a minimum. Low-cost in vitro hydroponic systems constructed using glass bottles, plastic film, and liquid nutrient media have been shown to increase growth, improve rooting efficiency, and shorten Chrysanthemum's rooting time. Meanwhile, using a simple cylindrical plastic container with cotton layers' substrate consisting of one or three stories has improved microtuberization and quality of Potato in vitro culture.

References

Alatorre-Cobos F, Calderón-Vázquez C, Ibarra-Laclette E, Yong-Villalobos L, Pérez-Torres C-A, Oropeza-Aburto A (2014) An improved, low-cost, hydroponic system for growing *Arabidopsis* and other plant species under aseptic conditions. BMC Plant Biol 14:69

Asao T (2012) Hydroponics: a standard methodology for plant biological researches. Intech, pp 67–93

Buddendorf-Joosten JMC, Woltering EJ (1994) Components of the gaseous environment and their effects on plant growth and development *in vitro*. Plant Growth Regul 15:1–16

Castañeda-Méndez O, Ogawa S, Medina A, Chavarriaga P, Selvaraj MG (2017) A simple hydroponic hardening system and the effect of nitrogen source on the acclimation of *in vitro* cassava (*Manihot esculenta* Crantz). In Vitro Cell Dev Biol Plant 53:75–85

Gericke WF (1937) Hydroponics-crop production in liquid culture media. Science 85(2198):177–178

Gopal J, Minocha JL, Dhaliwal HS (1998) Microtuberization in potato (*Solanum tuberosum* L.). Plant Cell Rep 17:794–798

Hahn EJ, Lee YB, Ahn CH (1996) A new method on mass-production of micropropagated chrysanthemum plants using microponic system in plant factory. Acta Hortic 440:527–532

Hoang NN, Kitaya Y, Morishita T, Endo R, Shibuya T (2017) A comparative study on growth and morphology of wasabi plantlets under the influence of the micro-environment in shoot and root zones during photoautotrophic and photomixotrophic micropropagation. Plant Cell Tissue Organ Cult 130(2):255–263

Hoang NN, Kitaya Y, Shibuya T, Endo R (2019) Development of an *in vitro* hydroponic culture system for wasabi nursery plant production—effects of nutrient concentration and supporting material on plantlet growth. Sci Hort 245:237–243

Jimerez E, Perez N, Feria de M, Barbon R, Capote A, Chavez M, Quiala E, Perez JC (1999) Improved production of potato microtubers using a temporary immersion system. Plant Cell Tissue Organ Cult 59:19–23

Jones JB (1982) Hydroponics: its history and use in plant nutrition studies. J Plant Nutr 5:1003–1030

Kumar K, Rao IU (2012) Morphophysiological problems in acclimatization of micropropagated plants in *ex vitro* conditions—a review. J Ornam Hortic Plants 2:271–283

Linh NV (2002) Flower growing technique. Agricultural Publishing Hourse, Hanoi, pp 81–125

Mathur A, Mathur AK, Verma P, Yadav S, Gupta ML, Darokar MP (2008) Biological hardening and genetic fidelity testing of micro-cloned progeny of *Chlorophytum borivilianum* Sant. et Fernand. Afr J Biotech 7:1046–1053

Maucieri C, Nicoletto C, van Os E, Anseeuw D, Van Havermaet R, Junge R (2019) Hydroponic technologies. In: Goddek S, Joyce A, Kotzen B, Burnell GM (eds) Aquaponics food production systems. Springer, New York, pp 77–110

Murashige T, Skoog F (1962) A revised medium for rapid growth and bio-assays with tobacco tissue cultures. Plant Physiol 1(3):473–497

Nhut DT, Dieu Huong NT, Khiem DV (2004) Direct microtuber formation and enhanced growth in the acclimatization of *in vitro* plantlets of taro (*Colocasia esculenta* spp.) using hydroponics. Sci Hortic 101:207–212

Nhut DT, Don NT, An TTT, Van TPT, Vu NH, Huyen PX, Khiem DV (2005) Microponic and hydroponic techniques in disease-free chrysanthemum (*Chrysanthemum* sp.) production. J Appl Hortic 7(2):67–71

Nhut DT, Nguyen NH, Thuy DTT (2006) A novel *in vitro* hydroponic culture system for potato (*Solanum tuberosum* L.) microtuber production. Sci Hort 110:230–234

Oh MM, Seo JH, Park JS, Son JE (2012) Physicochemical properties of mixtures of inorganic supporting materials affect growth of potato (*Solanum tuberosum* L.) plantlets cultured photoautotrophically in a nutrient-circulated micropropagation system. Hortic Environ Biotechnol 53: 497–504

Purohit S, Rawat JM, Pathak VK, Singh DK, Rawat B (2020) A hydroponic-based efficient hardening protocol for *in vitro* raised commercial kiwifruit (*Actinidia deliciosa*). In Vitro Cell Dev Biol Plant 57(3):541–550

Sargsyan E, Vardanyan A, Ghalachyan L, Bulgadaryan S (2011) Cultivation of thymus by *in vitro* and hydroponics combined method. World Acad Sci Eng Technol 80:129–132

Schwartz M (1995) Soilless culture management. Springer, New York, pp 107–112

Tung HT, Nam NB, Huy NP, Luan VQ, Hien VT, Phuong TTB, Dung LT, Loc NH, Nhut DT (2018) A system for large scale production of chrysanthemum using microponics with the supplement of silver nanoparticles under light-emitting diodes. Sci Hort 232:153–161

Valero-Aracama C, Kane ME, Wilson SB, Vu JC, Anderson J, Philman NL (2006) Photosynthetic and carbohydrate status of easy-and difficult-to-acclimatize sea oats (*Uniola paniculata* L.) genotypes during *in vitro* culture and *ex vitro* acclimatization. In Vitro Cell Dev Biol Plant 42(6):572–583

Vinterhalter D, Dragicevic I, Vinterhalter B (2008) Potato *in vitro* culture techniques and biotechnology. Potato I fruit. Veg Cereal Sci Biotech 2:16–45

Ziv M (1991) Morphological and physiological disorders of *in vitro* plantlets. In: Debergh PC, Zimmerman RH (eds) Micropropagation: technology and application. Kluwer Academic Publishers, Dordrecht, pp 45–69

Microponic Culture System in the Propagation of Some Plants

11

Hoang Thanh Tung, Ha Thi My Ngan, Truong Thi Bich Phuong, and Duong Tan Nhut

Abstract

A microponic culture system, which combines micropropagation and hydroponic techniques, inherits many advantages of both procedures and improves conventional plant propagation methods. Simple, inexpensive equipment and materials can be used to construct a microponic system to suit one's objective. This chapter introduces a microponic system using circular plastic containers and rectangular plastic boxes for the culture of *Chrysanthemum* explants. In addition, the positive effects of using the light-emitting diodes as a light source and silver nanoparticles in nutrient media in conjunction with the microponic system are also presented. In addition, the microponic system has also been applied on a number of plants such as carnation, strawberry, potato, and gloxinia. The results also show that the plants gave good growth and development in the microponic system. From the research results recorded, in the future microponic system promises to be applied on many plants and can produce plantlets on a large scale.

Keywords

Chrysanthemum · Carnation · Strawberry · Circular plastic container · Light-emitting diodes · Microponic system · Rectangular plastic box · Silver nanoparticles

H. T. Tung (✉) · H. T. M. Ngan · D. T. Nhut
Tay Nguyen Institute for Scientific Research, VAST, Dalat City, Vietnam

T. T. B. Phuong
Hue University of Sciences, Hue City, Vietnam

D. T. Nhut et al. (eds.), *Plant Tissue Culture: New Techniques and Application in Horticultural Species of Tropical Region*,
https://doi.org/10.1007/978-981-16-6498-4_11

11.1 Introduction

11.1.1 General Introduction of Microponic Method

The micropropagation method has revolutionized plant production, which can provide many uniform plantlets while retaining the genetic characteristics of plants. However, micropropagation methods can have adverse effects on the growth of plants in vitro. In the in vitro culture environment, factors such as high relative humidity, constant temperature, low photon flux intensity, and highly variable CO_2 concentration can negatively impact plant growth and development (Kozai 1991). These factors can lead to a relatively high mortality rate when plantlets are transferred from a closed-in vitro environment to the open nursery for further development.

In recent years, there have been studies to develop new approaches and propagation protocols in which the culture medium and the in vitro environment can be controlled. Photo-autotrophic in vitro culture system has been introduced to overcome some disadvantages of traditional methods, for example, limiting contamination by using sugar-free culture media (Kozai 1991; Kozai et al. 2005). Moreover, the photo-autotrophic method focuses on controlling physical conditions of the culture environment, such as CO_2 content, light quality, and photoperiod. These efforts result in high operating costs due to the automation of environmental controls and associated labor costs in handling cultures.

Microponics combines micropropagation and hydroponic techniques to develop high-quality plantlets generated from in vitro cultures. This approach was first reported by Hahn et al. (1996, 1998, 2000), and it overcomes many disadvantages of the traditional micropropagation procedures. Using a similar approach, Nhut and associates (Nhut et al. 2005; Tung et al. 2018) developed a microponic system that has been successfully applied to *Chrysanthemum* plantlet propagation. In their studies, the microponic system was constructed using simple materials and equipment, that is, tubes made of nylon film and plastic containers. The construction and operation of such a microponic system are detailed in this chapter.

11.1.2 The Advantages of the Microponic Method

The microponic method is best applied at the rooting stage of plant propagation. With simple sterilization procedures and sanatory controls, in conjunction with liquid nutrient media without a carbon source such as sucrose, fungal and bacterial contaminations can be kept to a minimum. Without the added sugar and agar in the culture medium, the microponic method significantly reduces operating costs. Furthermore, without a sterilization treatment, media preparation costs and associated labor costs for subcultures are also minimized. Thus, for the rooting of explants, the operating procedure is simplified, and the overall operational cost can be significantly reduced compared to traditional in vitro culture methods (Nhut et al. 2005).

Proper hardening of tissue-culture derived plants is essential to the successful commercialization of plant products. The microponic system is conducted under natural environmental conditions; this is much more favorable in plantlet acclimatization to the ex-vitro environment. With better ventilation and normal oxygen and carbon dioxide levels, plantlet growth becomes more normal and acclimatizes readily to greenhouse conditions. By combining rooting induction with the acclimatization stage, the overall time of plant propagation is significantly reduced, and the production cost is also minimized (Hahn et al. 1996, 1998, 2000).

The flexibility in design changes is an added advantage of using the microponic system. The system can be adjusted to include optimizing lighting conditions using light-emitting diodes and enhancing the growth of plantlets using silver nanoparticles (Tung et al. 2018).

11.1.3 Research on Microponic Methods

A microponic system was first described by Hahn et al. (1996) for the production of micropropagated *Chrysanthemum* plants. They aimed to see whether combing in vitro culture with hydroponics can simplify plant growth, rooting, acclimatization, with better results than traditional in vitro methods (Hahn et al. 1996, 1998, 2000). Using Rockwool, polyphenol resin, and polyurethane resin as supporting media, together with automatically controlled environmental conditions, *Chrysanthemum* plantlets in the microponic system had better and faster growth than those in in vitro systems (Hahn et al. 1996, 1998, 2000).

At present, two microponic research trends are noted, that is, (1) modernization and automation of equipment to optimize plant growth and production, and (2) implementation with simple, inexpensive equipment and materials but still ensuring good plant growth with large-scale production. In order to simplify the plant production process, we focus our efforts on the second trend. We investigate the impact of several factors on plants' growth in a microponic system, thereby proposing a large-scale plantlet production protocol. We aimed to generate disease-free and good quality plantlets, and also making further procedures easy to carry out, that is, simple in packing, distribution, and transportation and thereby reducing operating costs at the same time.

11.2 Establishment of a Microponic System for the Rooting Purposes

Rooting is usually the last stage of micropropagation before the plants are transferred to the greenhouse to acclimatize to the natural growing conditions. Rooting induction is generally conducted under sterile conditions. The study that in vitro rooting is possible under nonsterile conditions on *Chrysanthemum* is an important step in applying microponics in tissue culture. Nhut et al. (2005) demonstrated the microponic method's effectiveness compared with the in vitro propagation method in *Chrysanthemum* production. However, this earlier study serves only as a laboratory model; a scale-up design is presented in this chapter to grow a large number of plantlets and demonstrate the ease of operation.

11.2.1 Nylon Film Tube as Supporting Structure

Nylon films are used to prepare supporting tubes to house the shoot explants for root regeneration. The nylon film is the A4 cover (20 cm × 30 cm), made in Vietnam (Flex office, Thien Long Stationery, Vietnam); it is a common office supply used for binding books, documents, and records. Glossy A4 cover meets quality standards, that is, 1.5 mm in thickness, 100 sheets per package, transparent, and waterproof. It can be stored at 10–55 °C and 55–95% humidity.

11.2.2 Culture Systems

Three different microponic systems, MC, MR1, and MR2, are described as follows. The containers of the MC, MR1, and MR2 systems are made of polypropylene plastic, resistant to temperatures between −20 to 120 °C. Each container comes with a lid.

The MC system consists of a circular plastic container—12-cm-diameter at the top, 9-cm-diameter at the bottom, and 8.5-cm height (Dai Dong Tien Company, Vietnam). Each MC system has a 500 mL capacity and can house 15 nylon film tubes.

The MR1 system consists of a small rectangular plastic box with a height of 8.5 cm, 35 × 28 cm at the top, and 30 × 25 cm at the bottom (Tan Chi Thanh Company, Vietnam). Each MR1 system has a 5 L capacity and can house 300 nylon film tubes.

The MR2 system consists of a larger rectangular plastic box with a height of 16.1 cm, 31.8 cm in width, and 45.7 cm in length (Duy Tan Plastic Company, Vietnam). Each MR2 system has a 15 L capacity and can house 600 nylon film tubes.

11.2.3 Assembling a Microponic System

Wrap a nylon film 20 cm × 30 cm in size around a test tube (1.5 cm in diameter) and seal the nylon film using a heated metal rod. Next, remove the test tube and excise the excess nylon film to form a nylon tube of 20 cm long and 1.5 cm in diameter. Shorten the nylon tubes to 2 cm in height and 1.5 cm in diameter (Fig. 11.1).

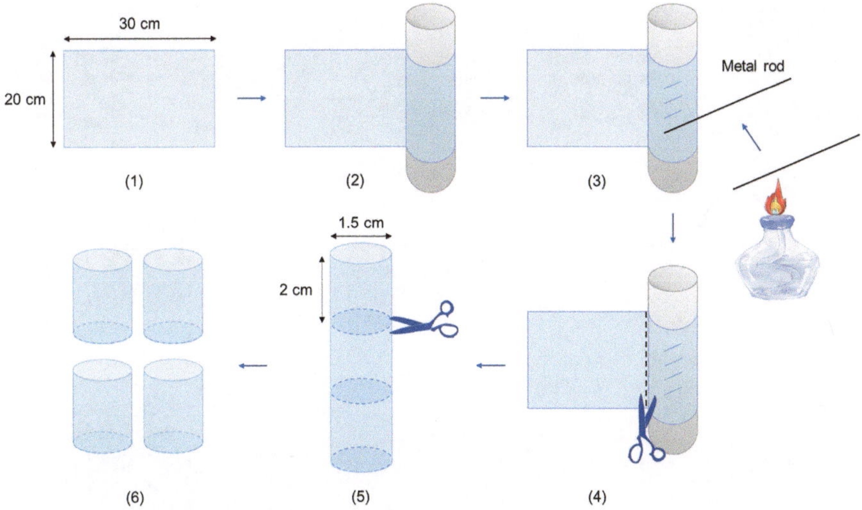

Fig. 11.1 Making nylon film tubes used in the microponic system. **(1):** Nylon film; **(2):** Wrap the film around a 1.5 cm diameter glass tube; **(3):** Seal the film by using a hot metal stick; **(4):** Cut off the excess film; **(5):** Cut into short tubes; **(6):** Nylon tubes

11.2.4 Processing of Chrysanthemum Plantlets for Rooting Using Microponics

Source of explants: Use *Chrysanthemum morifolium* Ramat cv. Jimba shoots (3 cm in length with 2 pairs of leaves) as explants. These shoots were obtained and excised from a mass of in vitro shoots maintained on MS medium (Murashige and Skoog, 1962) with 8 g L^{-1} agar and 30 g L^{-1} sucrose after 45 days of cultured (Nhut et al. 2005).

Pretreatment with IBA (Indole-3-butyric acid): Dip the shoots, 3 cm in length, with two pairs of leaves in the 500 ppm IBA solution for 20 min. Ensure the shoot tip is not in contact with the pretreatment solution (Fig. 11.2). After pretreatment, rinse the shoots with distilled water. Then, collect the shoots and place them into nylon tubes (Fig. 11.2).

Assembling the microponic systems: Place 15, 300, and 600 nylon tubes into the MC, MR1, and MR2 systems, respectively. Then add culture MS medium to the containers. Place the shoots collected after IBA pretreatment into the culture systems. The number of shoots depends on the number of nylon tubes.

11.3 The Rooting and Acclimatization of Chrysanthemum Plantlets Using Microponics: A Case History

Chrysanthemum is a popular potted and cut flower plant species with billions of stems sold each year and is favored for its rich color and a wide variety of flower shapes and sizes. *Chrysanthemum* accounts for 25–30% of the cut flowers in the

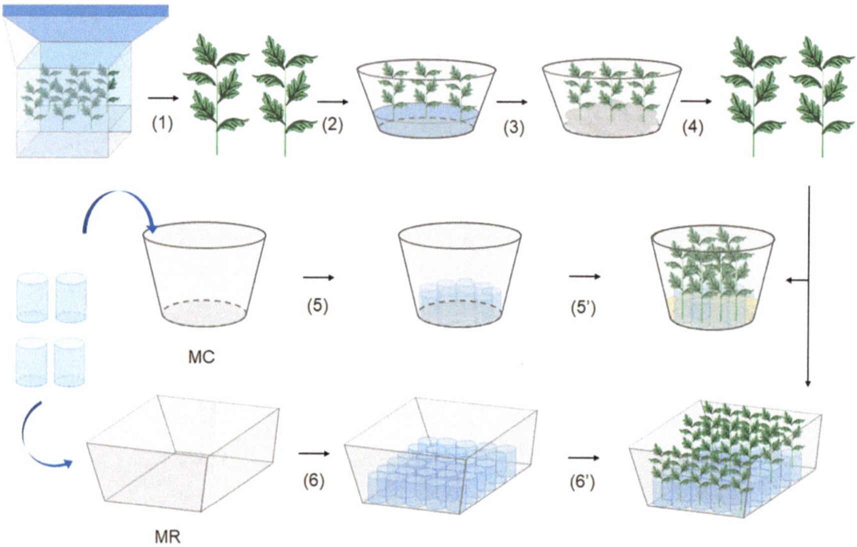

Fig. 11.2 Experimental steps. **(1):** In vitro shoots; **(2):** IBA pretreatment; **(3):** Wash with distilled water; **(4):** In vitro shoots after pretreatment; **(5, 6):** Culture systems; **(5′, 6′):** Shoots placed into the microponic system; *MC* Circular plastic box; *MR* Rectangular plastic box

world and Vietnam. Currently, the cultivation of *Chrysanthemum* and production of flowers in Vietnam in recent years have not met the quality and the demand needed by the farmers. Traditional methods give low propagation coefficient and poor plantlet quality; therefore, the economic efficiency is very low. Also, exporting *Chrysanthemum* plants to foreign countries still faces difficulties such as plantlets not being virus-free and poor acclimatization performance. Furthermore, the packaging is cumbersome as plantlets can only be maintained in blister foam or plastic trays, and it is difficult to transport the packaged material by air or sea for export. Hence, it is necessary to improve propagation methods that can overcome the above difficulties.

11.3.1 IBA Pretreatment

Pretreatment with IBA generates better quality plantlets and acclimatization characteristics. After 2 weeks of culture, the results showed that shoots pretreated with 500 ppm IBA gave 100% rooting, better acclimatization and growth as compared to pretreatment with distilled water (60%), or with 500 ppm IBA added directly to the medium (20%) (Fig. 11.3a). Also, the growth of plants (number of roots, root length, fresh weight, and dry weight) derived from IBA pretreatment was also better than the other treatments (Figs. 11.3b, c).

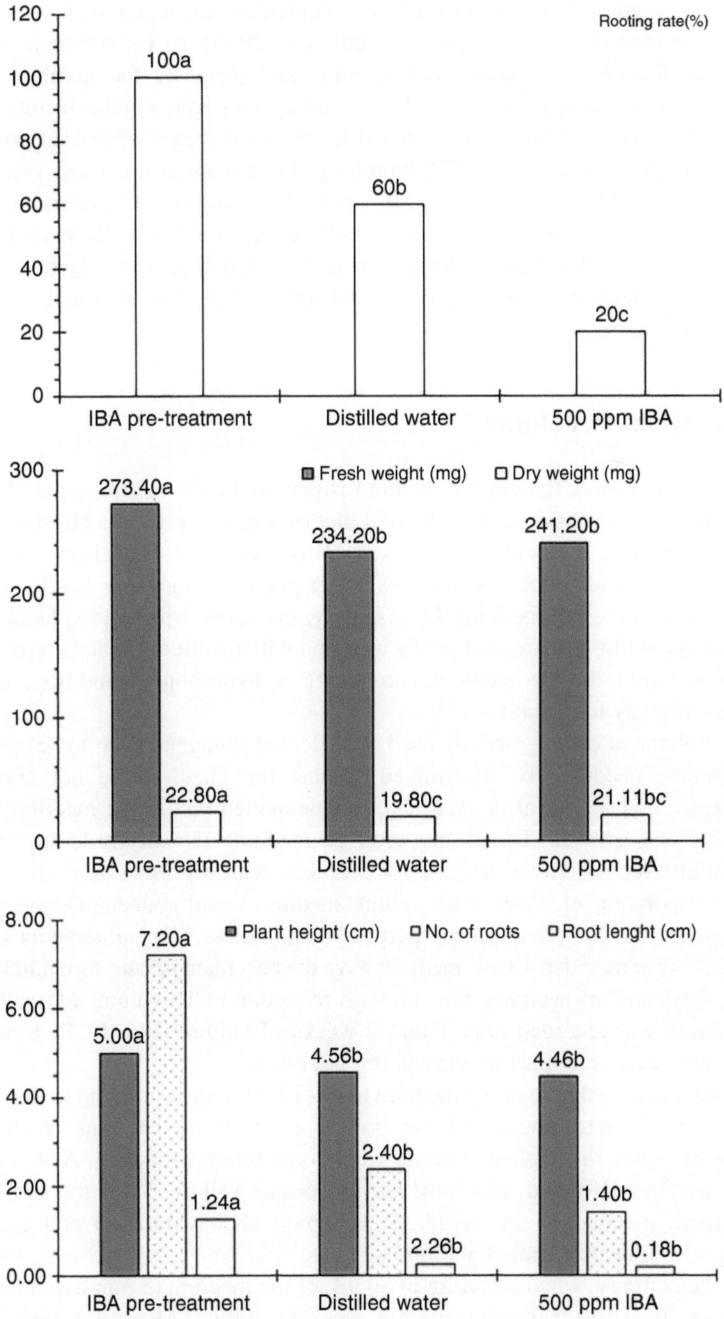

Fig. 11.3 Effects of 500 ppm IBA pretreatment on rooting of *Chrysanthemum* plants in the MC system after 2 weeks of culture

Root formation is a complex process dependent on many factors, including auxin's interaction with the explants (Saini et al. 2013). Auxin stimulates adventitious root formation, shortens rooting time, and increases the number of roots and root length (Saini et al. 2013), but not all species have similar results (Enders and Strader 2015). Auxin is only needed for the early stages of cell differentiation and organ initiation (Davies 1987); a prolonged treatment results in adverse effects. This is why shoots pretreated with 500 ppm IBA result in 100% of rooting; while adding 500 ppm IBA directly to the culture medium, only 20% rooting was observed. The results in the water treatment recorded approximately 60% rooting; however, the number of roots and the root length were less than that of the IBA pretreatment.

11.3.2 Medium Volume

In micropropagation, the culture medium is usually liquid or semi-solid. The most common gelling agent used in semi-solid media is agar. It provides physical support to the explant in the culture container. However, agar influences the culture medium's physical properties, such as water potential, moisture levels, available water, and solutes in a medium. In this study, the support for the explants is tubes made of nylon film that are chemically inert and will not affect explants' growth. The use of a liquid culture medium, simulating a hydroponic condition, provides nutrients directly to explants.

The volume of culture medium used in a selected container needs to be optimized. The volume needs to be determined so that the plantlets are not completely submerged, and sufficient oxygen level is maintained within the medium for root induction and growth. The evaporation of medium also needs to be regulated through different lid-cover designs, for example, with Millipore filter covered lids. Rapid evaporation of water from culture medium could concentrate the medium components altering its osmotic properties. In our studies, *Chrysanthemum*'s growth in the MC system with a 40 mL medium gave the best plant height, root number, root length, fresh and dry weights (Fig. 11.4). A reduction in the volume of media in the MC system was recorded after 1 and 2 weeks of culture (Fig. 11.5); however, a significant negative impact on growth did not occur.

Increasing the volume of the medium to 60–70 mL (corresponding to the medium depth of 11–13 mm) had a negative growth effect on the explants. With a large volume of liquid, the explants became submerged into the medium. As a result, the shoots became deformed, and most leaves became yellow. Since the shoots were submerged, a reduction in oxygen level would alter respiration and eventually leading to cell death (Choi et al. 2000).

On the contrary, when culturing in 30 mL of the medium (5 mm depth medium), the shoots began to root after about 1 week of culture. Moreover, only a small amount of medium remained, and plantlet growth became slow in the second week. This is likely due to the evaporation of water from the medium resulting in changes in the medium's osmotic condition. Therefore, for the MC system, a medium volume

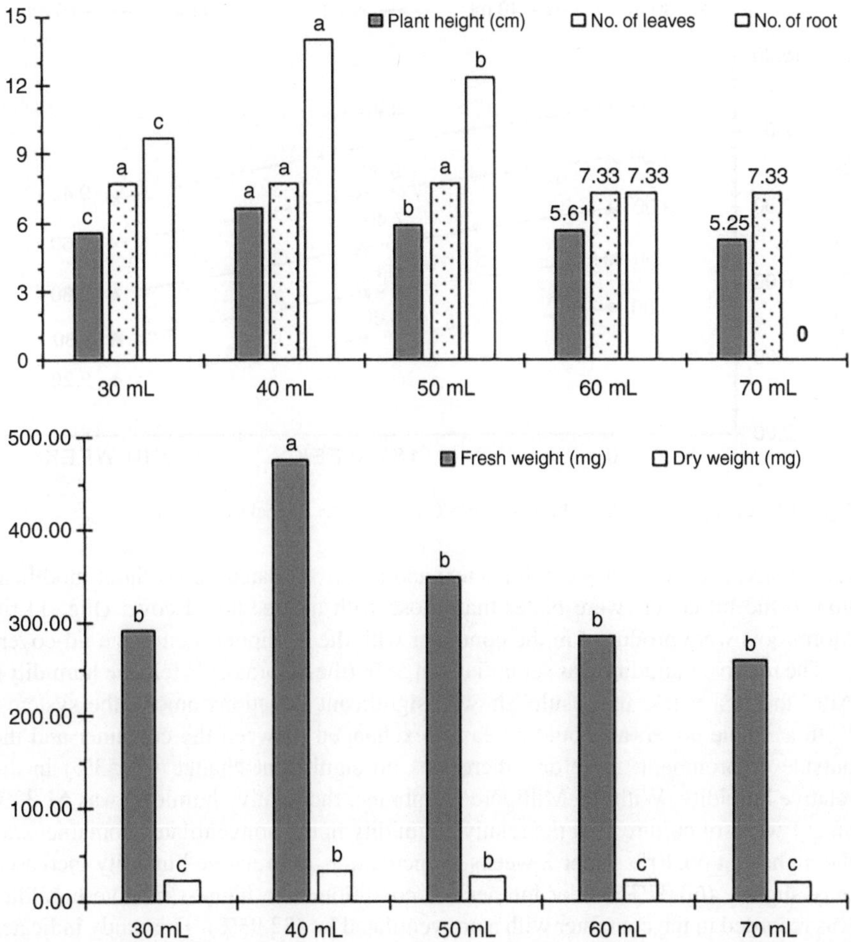

Fig. 11.4 Effect of volume medium on the growth of *Chrysanthemum* plants in MC system after 2 weeks of culture

of 40 mL (equivalent to the ambient depth of 7 mm) was best suited for proper growth of *Chrysanthemum* plantlets.

11.3.3 Ventilation Conditions

The effects of container ventilation were studied using modifications to the lid cover. The change of relative humidity in the MC system with different ventilation conditions after 2 weeks of culture was noted in Fig. 11.6. *Chrysanthemum* growth (plant height, number of roots, fresh and dry weights) with a Millipore membrane lid-cover [Millipore membrane with Milliseal™, pore size 0.5 μm (Nihon Millipore

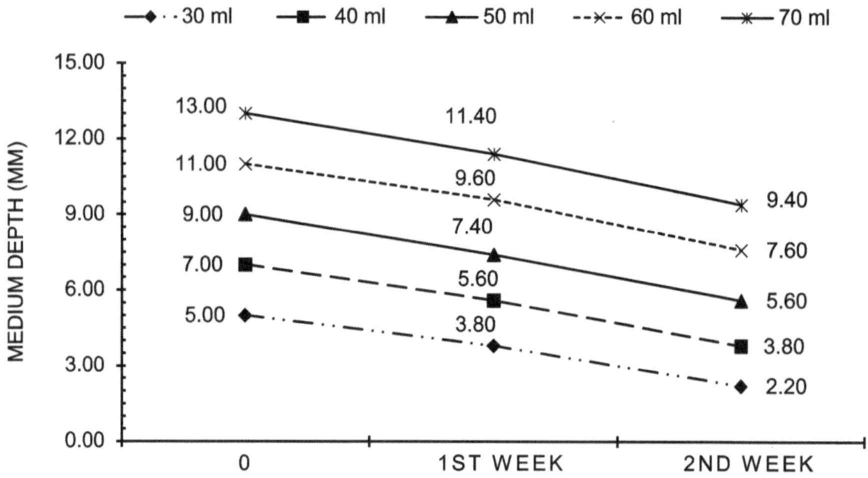

Fig. 11.5 Changes in media volume in the MC system after 2 weeks of culture

Ltd., Tokyo, Japan) and 2 cm diameter] and a nonventilated lid (without modification to the lid cover) were better than those with a three-holed cover (Fig. 11.6). More roots were produced in the container with the Millipore membrane lid-cover.

The relative humidity was set initially at 55% (the laboratory's relative humidity). After the first week, the results showed significant variations among the systems. With a 3-hole cover, air could be easily exchanged between the container and the outside environment; therefore, there was no significant change (57.53%) in the relative humidity. With the Millipore membrane, the relative humidity was 61.33% after 1 week of culture, and the relative humidity in the nonventilated container was the highest at 66.40%. After 2 weeks of incubation, the relative humidity increased only slightly (60.07%) under the aerated condition. The highest relative humidity was recorded in the container with a nonventilated lid (82.93%). This study indicates that proper ventilation is important to root organogenesis and the acclimatization process. Besides considering moisture levels, the air composition within containers must be analyzed to determine the beneficial components and their concentrations for plant development in a microponic system.

11.3.4 AgNPs in Reduced Microbial Contamination and Improved Growth

Plant shoots (3 cm) were cultured in the MR system supplemented with different AgNPs concentrations (0; 2.5; 5.0; 7.5 and 10 ppm). AgNPs were added directly to the culture medium and boiled to mix (Tung et al. 2018).

Results showed that by adding AgNPs into the culture medium, the plants grew well. Plant height, root length, leaf length, leaf width, fresh weight, and dry weight increased proportionally with the increase in AgNPs (0–7.5 ppm) concentration and

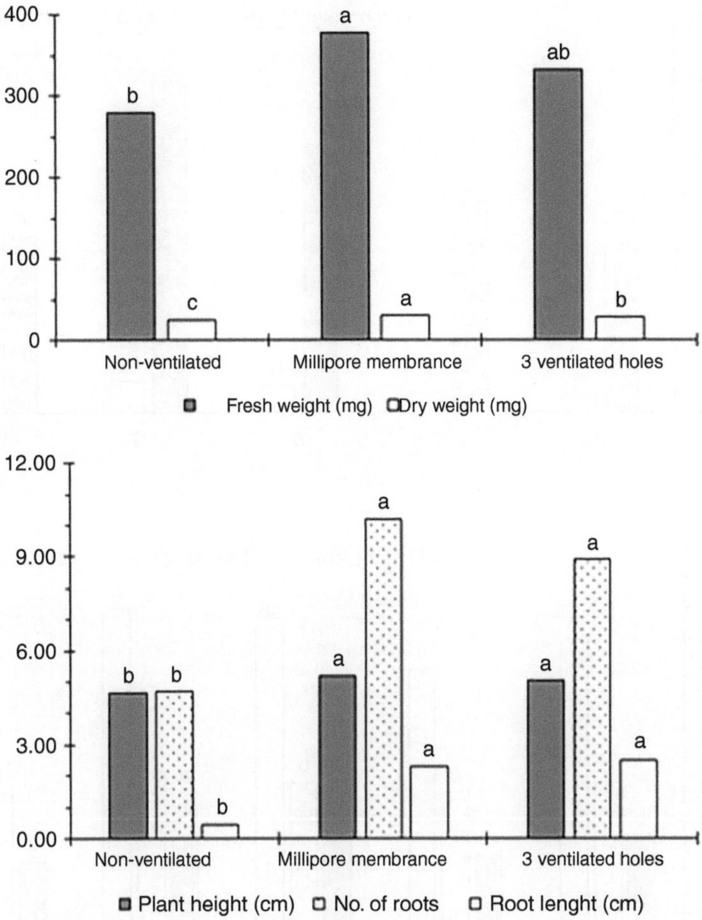

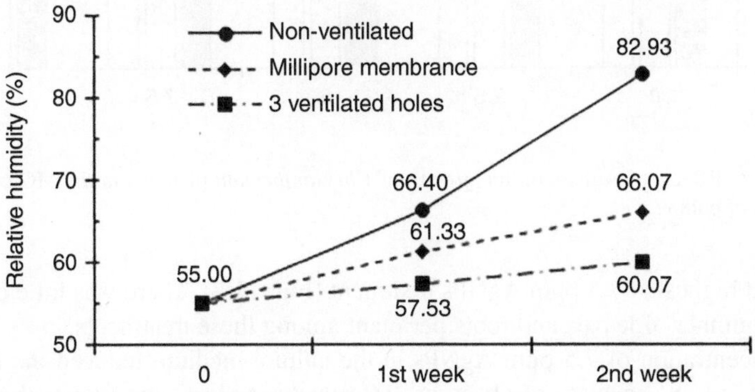

Fig. 11.6 Changes in relative humidity in the MC system with different ventilated conditions after 2 weeks of culture

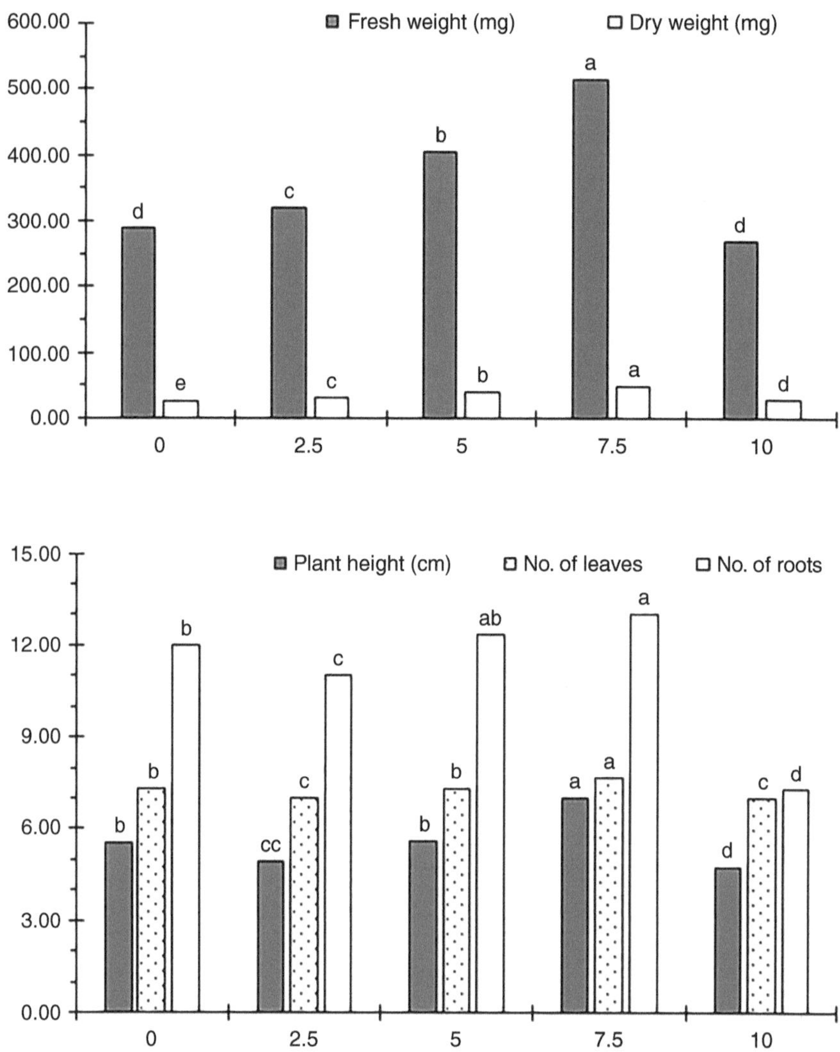

Fig. 11.7 Effects of AgNPs on the growth of *Chrysanthemum* plantlets in the MC system after 2 weeks of culture

reached highest at 7.5 ppm AgNPs treatment (Fig. 11.7). There was little difference in the number of leaves and roots per plant among these treatments.

Concentration of 7.5 ppm AgNPs in the culture medium reduced the microbial content of 8 species of bacteria (*Corynebacterium* sp., *Enterobacter* sp., *Arthrobacter* sp., *Agrobacterium* sp., *Xanthomonas* sp., *Pseudomonas* sp., *Bacillus* sp. and *Micrococcus* sp.) and 3 species of fungi (*Aspergillus* sp., *Fusarium* sp. and *Alterneria* sp.) (Tung et al. 2018). Although the effects of AgNPs have been reported on microorganisms and animal cells, fewer studies are available on plants (Krishnaraj et al. 2012; Salama 2012; Savithramma et al. 2012; Sharma et al. 2012). In plant tissue culture, AgNPs not only play a role in enhancing growth and development and in plant metabolic processes (Salama 2012; Sharma et al. 2012) but also enhance root formation ability and inhibit ethylene formation (Rezvani et al. 2012).

Most previous studies have focused on the effects of AgNPs on plants, such as explant disinfection (Mahna et al. 2013), germination (Rezvani et al. 2012), physiology, or morphology (Syu et al. 2014). To date, very few reports have focused on the effects of AgNPs on explant growth when transferred to greenhouse conditions. This study showed that the *Chrysanthemum* cultured in the MC system supplemented with different concentrations of AgNPs resulted in good growth. This indicates that the inclusion of AgNPs to the culture medium is effective for the growth of *Chrysanthemum*.

11.3.5 Lighting Conditions on Enhanced Growth of Chrysanthemum in MC system

Currently, LEDs are used in different fields of research such as chlorophyll synthesis (Tripathy and Brown 1995), photosynthesis (Tennessen et al. 1994), and morphogenesis (Hoenecke et al. 1992). Their effects on plants depend on the light intensity and quality (light spectrum, the direction of illumination) and illumination duration (Taiz and Zeiger 2007).

In this study, the effects of LEDs on the growth of *Chrysanthemum* shoots in the MR system were examined. Shoots were cultured under different LEDs, including Green—G (565 nm), Blue—B (450 nm), Red—R (660 nm), Yellow—Y (590 nm) (Steigerwald et al. 2002), and B combined with R of different ratios (10:90, 20:80, 30:70, 40:60, 50:50 and 60:40) at 40–45 μmol m^{-2} s^{-1}. Fluorescent lamps (FL) were used as the control. The microponic systems were maintained at 25 ± 2 °C with a humidity of 55–60% and a photoperiod of 16 h/day under LEDs and FL.

Figure 11.8 presents the data on the effectiveness of different lighting conditions on the growth of *Chrysanthemum* plantlets. The chlorophyll content of the leaves increased with the addition of Blue and Red combinations. The best results were obtained when 30B + 70R were combined (Fig. 11.8).

Under Green and Yellow light conditions, the plants die. This is because the wavelengths of Y (570–590 nm) and G (490–570 nm) do not match the maximum absorption spectral region of chlorophyll a (662 nm) and chlorophyll b (645 nm) (Lichtentaler and Wellburn 1985). Under R light, leaves tend to be elongated, and the chlorophyll content decreased. Strawberry, *Eucalyptus*, *Cymbidium*, orchid, and banana had elongated stems under R light, while lilies usually developed well under these conditions (Nhut 2002). In this study, the *Chrysanthemum* shoots under R lighting conditions had slender, yellowish stems; chlorophyll a, chlorophyll b, and chlorophyll (a + b) were low, and the fresh weight was also lower than plants grown

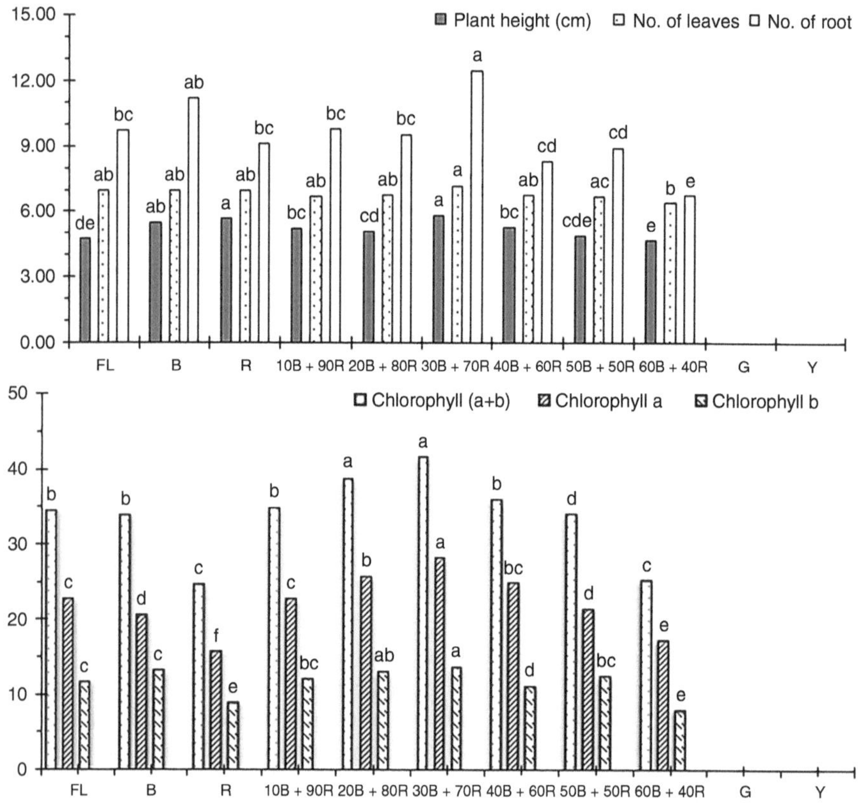

Fig. 11.8 The growth of *Chrysanthemum* leaves in the MC system under different lighting conditions after 2 weeks of culture

under B combined with R [10B: 90R, 20B: 80R, 30B: 70R, 40B: 60R and 50B: 50R].

Besides serving as a light source, the use of an LED lighting system has additional benefits. In commercial tissue culture rooms, the lighting system commonly used for clonal propagation are fluorescent lamps. Although different types of fluorescent lamps are available for growing plants, all these lamps emit heat. Therefore, a cooling system is needed to stabilize the heat in a culture room. There are two main components related to the cost of electricity in tissue culture rooms: 65% of the electricity is needed to light the room, and 25% of the electricity is used to cool the room. Therefore, developing nonheat lighting systems in growth rooms will provide significant benefits in cost reduction in micropropagation (Standaert-De Metsenaere 1991).

11.4 The Effectiveness of Microponics and Micropropagation

11.4.1 MC and MR Systems on Growth

After 2 weeks of culture, the results showed that plants cultured using the MR systems were not significantly different from those in the MC system. Moreover, more plants can be grown using the MR systems than the MC system per unit space (Tung et al. 2018). The MR systems are better use of resources and physical space associated with plant propagation when compared to the MC system.

To set up MR1 and MR2 systems, it is easy and less expensive to buy items on the market. They are made from polypropylene plastic material; hence, they are light and ensure durability. Importantly they are transparent so that light can easily penetrate to reach the plantlets. Both of these systems have sealed lids, and ventilation holes can be made easily. They are readily arranged in a culture room and transported long distances. It is important to note that using the traditional method of micropropagation, the same physical space occupied by MR systems can house only a few glass bottles (500 mL) with only 10 plants each.

11.4.2 Evaluate the Effectiveness of Microponics and Micropropagation

The microponic system we described is a propagation system that combines micropropagation and hydroponics, a potential method for producing plantlets. The plantlets were cultured on nonsterile conditions, sugar-free, and agar-free. This is a simple and effective method for large-scale seedling production. This method can completely replace the current propagation method, and commercially produce plantlets of many different plants. The microponic system has the following advantages over traditional methods. (1) Microponics is an open system—this method can shorten the micropropagation process by combining the rooting stage with the acclimatization stage, eliminating a transfer-step, that is, from in vitro rooting to greenhouse acclimatization. As a result, the plantlet survival rate is high (100%). (2) Cost savings—Microponics can save costs by not using sugar, agar, and alcohol in the culture process and save energy by not needing autoclave and culture under LED lighting conditions. (3) Transport and stacking—The rooted plantlets can be easily packed in small glass containers and transported because of their small mass (0.5 kg/box). They are 20 times more efficient than conventional shipping methods since the traditional containers weigh about 10 kg (about 10 blisters of 60 plants/blister). (4) Simple and easy to design—The rectangular plastic boxes and nylon film holders are commonly available, easy to prepare, durable, lightweight, and most importantly, reusable.

Improving plantlets quality*:* Simple modifications can enhance the usefulness of microponic systems. Plantlet quality is readily enhanced by adding AgNPs to the culture medium and the use of an LED lighting system. Using these combinations, the growth and quality of plantlets improved together with a reduction of microbial

Fig. 11.9 The growth and development of *Chrysanthemum* plantlet in MR2 system and packed plastic box. (**a, b**) *Chrysanthemum* growth and development in MR2 system after 2 weeks of culture; (**c**)**:** Plants packed plastic box

growth. With creative designs, large-scale production of plantlets is possible, as shown in the MR2 system (Fig. 11.9).

11.5 Planting in Greenhouse and Field Trials—Chrysanthemum Plantlets Culture in MR System Until Flowering Stages

Plants cultured on the microponic system grew better with better plant height than those obtained from micropropagation after being transferred to the greenhouse (Fig. 11.10). After 12 weeks, plants cultured on the microponic system began to flower. At week 13, flower buds were visible. The results of this experiment showed that the growth potential of the plants cultured on microponic system is more optimal than those obtained from micropropagation, at the flowering stage.

The propagation efficiency of the microponic method has been proven to be better than the micropropagation method. The ability to produce flower buds (number of buds/plant), plant height, stem diameter, and flower size derived from microponic culture was also better than micropropagated plants. The *Chrysanthemum* plantlets cultured with microponics have beautiful flowers and have the ability to flower earlier than micropropagated plants. The results of our studies suggest that the microponic method is very useful in multiplying *Chrysanthemum*, in obtaining superior quantity plants with better adaptability and good growth at the nursery stage, giving high quantity, good quality, adaptability, and good growth at the nursery stage.

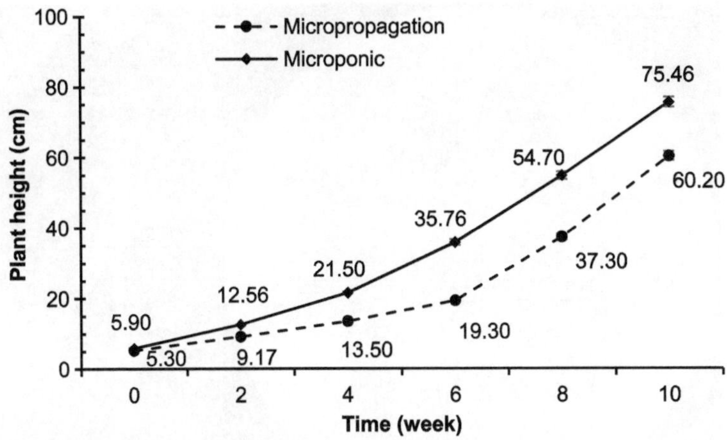

Fig. 11.10 Plant height of in vitro and microponic plants transferred into a greenhouse

11.6 Application of Microponic System on Some Other Plants

Apical shoots (1.5 cm) with 4 leaves of carnation in vitro were cultured on ½ MS medium supplemented with 1.0 mg/L NAA, 30 g/L sucrose, 8 g/L agar (Kharrazi et al. 2011) for in vitro and microponic culture (without sucrose) systems supplemented with AgNPs at different concentrations (0.5; 1; 2; 3; 5 mg/L). Controls were shoots grown on medium without AgNPs.

The effectiveness of microponic culture systems under aeration conditions and 2 mg/L AgNPs (Fig. 11.11) has shown their positive effects on carnation growth and development in vitro after 3 weeks of culture. This study shows that the microponic system is an effective system for the rooting stage of carnations because the aeration conditions along with the liquid media facilitate the absorption of nutrients from the culture medium by enhancing transpiration, reducing hydration, helping plants grow better and increase plant quality. Relative humidity (65–70%) in the microponic system is equivalent to that of the greenhouse, which helps the plants to acclimatization easily, with a high survival rate (Tung et al. 2018).

In addition, the growth of carnations as well as the ability to form new plants of plants cultured in the microponic system with/without the addition of 2 mg/L AgNPs was also better than in the in vitro culture system at the acclimatization stage after 2 months (Fig. 11.12).

Besides carnations, this system has also been applied to strawberry, potato, and gloxinia plants (Figs. 11.13, 11.14, 11.15, and 11.16). Subsequent growth of microponic derived gloxinia plantlets in handmade hydroponic systems (Fig. 11.17).

Fig. 11.11 The growth and development of carnation plantlet in microponic system after 3 weeks of culture. (**a, b**): Plant in microponic system without 2 mg/L AgNPs; (**c, d**): Plant in microponic system without AgNPs

11.7 Conclusion

This chapter aims to provide an example of setting up a microponic system and evaluating the effects of some factors on the growth of *Chrysanthemum* plantlets and the explants' ability to acclimatize, grow, and flowering in the greenhouse. *Chrysanthemum* shoots were pretreated with 500 ppm IBA in a circular plastic container

Fig. 11.12 The growth and development of carnation plantlet in in vitro and microponic system supplemented without/with 2 mg/L AgNPs (left to right) after 3 weeks of culture

Fig. 11.13 The growth and development of strawberry plantlet in microponic system

system with a Millipore membrane cap containing 40 mL of ½MS medium, at a density of 15 stems cuttings per container. A concentration of 7.5 ppm AgNPs in the microponic medium reduced the microbial contamination of 8 species of bacteria and 3 species of fungi. The lighting condition with a ratio of 70% Red LED and 30% Blue LED gave optimal growth of plantlets and acclimatization in the greenhouse. The rectangular plastic box-1 (300 plants/box) and box-2 (600 plants/box) systems

Fig. 11.14 The growth and development of potato plantlet in microponic system

Fig. 11.15 The growth and development of gloxinia plantlet in microponic system

gave uniform and healthy plants, and these systems are suitable for large-scale commercial plant production. *Chrysanthemum* plants cultured on the microponic system flowered faster with large size than in the traditional micropropagation method. In addition, the microponic system has also been applied on a number of plants such as carnation and strawberry.

Fig. 11.16 Comparison of microponic and traditional culture (bottle) systems for gloxinia growth and development. (**a, c**): Microponic system; (**b, d**) Traditional culture system

Fig. 11.17 Subsequent growth of microponic derived gloxinia plantlets in handmake hydroponic systems

References

Choi SM, Son SH, Yun SR, Kwon OW, Seon JH, Paek KY (2000) Pilot-scale culture of adventitious roots of ginseng in a bioreactor system. Plant Cell Tissue Organ Cult 62:187–193

Davies PJ (1987) Plant hormones and their role in plant growth and development. Martinus Nijhoff Publishers, Dordrecht, The Netherlands, pp 1–12

Enders TA, Strader LC (2015) Auxin activity: past, present, and future. Am J Bot 102(2):180–196

Hahn EJ, Lee YB, Ahn CH (1996) A new method on mass-production of micropropagated Chrysanthemum plants using microponic system in plant factory. Acta Hortic 440:527–532

Hahn EJ, Bae JH, Lee YB, Beom Y (1998) Growth and leaf-surface characteristics of Chrysanthemum plantlets between hydroponic and microponic system. J Korean Soc Hort Sci 39(6):838–842

Hahn EJ, Bea JH, Lee YB (2000) Growth and photosynthetic characteristics of Chrysanthemum plantlets as affected by pH and EC of nutrient solution in microponic culture. J Korean Soc Hort Sci 41(1):12–15

Hoenecke ME, Bula RJ, Tibbitts TW (1992) Importance of blue photon levels for lettuce seedlings grow under red light-emitting diodes. Hortic Sci 27:427–430

Kharrazi M, Nemati H, Tehranifar A, Bagheri A, Sharifi A (2011) In vitro culture of carnation (*Dianthus caryophyllus* L.) focusing on the problem of vitrification. J Biol Environ Sci 5(13):1–6

Kozai T (1991) Micropropagation under photoautotropic conditions. In: Debergh PC, Zimmerman RH (eds) Micropropagation, technology and application. Kluwer Academic Publishers, Dordrecht, The Netherlands, pp 447–469

Kozai T, Afreen F, Zobayed SMA (2005) Photoautotrophic (sugar-free medium) micropropagation as a new micropropagation and transplant production system. Springer, Dordecht, The Nertherlands, p 315

Krishnaraj C, Jagan EG, Ramachandran R, Abirami SM, Mohan N, Kalaichelvan PT (2012) Effect of biologically synthesized silver nanoparticles on *Bacopa monnieri* (Linn.) Wettst, plant growth metabolism. Proc Biochemist 47(4):651–658

Lichtentaler HK, Wellburn AR (1985) Determination of total carotenoids, chlorophyll a and b of leaf in different solvents. Biochem Soc Trans 11:591–592

Mahna N, Vahed SZ, Khani S (2013) Plant in vitro culture goes nano: nanosilver-mediated decontamination of ex vitro explants. J Nanomed Nanotech 4(161):1

Murashige T, Skoog F (1962) A revised medium for rapid growth and bio assays with tobacco tissue culture. Physiol Plant 15:473–497

Nhut DT (2002) In vitro growth and physiological aspects of some horticultural plantlets cultured under red and blue light-emitting diodes (LEDs). Doctoral thesis. Kagawa University, Japan, pp 1–18

Nhut DT, Don NT, An TTT, Van TPT, Vu NH, Huyen PX, Khiem DV (2005) Microponic and hydroponic techniques in disease-free Chrysanthemum (*Chrysanthemum* sp.) production. J Appl Hortic 7(2):67–71

Rezvani N, Sorooshzadeh A, Farhadi N (2012) Effect of nano-silver on growth of saffron in flooding stress. World Acad Sci Eng Tech 1:517–522

Saini S, Sharma I, Kaur N, Pati PK (2013) Auxin: a master regulator in plant root development. Plant Cell Rep 32(6):741–757

Salama HMH (2012) Effects of silver nanoparticles in some crop plants, common bean (*Phaseolus vulgaris* L.) and corn (*Zea mays* L.). Inter Res J Biotech 3(10):190–197

Savithramma N, Ankanna S, Bhumi G (2012) Effect of nanoparticles on seed germination and seedling growth of *Boswellia ovalifoliolata* an endemic and endangered medicinal tree taxon. Nano Vis 2:61–68

Sharma P, Bhatt D, Zaidi MG, Saradhi PP, Khanna PK, Arora S (2012) Silver nanoparticle-mediated enhancement in growth and antioxidant status of *Brassica juncea*. Appl Biochem Biotech 167:2225–2233

Standaert-De Metsenaere REA (1991) Economic considerations. In: Debergh PC, Zimmermann RH (eds) Micropropagation: technology and application. Kluwer Academic Publish, Dordrecht, The Netherlands, pp 123–140

Steigerwald DA, Bhat JC, Collins D, Fletcher RM, Holcomb MO, Ludowise MJ, Martin PS, Rudaz SL (2002) Illumination with solid state lighting technology. IEEE J Sel Top Quantum Electron 8(2):310–320

Syu YY, Hung JH, Chen JC, Chuang HW (2014) Impacts of size and shape of silver nanoparticles on *Arabidopsis* plant growth and gene expression. Plant Physiol Biochem 83:57–64

Taiz L, Zeiger E (2007) Plant physiology, vol 115. Benjamin Cummings Publishing Company, New York, p 575

Tennessen DJ, Singsaas EL, Sharkey TD (1994) Light-emitting diodes as a light source for photosynthesis research. Phot Res 39:85–92

Tripathy BC, Brown CS (1995) Root-shoot interaction in the greening of wheat seedlings grown under red light. Plant Physiol 107:407–411

Tung HT, Nam NB, Huy NP, Luan VQ, Hien VT, Phuong TTB, Dung LT, Nhut DT (2018) A system for large scale production of Chrysanthemum using microponics with the supplement of silver nanoparticles under light-emitting diodes. Sci Hortic 232:153–161

The Application of Thin Cell Layer Culture Technique in Plant Regeneration and Micropropagation: Latest Achievements

12

Hoang Thanh Tung, Tran Hieu, Truong Hoai Phong, Hoang Dac Khai, Nguyen Thi My Hanh, K. Tran Thanh Van, and Duong Tan Nhut

Abstract

The thin cell layer (TCL) technique is a method that uses "thin" specimens cut from different plant organs (leaf, root, bud, flower stalk, cotyledon, hypocotyl, etc.) for micropropagation. Thin cell layers excised longitudinally (lTCLs) or transversely (tTCLs) from organs have many advantages as explant sources. Because of the thickness of the explants, cells and tissues are in direct contact with the medium, and the endogenous hormone amount in the explants is low. Thus, the explants can respond quickly to medium conditions and regenerate organs and somatic embryos readily. As a result, the TCL technique has been successfully applied to plant micropropagation. Also, this technique has a high success rate on species considered difficult to cultivate, such as wheat, bamboo, and pine. In this chapter, the successful applications of the TCL technique using a fruit plant (*Passiflora* sp.), a flower (*Begonia* sp.), and a medicinal plant (*Panax vietnamensis*) are documented.

Keywords

Thin cell layer · Panax vietnamensis · Passiflora · Begonia

H. T. Tung · T. Hieu · T. H. Phong · H. D. Khai · D. T. Nhut (✉)
Tay Nguyen Institute for Scientific Research, VAST, Dalat City, Vietnam

N. T. M. Hanh
Dalat Vocational Training College, Dalat City, Vietnam

K. T. T. Van
International Centre for Interdisciplinary Science and Education, Quy Nhon City, Vietnam

D. T. Nhut et al. (eds.), *Plant Tissue Culture: New Techniques and Application in Horticultural Species of Tropical Region*,
https://doi.org/10.1007/978-981-16-6498-4_12

12.1 Introduction

Almost 50 years since the thin cell layer (TCL) concept (Tran Thanh Van 1973) was proposed, over 77 different plant species or hybrids have been successfully micropropagated using the TCL method (Teixeira da Silva and Dobránszki 2013; Teixeira da Silva 2013). The species studied consisted of 50% ornamentals, 25% agricultural or vegetable crops, 8% medicinal plants, etc. The concept, successes, and the potential of the TCL method have been discussed (Teixeira da Silva and Dobránszki 2013, 2015; Teixeira da Silva 2013, 2015). The culture of "thin" explants has many advantages. A sampling of appropriate organs and tissues is quickly done after surface decontamination. This allows the selection of which organ is most suitable for regeneration. The optimum section thickness, location, and age of explants can be determined by culturing explants along the axial axis. For thin transverse sections (tTCL), the cell types and tissues responsible for regeneration can be studied using careful histological methods. The effects of the axial polarity, if any, on regeneration can be investigated using longitudinally excised thin sections (lTCL). Furthermore, da Silva et al. (2015) showed that "untapped potential of plant thin cell layers" as "marker molecules/genes of differentiation can be easily localized in situ in the target/responsible cells".

12.2 Recent Successes Using the TCL Technique

The TCL technique has been successful in the morphogenesis and regeneration of many previously reported plant species in many orchids, medicinal plants, woody plants, field crops, and vegetables (Teixeira da Silva and Dobránszki 2015), and it continues to show superiority in a variety of plant species (Table 12.1).

This technique can also apply to the study of developmental processes and transformation studies. The tTCL-stem of *Tanacetum cinenariifolium* was induced for shoot regeneration, and all hypothetical transgenic lines regenerated from TCLs were GFP (green fluorescent protein gene) positive (Mao et al. 2014). The study of nuclear DNA content, hydrogen peroxide, and antioxidant enzyme activities during organogenesis was carried out through the transverse thin cell layer of *Malaxis wallichii* (Bose et al. 2017). Betti et al. (2019) showed that controlling antagonism between ethylene and jasmonate indicated in adventitious rooting of *Arabidopsis thaliana* thin cell layers.

In summary, the TCL technique has been successfully applied in plant cell, tissue, and organ culture. The low variability and fast reaction times facilitate the rapid multiplication of plant varieties. This method also achieved high success rates on subjects considered difficult to cultivate, such as cereals (wheat and rice) and forest trees (bamboo, melaleuca, etc.). In this chapter, the successful applications of the TCL technique using a fruit plant (*Passiflora* sp.), a flower (*Begonia* sp.), and a medicinal plant (*Panax vietnamensis*) are documented.

Table 12.1 The recent applications of the thin cell layer culture technique in some plant species

Species/hybrids	References
Stevia rebaudiana Bertoni	Ramírez-Mosqueda and Iglesias-Andreu (2016)
Agave fourcroydes Lem.	Monja-Mio and Robert (2016)
Panax vietnamensis Ha et Grushv.	Hien et al. (2016)
Jatropha curcas L.	Loan et al. (2016), Cristian and Anne (2021)
Scutellaria ocmulgee	Vaidya et al. (2016)
Bacopa monnieri L. *Wettst.*	Croom et al. (2016)
Rubus sanctus	Sabooni and Shekafandeh (2017)
Rubus hirtus	Sabooni and Shekafandeh (2017)
Malaxis wallichii	Bose et al. (2017)
Crocus sativus L.	Azadi et al. (2017)
Dendrobium aqueum	Parthibhan et al. (2018)
Passiflora edulis Sims.	Hieu et al. (2018a, b, 2019)
Withania coagulans Dunal	Tripathia et al. 2018
Paphiopedilum callosum var. sublaeve	Wattanapan et al. 2018
Hedychium coronarium	Tu et al. (2018)
Dendrobium aphyllum Roxb	Bhattacharyya et al. (2018)
Pinus patula Schl. Et Cham	Ramírez-Mosqueda et al. (2019)
Cattleya forbesii Lindl.	Ekmekçigil et al. (2019)
Gerbera jamesonii (H. bolus ex bolus f.)	Winarto et al. (2019)
Hadrolaelia grandis	Vudala et al. (2019)

12.3 Regeneration Studies of *Passiflora* Sp. Using the TCL Technique

12.3.1 Plant Materials

In vitro shoots served as the explant source for this study. The shoots were obtained from in vitro culture of ex vitro axillary buds from plants maintained in the greenhouse (Tay Nguyen Institute for Scientific Research, Dalat, Vietnam). The leaves and stems were excised from in vitro shoots (1.5 months old) of *Passiflora edulis* Sims. and *P. edulis* f. *flavicarpa* and used as materials in this study.

12.3.2 Culture Medium

The shoot regeneration medium used in the experiments was MS medium (Murashige and Skoog 1962) supplemented with 30 g/L sucrose, 8 g/L agar and different concentrations of BA (6-benzyladenine) (0, 0.5, 1.0, 1.5, and 2.0 mg/L) or combined with NAA (α-Naphthaleneacetic acid) (0, 0.5, and 1.0 mg/L). The medium was adjusted to pH 5.7–5.8 and then sterilized (autoclaved for 30 min at 121 °C and 15 psi).

12.3.3 Establishment of Longitudinal and Transverse Thin Cell Layer Explants

In this study, the in vitro leaf (10 mm × 10 mm) was cut longitudinally (10 mm × 2.5 mm) or transversally (2.5 mm × 10 mm) yielded four thin cell layer explants (lTCL-L or tTCL-L). The in vitro stem (1 mm × 10 mm) was cut longitudinally yielded two lTCL-S explants (10 mm × 0.5 mm) or transversally yielded four tTCL-S explants (2.5 mm × 1 mm) (Fig. 12.1).

12.3.4 Shoot Regeneration of tTCL and lTCL Derived from Leaf Explants

For *P. edulis* Sims., BA treatment alone gave a high shoot regeneration efficiency from tTCL-L and lTCL-L (Fig. 12.2). The tTCL-L explants in all treatments supplemented with BA gave the shoot regeneration rate more than 50%, and the shoot regeneration rate was the highest (100%) at 1.0 mg/L BA after 8 weeks of culture (Fig. 12.2). The number of shoots (4.00 shoots) and shoot height (1.53 cm) were also the highest in this treatment (Figs. 12.3 and 12.4). The lTCL-L explants in the treatment with only BA showed a lower shoot regeneration rate than tTCL-L and reached the highest rate of 56.67% at 1.5 mg/L BA (Fig. 12.2). Meanwhile, in the combination of BA (0.5, 1.0, 1.5 and 2.0 mg/L) and NAA (0.5 and 1.0 mg/L), tTCL-L and lTCL-L explants were necrosis after 8 weeks of culture (data not show).

For leaf explants of *P. edulis* f. *flavicarpa*, there was no regeneration including just BA or BA combined with NAA in both tTCL-L and lTCL-L, all explants were browning or necrosis after 8 weeks of culture. In studies of the micropropagation of Passiflora species, cytokinin was used in a wide range of concentrations from 0.5 to 10.0 mg/L BA, with or without in combination with auxin (Ozarowski and Thiem 2013). However, concentrations of just BA or combination with NAA in this experiment may not be appropriate in terms of concentration for the regeneration from TCL explants of *P. edulis* f. *flavicarpa*.

Da Silva et al. (2011) indicated that most of the plant regeneration cases were via the organogenesis pathway through the influence of BA with different concentrations. Cytokinin is essential for in vitro regeneration of *Passiflora* species and is effective in inducing direct or indirect shoot regeneration. Furthermore, the stimulation of adventitious shoot is often dependent on the combination of auxin and cytokinin. Besides, the previous studies obtained shoot regeneration from disc-shaped leaf samples with different diameters (2 and 8 mm) (Dornelas and Vieira 1994; Lombardi et al. 2007). Lombardi et al. (2007) studied shoot regeneration from disc-shaped leaves (5 mm) on *P. cincinnata*; the results indicated that leaves cultured on MS medium supplemented with 1.0 mg/L BA showed indirect shoot regeneration rate via callus reached only 41.33%, and the number of shoots was 2.32 shoots. Similarly, in the *P. edulis* f. *flavicarpa*, the direct shoot regeneration from the disc-shaped leaf sample (8 mm) on MS medium supplemented with 1.0 mg/L BA for the shoot regeneration rate was only 41.7% after 28 days of culture (Dornelas and Vieira

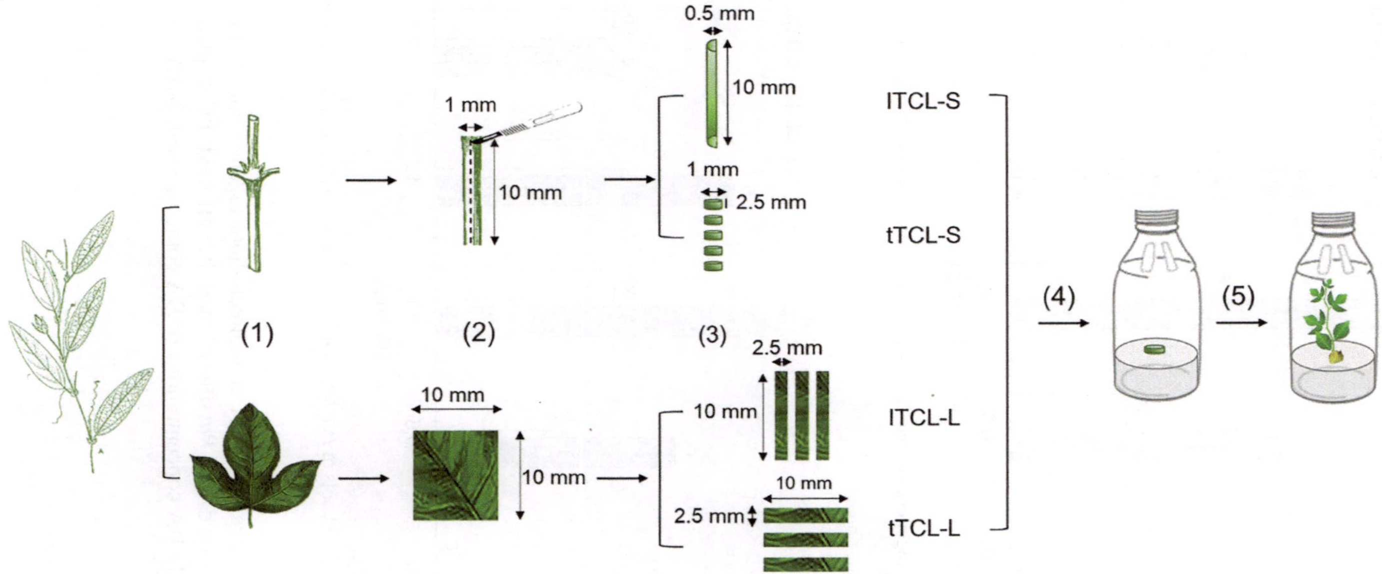

Fig. 12.1 Diagram of establishing a TCL system for shoot regeneration of *P. edulis* Sims. and *P. edulis* f. *flavicarpa*; **(1)** leaf and stem; **(2)** cut leaf into square pieces (10 mm × 10 mm) and stem (1 mm × 10 mm); **(3)** cut lTCL and tTCL of stem and leaf; **(4)** transfer TCL-L and TCL-S to shoot regeneration medium; and **(5)** shoots were recorded after 8 weeks of culture

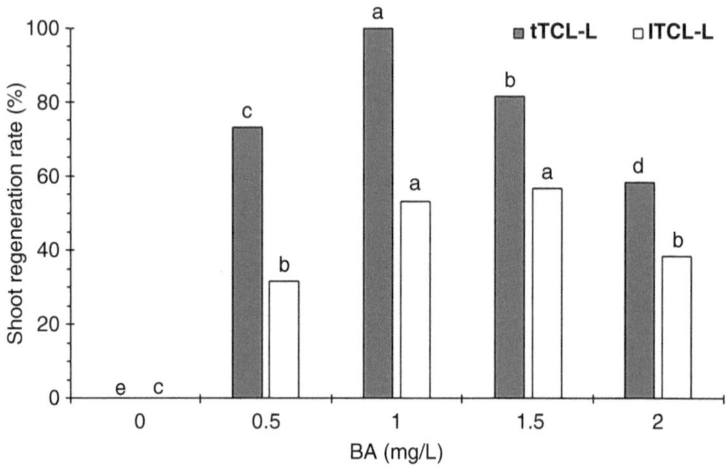

Fig. 12.2 The shoot regeneration rate of *P. edulis* Sims. derived from tTCL-L and lTCL-L after 8 weeks of culture

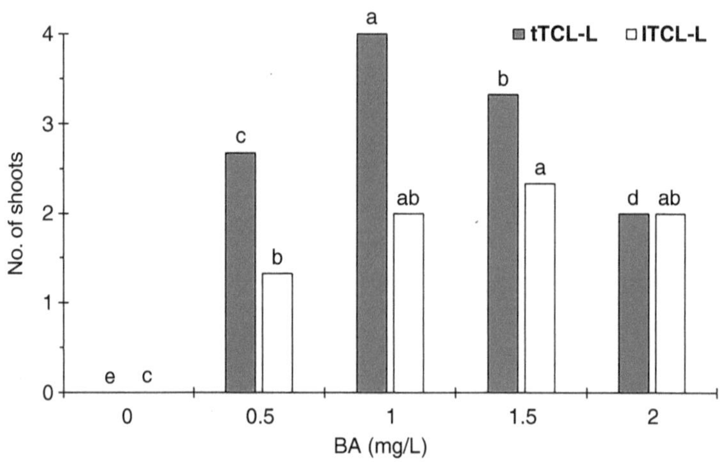

Fig. 12.3 The number of shoots of *P. edulis* Sims. derived from tTCL-L and lTCL-L after 8 weeks of culture

1994). Thus, differences in direct or indirect shoot regeneration in this study and those previously noted on *Passiflora* may be affected by explants cuttings, genotypes, and only BA concentrations or BA combined with NAA.

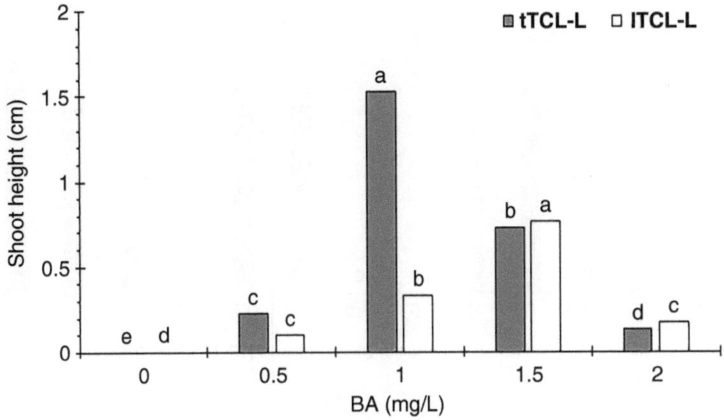

Fig. 12.4 The shoot height of *P. edulis* Sims. derived from tTCL-L and lTCL-L after 8 weeks of culture

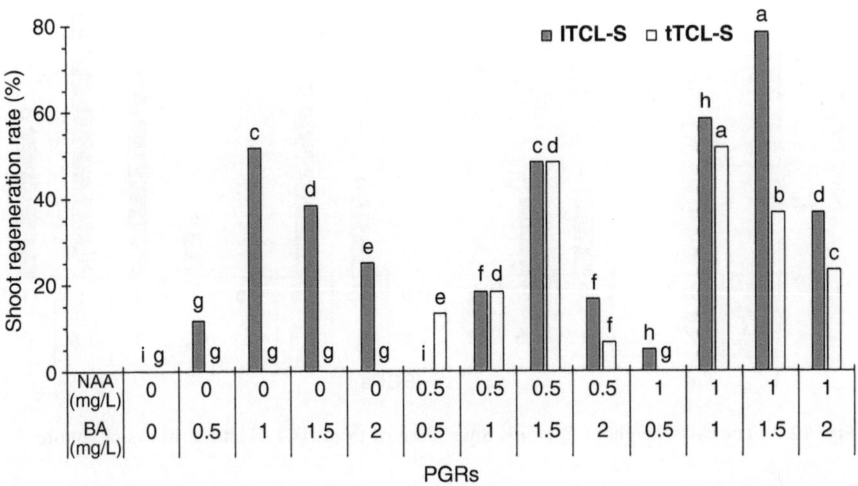

Fig. 12.5 The shoot regeneration rate of *P. edulis* Sims. derived from tTCL-S and lTCL-S after 8 weeks of culture

12.3.5 Shoot Regeneration of tTCL and lTCL Derived from Stem Explants

The shoots induced from lTCL-S and tTCL-S of *P. edulis* Sim. (Figs. 12.5, 12.6, 12.7, and 12.8) and *P. edulis* f. *flavicarpa* (Figs. 12.9, 12.10, 12.11, and 12.12) on MS medium supplemented with BA alone or a combination of NAA was observed after 8 weeks of culture.

For *P. edulis* Sims, only BA or BA combined with NAA is suitable for shoot regeneration. The highest shoot regeneration rate (78.33%) was obtained from

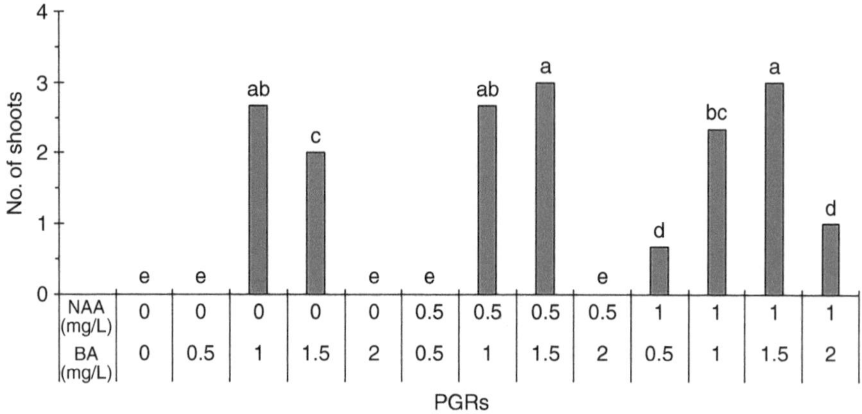

Fig. 12.6 The number of shoots of *P. edulis* Sims. derived from lTCL-S after 8 weeks of culture

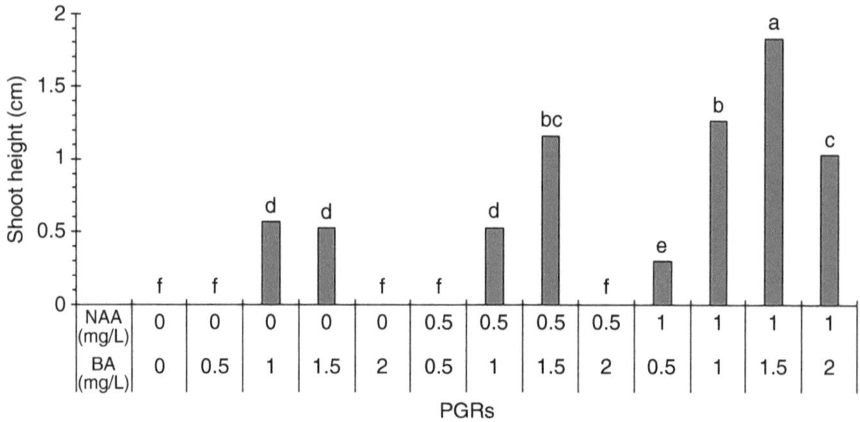

Fig. 12.7 The shoot height of *P. edulis* Sims. derived from lTCL-S after 8 weeks of culture

lTCL-S cultured on MS medium supplemented with 1.5 mg/L BA combined 1.0 mg/L NAA (Fig. 12.5), and the number of shoots (3.00 shoots) and shoot height (1.83 cm) were also highest at this concentration (Figs. 12.6 and 12.7).

For *P. edulis* f. *flavicarpa*, lTCL-S cultured on MS medium supplemented with 1.5 mg/L BA for shoot regeneration rate (66.67%) (Fig. 12.9), number of shoots (2.33 shoots), and shoot height (0.43 cm) (Figs. 12.10 and 12.11) higher than lTCL-S cultured on MS medium supplemented with others BA (Fig. 12.12).

For lTCL-S explant, shoot regeneration rate (71.67%), the number of shoots (3.00 shoots), and shoot height (1.13 cm) are the highest in medium containing 1.5 mg/L BA and 1.0 mg/L NAA (Fig. 12.9). In this study, tTCL-S did not shoot regeneration at any concentration of only BA or BA plus NAA. Only BA or BA combined with

Fig. 12.8 Effects of BA combined with NAA on shoot regeneration of *P. edulis* Sims derived from tTCL-S and lTCL-S after 8 weeks of culture. (**a, c**) tTCL-S and lTCL-S (0.5, 1.0, 1.5, and 2.0 mg/L BA combined with 0.5 mg/L NAA, left to right); (**b, d**) tTCL-S and lTCL-S (0.5, 1.0, 1.5, and 2.0 mg/L BA plus 1.0 mg/L NAA, left to right)

1.0 mg/L NAA, tTCL-S was necrosis after 8 weeks of culture; however, BA combined with 0.5 mg/L NAA, 100.00% tTCL-S induced callus (Fig. 12.12).

This study compares the effectiveness of shoot regeneration from lTCL-S and tTCL-S of *P. edulis* Sims. and *P. edulis* f. *flavicarpa* cultured on MS medium supplemented with only BA or BA plus NAA. The results showed that lTCL-S showed higher efficiency in shoot regeneration than tTCL-S in both varieties. MS medium supplemented with 1.5 mg/L BA plus 1.0 mg/L NAA was the most suitable medium for induction of shoot regeneration from lTCL-S of both purple and yellow passion fruit varieties (Figs. 12.2 and 12.8).

Explant sources are one of the key elements in an in vitro propagation process. Several biological factors include the genotype, tissue, or organ used as the sample source and the age, size, and shape of the explants (tTCL or lTCL), all of which can influence morphogenesis (Teixeira da Silva and Dobránszki 2015). tTCL-S or lTCL-S was often used to study shoot regeneration during micropropagation, such as *Sesamum indicum* L. (Chattopadhyaya et al. 2010) and *Vanilla planifolia* (Jing et al. 2014).

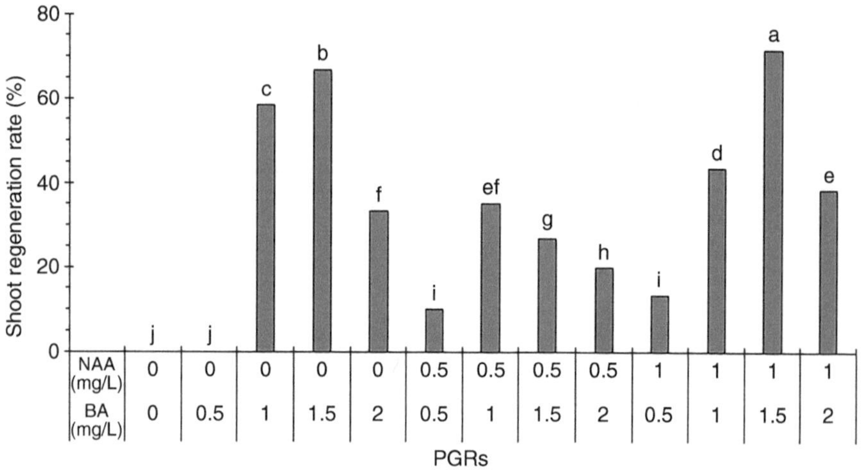

Fig. 12.9 The shoot regeneration rate of *P. edulis* f. *flavicarpa* derived from lTCL-S after 8 weeks of culture

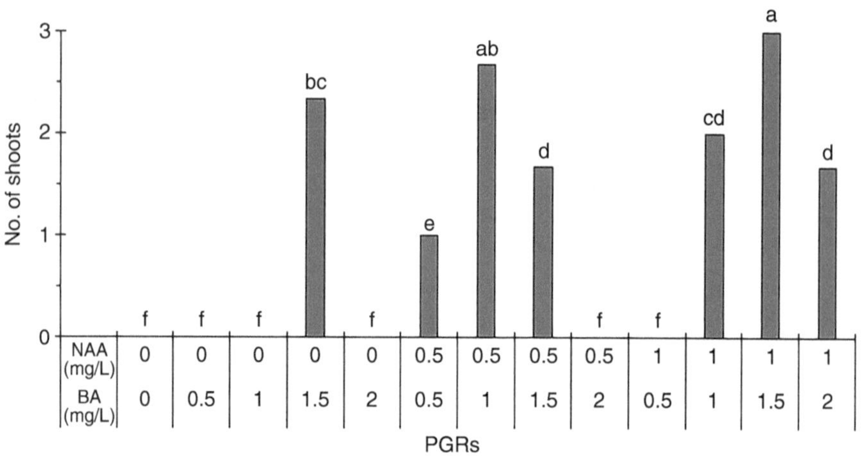

Fig. 12.10 The number of shoots of *P. edulis* f. *flavicarpa* derived from lTCL-S after 8 weeks of culture

The original selection of explant sources was important for the efficiency of shoot regeneration using the TCL technique in micropropagation (Chattopadhyaya et al. 2010). Currently, plantlet regeneration via organogenesis uses explant sources such as stem, internode, leaf, cotyledon, and shoot tip; however, the type of explants suitable for success in micropropagation depends on different varieties and species (Ozarowski and Thiem 2013).

The effectiveness of shoot regeneration from TCL fragments was assessed by shoot regeneration coefficients. Shoot regeneration coefficients were compared by:

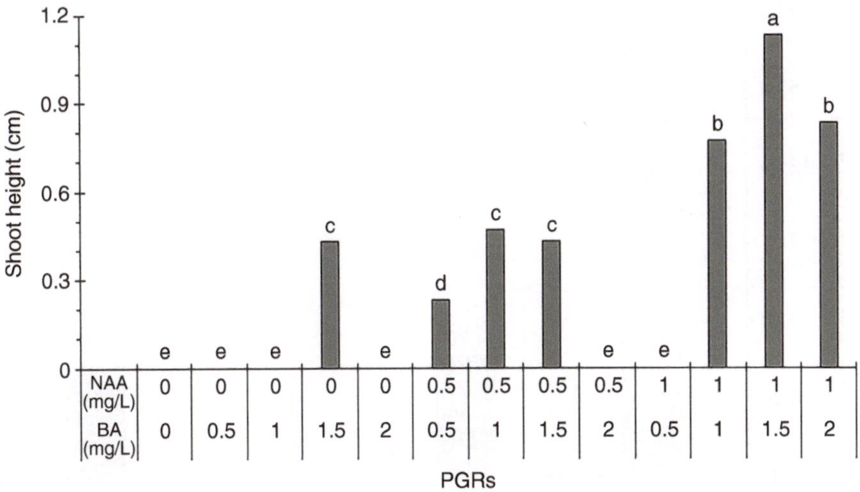

Fig. 12.11 The shoot height of *P. edulis* f. *flavicarpa* derived from lTCL-S after 8 weeks of culture

Fig. 12.12 Effects of BA combined with NAA on shoot regeneration of *P. edulis* f. *flavicarpa* derived from tTCL-S and lTCL-S after 8 weeks of culture. (**a, c**) tTCL-S and lTCL-S (0.5, 1.0, 1.5, and 2.0 mg/L BA combined with 0.5 mg/L NAA, left to right); (**b, d**) tTCL-S and lTCL-S (0.5, 1.0, 1.5, and 2.0 mg/L BA plus 1.0 mg/L NAA, left to right)

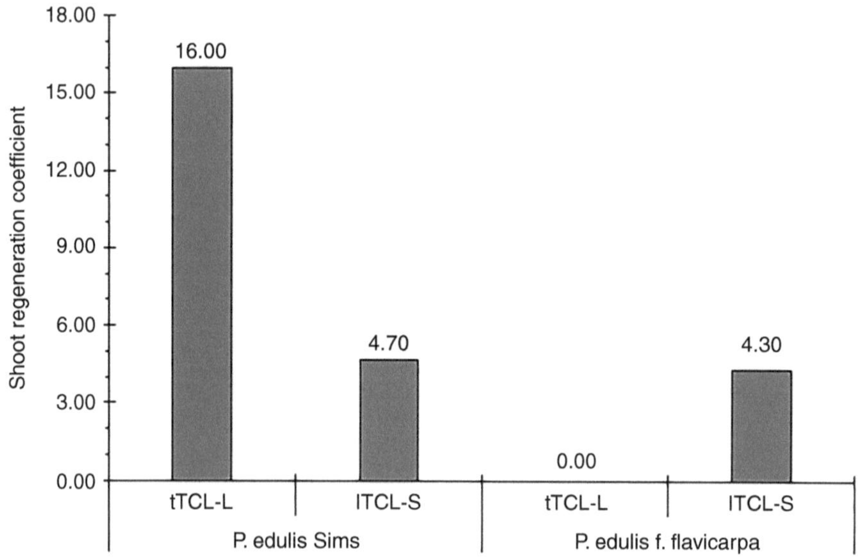

Fig. 12.13 Shoot regeneration coefficient from tTCL-L and lTCL-S explants of *P. edulis* Sims and *P. edulis* f. *flavicarpa* after 8 weeks of culture

$S_c = R_s\ (\%) \times S \times N$ (Hieu et al. 2018a, b). Where S_c: Shoot regeneration coefficient; R_s: Shoot regeneration rate (%); *S*: Number of leaf or internodal fragments cut longitudinally or transversally; *N*: Average of shoots/explant.

Comparing shoot regeneration coefficients between leaf and stem for TCLs culture, the results showed that the shoot regeneration coefficient of tTCL-L explant (16.00) was higher (3.40 times) as compared with lTCL-S (4.70) of *P. edulis* Sims. In comparison, the shoot regeneration coefficient of lTCL-S explants was 4.30 times that of the tTCL-L explants or lTCL-L explants (1.00) of *P. edulis* f. *flavicarpa* (Fig. 12.13).

12.4 Regeneration Studies of *Begonia* Sp. Using the TCL Technique

12.4.1 Plant Material

The petiole (P), flower stalk (F), and stem (S) explants of 3-month-old ex vitro *B. tuberous* plant at the Tay Nguyen Institute for Scientific Research (VAST, Vietnam) were used for materials. The explants were collected and cut transverse thin cell layer (tTCL) with 1 mm × 10 mm in size (Fig. 12.14).

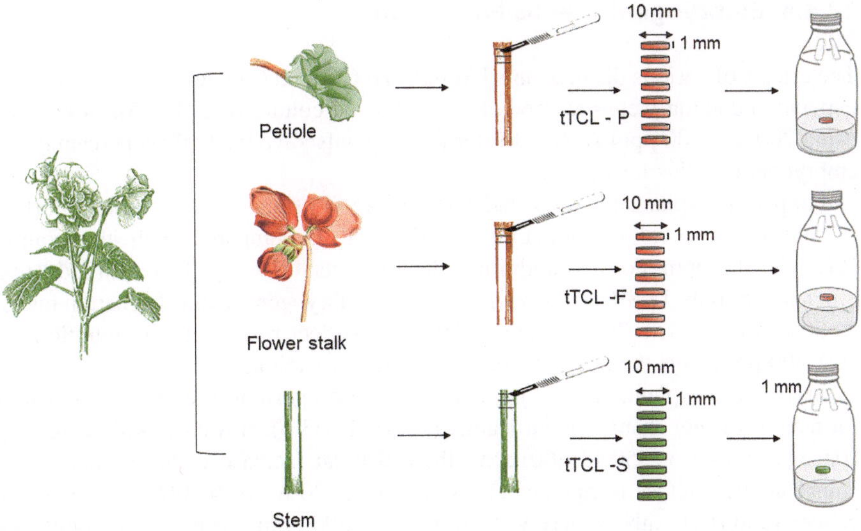

Fig. 12.14 Diagram describes the t-TCL culture of different explant types

12.4.2 Silver Nanoparticles Solution

AgNPs with a size of less than 20 nm were used as follows: $AgNO_3$, 750–1000 ppm; β–chitosan, 250–300 ppm; $NaBH_4$, 200 ppm; Mole ratio $NaBH_4/AgNO_3$, ¼; and a drip rate of $NaBH_4$ is 10–12 drops per min (Chau et al. 2008). AgNP solution was mixed in five concentrations of 0, 100, 200, 300 and 400 ppm.

Explants of the petiole, flower stalk, and stem of *B. tuberous* were sterilized with disinfectants. Using AgNP solution with different concentrations (50, 100, 200, 300, and 400 ppm) for 20 min as an alternative bactericidal agent for the common disinfectant such as 0.1% $HgCl_2$ in 6 min or 60 g/L $Ca(ClO)_2$ in 10 min and the control using distilled water as an antiseptic agent.

12.4.3 TCLs Culture

The tTCL explants (tTCL-P, tTCL–F, and tTCL-S) were cultured on MS medium supplemented with 0.2 mg/L NAA, 0.1 mg/L TDZ (Thidiazuron), 30 g/L sucrose, and 8 g/L agar (Nhut et al. 2005) to evaluate the embryogenic callus after 28 days of culture.

Twenty-eight-day-old embryogenic calli derived from petiole, flower stalk, and stem explants were cut into 0.5 × 0.5 cm. The calli were cultured on ½ MS medium supplemented with 0.2 mg/L TDZ, 0.2 mg/L NAA, 30 g/L sucrose, and 8 g/L agar (Nhut et al. 2005) to evaluate the somatic embryogenesis after 42 days of culture.

12.4.4 Embryogenic Callus Formation

The effects of various disinfectants [AgNPs, $HgCl_2$, and $Ca(ClO)_2$] on the ability to explants induction were recorded after 28 days of culture (Fig. 12.15). In general, using 200 and 300 ppm AgNPs to sterilize explants gave the highest percentage of embryogenic callus formation.

For petiole explant, 100, 200, and 300 ppm AgNPs showed the highest proportion of embryogenic callus formation (35.53–37.8%), compared with the control (11.13%) at 0 ppm. For two traditional disinfectants, 0.1% $HgCl_2$ had an induction efficiency of only 17.8%, and $Ca(ClO)_2$ had an embryogenic callus formation rate of 22.20% (Fig. 12.15). Thus, using AgNPs to disinfect petiole at concentrations of 100–300 ppm gave the best embryogenic callus formation.

In the stem explant, at 0–50 ppm AgNPs, the proportion of embryogenic callus formation was not changed significantly (13.33–15.53%). However, with increasing concentrations from 50 to 200 ppm, there was an increase in the proportion of embryogenic callus formation (13.33% to 22.20% to 40.00%). When the concentrations of AgNPs increased from 200 to 300 ppm, there was no significant change in the rate of embryogenic callus formation (40.00% and 42.20%, respectively). These two AgNP concentrations were the most effective in embryogenic callus induction of explants (Fig. 12.16). For the two traditional disinfectants, the rate of embryogenic callus formation when disinfected with 60 g/L $Ca(ClO)_2$ had an induction rate of 31.13%, which was higher than the 0.1% $HgCl_2$ treatment (20%). Thus, 200 and 300 ppm AgNP treatments were the most effective in embryogenic callus induction.

For flower stalk explants, the proportion of embryogenic callus formation when using AgNPs was similar to tTCL-S. The proportion of embryogenic callus

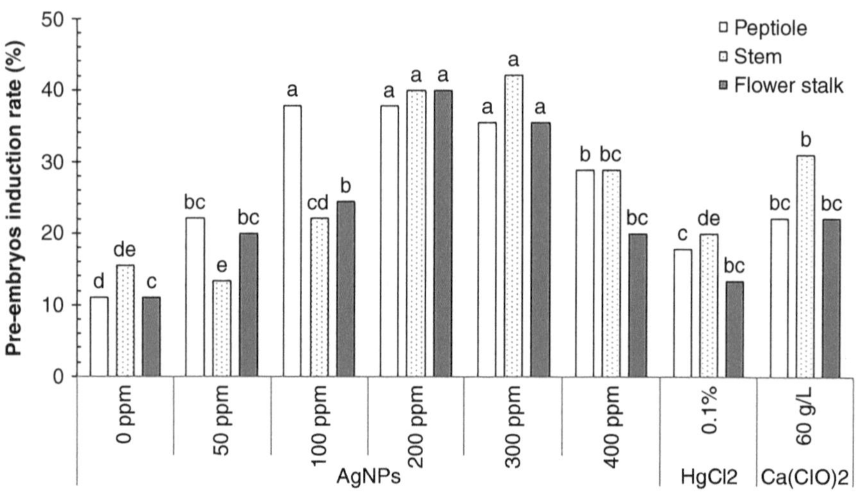

Fig. 12.15 The effect of AgNPs, $HgCl_2$, and $Ca(ClO)_2$ on the embryogenic callus formation after 28 days of culture

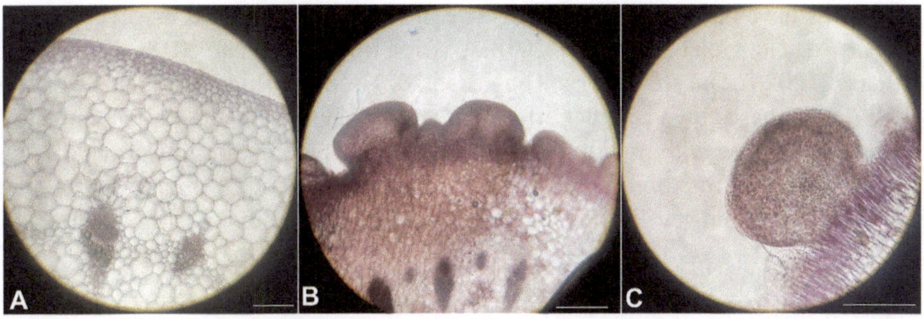

Fig. 12.16 The morphology of embryogenic callus derived from AgNPs treatment. (**a**) Original explant at day 0; (**b**) Embryogenic callus at day 14; (**c**) Globular embryos at day 28 (bars: 1 mm)

formation from 11.13% to 40.00% increased proportionally to an increase in the concentration of AgNPs from 0 to 200 ppm. Increasing the concentration of AgNPs from 200 to 300 ppm, the results showed a negligible change in the proportion of embryogenic callus formation (40.00% and 35.53%). At the 400 ppm AgNPs, the proportion of embryogenic callus formation was not high (20.00%). The efficiency of tTCL-F induction when disinfected with 60 g/L $Ca(ClO)_2$ and 0.1% $HgCl_2$ was much lower than using 200 and 300 ppm AgNPs as sterilants. Thus, AgNPs at concentrations of 200 and 300 ppm gave the best embryogenic callus formation in t-TCL-F.

The results noted above showed that AgNPs are effective for disinfection of samples and explant induction. However, each type of explant responds differently to the appropriate concentration of AgNPs. In general, using AgNPs at concentrations of 200 and 300 ppm gives the best induction effect in all three explant types (petiole alone can use 100–300 ppm AgNPs). If the concentration is too high (400 ppm) or too low (0 and 50 ppm), the disinfection and induction effect is not high. Meanwhile, 60 g/L $Ca(ClO)_2$ and 0.1% $HgCl_2$ have lower induction rate than AgNPs.

AgNPs have been shown to have high antibacterial properties. The effectiveness of AgNPs in general and nano solutions, in particular, depends on the combination of chemistry, size, surface coverage, chemical interactions, concentrations, and the plants used (Syu et al. 2014). The application of AgNPs in plant tissue culture to prevent microbial contamination was first reported by Abdi et al. (2008). Tung et al. (2018) used AgNPs in microponic culture, and the results showed that AgNPs played a role in increasing plant growth as well as reducing the number of microorganisms (eight bacterial species and three fungal species) in the microponic medium.

Disinfecting ex vitro leaves, flower stalks, and stems with $HgCl_2$ disinfectant, the recorded results showed that the explant induction only reached about 30–35% (Nhut et al. 2005). Meanwhile, this study indicates that AgNPs have higher disinfection efficiency (40–42%) at 200–300 ppm AgNPs in all three different types of explant sources.

AgNPs have been shown to be effective in disinfecting cultures on several crops, such as African violet (Nhut et al. 2018), passion fruit (Hieu et al. 2018a, b), seaweed (Mo et al. 2020), and strawberry (Tung et al. 2021). In this study, sterilizing explants with AgNPs has shown a higher effect on the survival rate and the explant induction compared to those with traditional disinfectants. The recorded results provide a new direction in plant micropropagation through this investigation, which is to optimize the ex vitro explant sterilization stage by using AgNP as a disinfecting agent.

12.4.5 Somatic Embryogenesis

The different stages of embryos derived from AgNPs and $HgCl_2$ treatment were recorded after 42 days of culture (Figs. 12.17 and 12.18). After about 1 week of culture, all embryogenic callus formed somatic embryos (100%). However, there are differences in the number of embryos at different stages (Fig. 12.17).

The total number of embryos of embryogenic calli derived from the stalk, petiole, and stem segments sterilized with AgNPs was higher than those sterilized with $HgCl_2$ (Fig. 12.17). Then, internode and flower stalk-derived explants sterilized with AgNPs were highest in the total number of embryos (34.33 and 36.33 embryos, respectively) as well as the cotyledon stage (12.33 and 14.33 embryos, respectively) (Fig. 12.17).

This is the first study to describe somatic embryogenesis with full stages such as globular, heart, torpedo, and cotyledon. Research by Nhut et al. (2005, 2010) only described the process of shoot regeneration from different explant sources. The

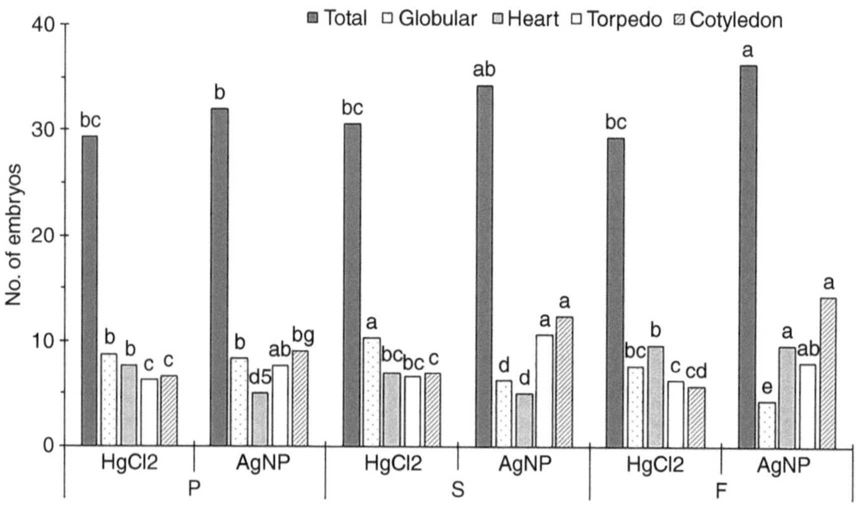

Fig. 12.17 The different stages of embryos derived from AgNPs and $HgCl_2$ treatment after 42 days of culture (the total number of embryos was the sum of the number of embryos at all stages [Globular, Heart, Torpedo, and Cotyledon])

Fig. 12.18 The effect of AgNPs and $HgCl_2$ on embryogenesis from different explants after 42 days of culture. (**a**) AgNPs, (**b**) $HgCl_2$, (**P**) Peptiole, (**S**) Stem, (**F**) Flower stalk (bars: 1 cm)

results of this study are significant in demonstrating the morphogenesis of the explants via the embryogenesis pathway and the role of AgNP in increasing the efficiency of embryogenesis and the short time of embryogenesis with explant sources sterilized by common disinfectant.

12.5 Regeneration Studies of *Panax Vietnamensis* Using the TCL Technique

12.5.1 Plant Material

Three-month-old leaves of in vitro *P. vietnamensis* plants were used as the source of explants. The selected plants were vitrification-free with healthy leaves and shoots.

12.5.2 Culture Medium

The tTCL-L (1 mm × 10 mm) explants were cut from in vitro leaves as initial explants and culture on MS medium supplemented with plant growth regulators (PGRs) including NAA (0.1–2.0 mg/L), 2,4-D (0.1–2.0 mg/L), BA (0.1–2.0 mg/L), TDZ (0.01–1.0 mg/L), and in combination for callus induction experiments.

For callus proliferation, explants were cultured in MS media supplemented with 0.2 mg/L TDZ and different concentrations of the auxins, that is, 2,4-D, IBA, and NAA at different concentrations (0.5, 1.0, 2.0, 3.0, and 5.0 mg/L) in a 16-h photoperiod. After 8 weeks of culture, the white calli were used as primary explants to establish embryogenic cultures.

White calli derived from tTCL-L were cut into small pieces (1.0 cm × 1.0 cm) and placed on MS media containing 1.0 mg/L 2,4-D, 0.5 mg/L NAA, and TDZ at various concentrations (0.01, 0.1, 0.2, and 0.5 mg/L) for embryogenesis experiments.

12.5.3 Callus Induction

The tTCL-L explants cultured on PGR-free medium and media containing different concentrations of TDZ or BA under either 16-h photoperiod or darkness were necrotic (data not shown). Calli stemming from the edges of the explants were obtained from tTCLs cultured on media supplemented with different concentrations of 2,4-D (0.2–2.0 mg/L) or NAA (1.0–2.0 mg/L) (Fig. 12.19).

Soft friable and hard non-friable calli were obtained on media supplemented with 2,4-D and NAA, respectively. The callogenesis rate (66.7%) was highest in the medium supplemented with 2.0 mg/L 2,4-D under darkness. 2,4-D is the most effective auxin for callus induction of *Panax* sp. (Choi et al. 1994). Our result also showed that 2,4-D was optimal to callus induction after 8 weeks of culture. NAA also induced callus formation, while media containing TDZ and BA resulted in necrotic explants. tTCL-L explants regardless of light or dark conditions cultured on media supplemented with 2,4-D combined with BA induced callus formation under both 16-h photoperiod and darkness after 8 weeks of culture. Initial callus tissue emerged from the edges of explants followed by the surface. At a 16-h photoperiod, callogenesis rates were similar to those under darkness, and 6 out of 18 treatments gave callogenesis rates of 100% (Fig. 12.20).

Explants cultured under 16-h photoperiod induced green hard calli, while milky, transparent-white, and brownish-yellow friable calli were observed when explants

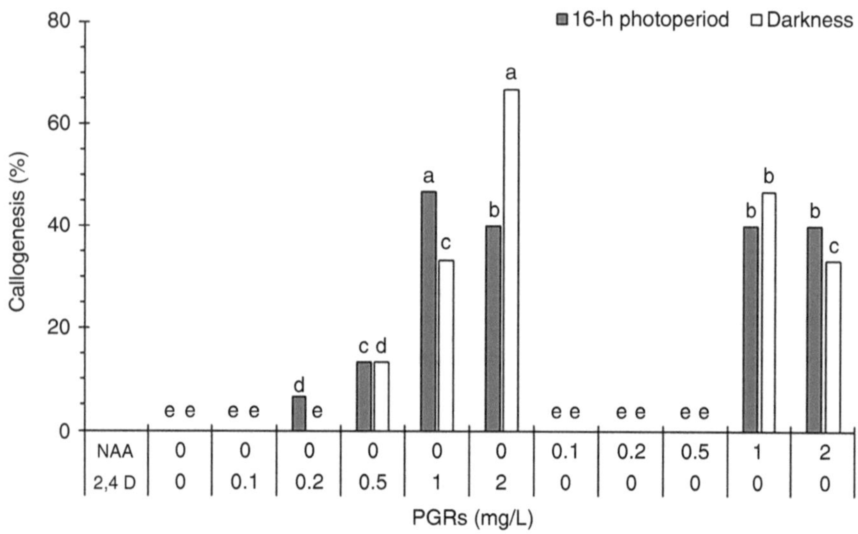

Fig. 12.19 Effects of PGRs on the callogenesis of *P. vietnamensis* tTCL-L after 8 weeks of culture

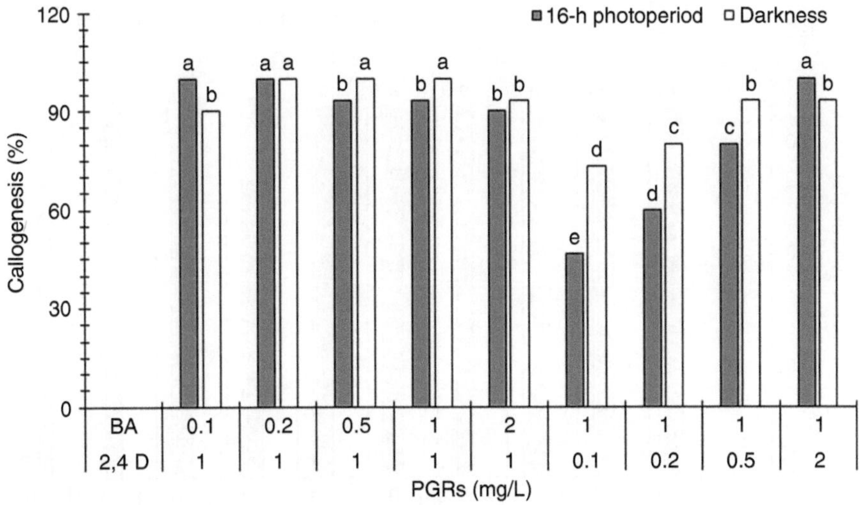

Fig. 12.20 Combinatorial effects of 2,4-D and BA on the callogenesis of *P. vietnamensis* tTCL-L after 8 weeks of culture

were maintained under dark conditions. Eleven of the 18 media treatments supplemented with 1.0 mg/L 2,4-D in combination with TDZ (0.01–1.0 mg/L) under 16-h photoperiod and in the darkness with TDZ (0.01–0.5 mg/L) gave callogenesis rates of 100% (Fig. 12.21).

In comparison with media containing 2,4-D and BA, media supplemented with 2,4-D, and TDZ promoted greater callus induction (data not shown). Darkness was as suitable as a light condition for callogenesis; however, calli produced under 16-h photoperiod were green and hard, while explants were white, yellow, and friable calli in the darkness. Under total darkness, medium containing 1.0 mg/L 2,4-D and 0.1 mg/L TDZ yielded milky white friable calli emerging from the edges and was the most suitable for callogenesis with the maximum callus induction.

NAA combined with BA was less effective at inducing callogenesis than 2,4-D and BA or 2,4-D and TDZ. Explants were necrotic in 6 of 18 treatments, and callogenesis was not observed in two other treatments even though the explants were still green (Fig. 12.22).

Total darkness was more suitable to callus formation than the 16-h photoperiod (Fig. 12.22), and explants cultured on media supplemented with 2.0 mg/L NAA and 1.0 mg/L BA under total darkness gave the best rate of callogenesis (100%), while 60% was achieved on the same media formulation under a 16-h photoperiod. Calli emerged from the edges of explants and were few in number.

In this study, the MS media supplemented with 2,4-D combined with TDZ was the most suitable combination for callus formation. This result is consistent with callus formation *in P. ginseng,* and *P. quinquefolius*, which was most successful in MS media supplemented with 2,4-D combined with kinetin (KIN) or with BA (Furuya et al. 1986; Jiu 1992).

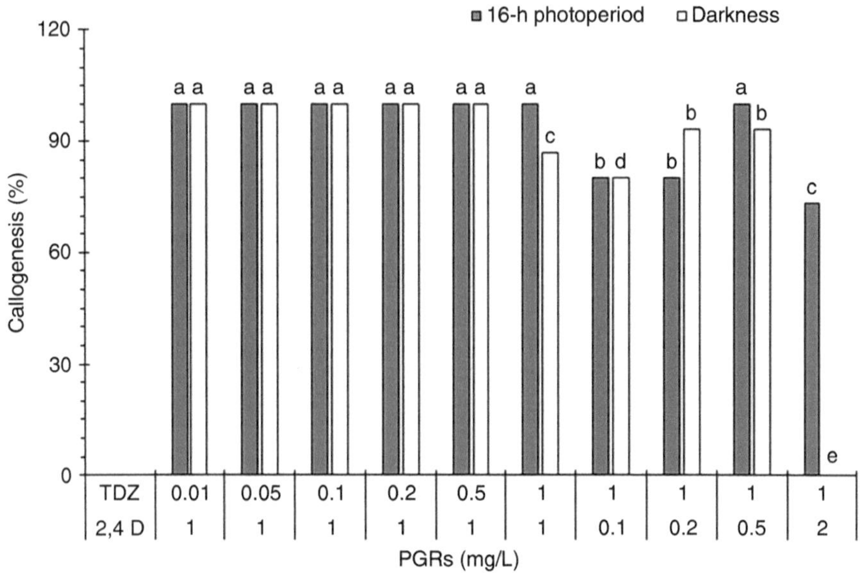

Fig. 12.21 Combinatorial effect of 2,4-D and TDZ on the callogenesis of *P. vietnamensis* tTCL-L after 8 weeks of culture

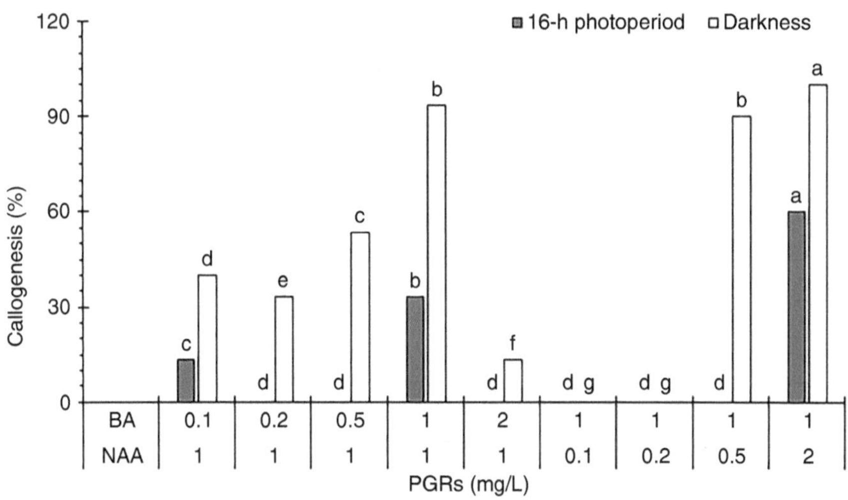

Fig. 12.22 Combinatorial effect of NAA and BA on the callogenesis of *P. vietnamensis* tTCL-L after 8 weeks of culture

12.5.4 Callus Proliferation

The media containing 2,4-D stimulated higher callus induction than those with IBA or NAA (Fig. 12.23). Calli growth in the medium supplemented with 1.0 mg/L 2,4-D with approximately fourfold fresh weight increases after 4 weeks of culture (Fig. 12.23).

The results also showed that TDZ combined with auxins (especially, 2,4-D) significantly improved callus growth. Guo et al. (2011) showed that TDZ had both auxin and cytokinin-like effects to induce and maintain a number of biological events in cells. It is thought that TDZ enhances the accumulation and transport of auxin in cultured tissues.

12.5.5 Embryogenesis

The most commonly used PGRs for induction of embryogenesis were 2,4-D, dicamba, and picloram (Roostika and Mariska 2003). Among all the PGRs, 2,4-D was optimal for callus and somatic embryo formation in Panax ginseng (Zhong and Zhong 1992; Arya et al. 1993).

Somatic embryogenesis could be further improved in combination between 2,4-D and KIN (Furuya et al. 1986; Lee et al. 1989) or NAA (Wang et al. 1999). In this study, the combinations of 1 mg/L 2,4-D with 0.5 mg/L NAA and 0.2 mg/L TDZ were suitable for somatic embryogenesis. All the somatic embryo development

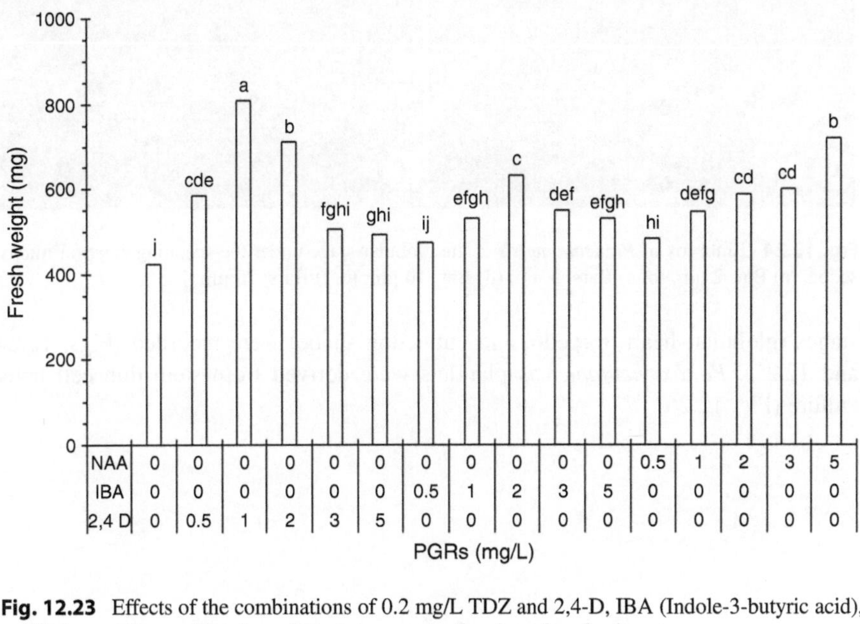

Fig. 12.23 Effects of the combinations of 0.2 mg/L TDZ and 2,4-D, IBA (Indole-3-butyric acid), or NAA on callus proliferation of *P. vietnamensis* after 4 weeks of culture

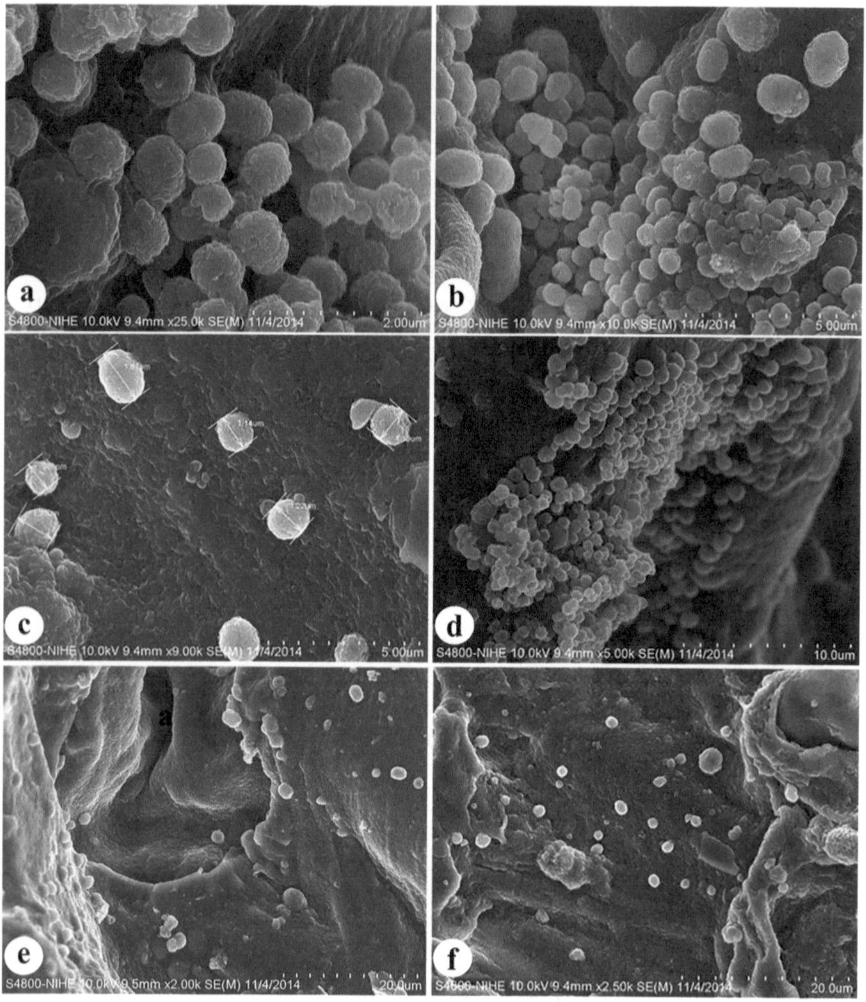

Fig. 12.24 Embryos of *P. vietnamensis* at the globular stage under the scanning electron microscope. (a) Bar: 2 μm; (b, c) Bars: 5 μm; (d) Bar: 10 μm; (e, f) Bars: 20 μm

stages (globular, heart, torpedo, and cotyledon shape) were recorded (Figs. 12.24 and 12.25). *Panax vietnamensis* plantlets were derived from stem thin cell layer culture (Fig. 12.26).

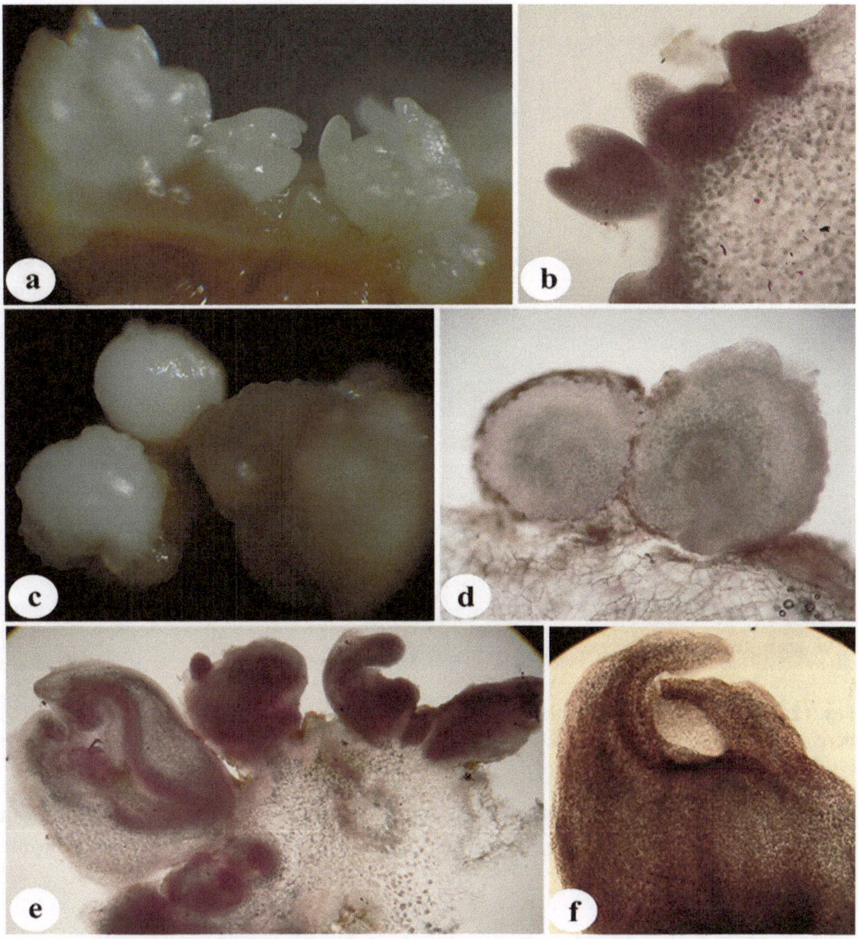

Fig. 12.25 All stages of somatic embryos development of *P. vietnamensis*. (**a, b**) Heart stage; (**c, d**) Globular stage; (**e**) Cotyledon stage; (**f**) Torpedo stage

12.6 Conclusion

The TCL approach has been successful in the morphogenesis and regeneration of some previously reported plant species, and it continues to demonstrate dominance in many plant species such as fruit trees (*Passiflora* sp.), flowers (*Begonia* sp.), and medicinal plants (*Panax vietnamensis*). The response of each plant species varies depending on many factors such as size and origin. The optimal TCL explant, in combination with other controllable factors, has shown this system to be superior to the use of conventional cultures. Hence, depending on the research target, we can choose the appropriate morphogenesis type. With its application potential, based on

Fig. 12.26 *Panax vietnamensis* plantlets were derived from stem thin cell layer culture. (**a**) l-TCL-derived somatic embryo, (**b**) t-TCL-derived somatic embryo, (**c**) somatic embryo-derived plantlet

the advantages in improving the efficiency of regeneration in both coefficient and quality, the TCL technique will also be applied on many other crops of economic value in the future.

References

Abdi G, Salehi H, Khosh-khuri M (2008) Nano silver: a novel nanomaterial for removal of bacterial contamination in *valerian* (*V. officinalis*) tissue culture. Acta Physiol Plant 30:709–714

Arya S, Arya ID, Eriksson T (1993) Rapid multiplication of adventitious somatic embryos of *Panax ginseng*. Plant Cell Tissue Organ Cult 34:157–162

Azadi P, Bagheri K, Gholami M, Mirmasoumi M, Moradi A, Sharafi A (2017) Thin cell layer, a suitable explant for in vitro regeneration of saffron (*Crocus sativus* L.). J Agric Sci Technol 19: 1429–1435

Betti C, Rovere FD, Ronzan M, Fattorini L (2019) EIN2 and COI1 control the antagonism between ethylene and jasmonate in adventitious rooting of *Arabidopsis thaliana* thin cell layers. Plant Cell Tissue Organ Cult 138:41–51

Bhattacharyya P, Paul P, Kumaria S, Tadon P (2018) Transverse thin cell layer (t-TCL)-mediated improvised micropropagation protocol for endangered medicinal orchid *Dendrobium aphyllum* Roxb: an integrated phytomolecular approach. Acta Physiol Plant 40:137

Bose B, Kumaria S, Choudhury H, Tandon P (2017) Insights into nuclear DNA content, hydrogen peroxide and antioxidative enzyme activities during transverse thin cell layer organogenesis and ex vitro acclimatization of Malaxis wallichii, a threatened medicinal orchid. Physiol Mol Biol Plants 23:955–968

Chattopadhyaya B, Banerjee J, Basu A, Sen SK, Maiti MK (2010) Shoot induction and regeneration using internodal transverse thin cell layer culture in *Sesamum indicum* L. Plant Biotech Rep 4(2):173–178

Chau H, Bang L, Buu N, Dung T, Ha H, Quang D (2008) Some results in manufacturing of nanosilver and investigation of its application for disinfection. Adv Natl App Sci 9(2):241–248

Choi KT, Lee CH, Ahn IO, Lee JH, Park JC (1994) Characteristics of the growth and ginsenosides in the suspension–cultured cells of Korean ginseng (*Panax ginseng* C.a. Meyer). In Proceedings of the international ginseng conference, Vancouver, Canada, pp. 259–268

Cristian NMP, Anne KHE (2021) A novel genotype-independent technique for successful induction of somatic embryogenesis of adult plants of *Jatropha curcas* L. using petiole transverse thin cell layer (TCL). Afr J Biotechnol 20(2):85–91

Croom L, Jackson C, Vaidya B, Parajuli P, Joshee N (2016) Thin cell layer (TCL) culture system for herbal biomass production and genetic transformation of *Bacopa monnieri* L. Wettst. Am J Plant Sci 7:1232–1245

Da Silva CV, de Oliveira LS, Loriato VAP, Da Silva LC, De Campos JMS, Viccini LF, De Oliveira EJ, Otoni WC (2011) Organogenesis from root explants of commercial populations of *Passiflora edulis* Sims and a wild passion fruit species, *P. cincinnata* masters. Plant Cell Tiss Org Cult 107(3):407–416

Dornelas MC, Vieira MLC (1994) Tissue culture studies on species of *Passiflora*. Plant Cell Tissue Organ Cult 36(2):211–217

Ekmekçigil M, Bayraktar M, Akkuş O, Gürel A (2019) High-frequency protocorm-like bodies and shoot regeneration through a combination of thin cell layer and RITA® temporary immersion bioreactor in *Cattleya forbesii* Lindl. Plant Cell Tissue Organ Cult 136:465–466

Furuya T, Yoshikawa T, Ushiyama K, Oda H (1986) Formation of plantlets from callus cultures of ginseng (*Panax ginseng*). Experientia 42(2):193–194

Guo B, Abbasi BH, Zeb A, Xu LL, Wei YH (2011) Thidiazuron: a multi–dimensional plant growth regulator. Afr J Biotech 10(45):8984–9000

Hien VT, Huy NP, Vinh BVT, Chien HX, Tung HT, Nam NB, Luan VQ, Nhut DT (2016) Somatic embroygenesis from leaf transverse thin cell layer derived-callus of Vietnamese ginseng (*Panax vietnamensis* Ha et Grushv.). J Biotech Vietnam 14(1):63–73

Hieu T, Linh NTN, Tung HT, Bao HG, Nguyen CD, Nhut DT (2018a) Stimulation of shoot regeneration through leaf thin cell layer culture of *Passiflora edulis Sims*. J Biotech Vietnam 16(4):669–677

Hieu T, Tung HT, Nguyen CD, Nhut DT (2018b) Establishing aseptic explant source for *Passiflora edulis* Sims. And *Passiflora edulis* f. Flavicarpa. J Sci Hue Uni: Nat Sci 127(1C):71–84

Hieu T, Tung HT, Nguyen CD, Nhut DT (2019) Efficiency of shoot regeneration and micropropagation of purple passion fruit (*Passiflora edulis* Sims.) via internodal longitudinal thin cell layer culture. J Biotech Vietnam 17(4):699–708

Jing GF, Razak WNA, Rahman ZAB, Subramaniam S (2014) The effect of thin cell layer system in *Vanilla planifolia* in vitro culture. Curr Bot 5(1):22–25

Jiu SY (1992) Plant regeneration from adventitious root-derived calli of ginseng (*Panax ginseng* C.a. Meyer). J Agr Assoc China New Series 0:41–48

Lee HS, Lee KW, Yang SG, Jeon JH, Liu JR (1989) Plant regeneration through somatic embryogenesis from mature zygotic embryos of ginseng (*Panax ginseng* C. A. Meyer) and flowering of plantlets. Korean J Bot 32:145–150

Loan NTK, Tuan TT, Hien NTT, Huy NP, Du TX, Nhut DT (2016) Primary and secondary somatic embryogenesis in *Jatropha curcas* L. from leaf transverse thin cell layers. J Biotech Vietnam 14(4):661–671

Lombardi SP, Passos IS, Nogueira MCS, da Glória BA (2007) In vitro shoot regeneration from roots and leaf discs of *Passiflora cincinnata* mast. Brazilian Arch Biol Tech 50(2):239–247

Mao J, Zhou Y, Guo CX, Xu HL, Wang CY (2014) An agrobacterium-mediated transformation system from thin cell layer explants of pyrethrum (*Tanacetum cinenariifolium*). Acta Hortic 1035:163–170

Mo VT, Cuong LK, Tung HT, Huynh TV, Nghia LT, Khanh CM, Lam NN, Nhut DT (2020) Somatic embryogenesis and plantlet regeneration from the seaweed *Kappaphycus striatus*. Acta Physiol Plant 42:1–11

Monja-Mio KM, Robert ML (2016) Direct somatic embryogenesis of *Agave fourcroydes* Lem. Through thin cell layer culture. In vitro Cell Dev Biol-Plant 52:212

Murashige T, Skoog F (1962) A revised medium for rapid growth and bioassays with tobacco tissue cultures. Physiol Plant 15(3):473–497

Nhut DT, Ha NTT, Huyen PX, Anh TTL, Hai NT, Thi NN, Tram NT, Binh NV, Luan VQ (2005) Influence of the position of explants on morphogenesis of thin cell layers of *begonia tuberous*. Proc Nat Sci Conf 2005: Biotech Basic Res 8(2):279–283

Nhut DT, Hai NT, Phan MX (2010) A highly efficient protocol for micropropagation of *begonia tuberous*. In: Jain SM, Ochatt S (eds) Protocols for in vitro propagation of ornamental plants. Humana Press, United States, pp 15–20

Nhut DT, Trinh DB, Cuong DM, Tung HT, Huy NP, Hien VT, Luan VQ, Hien LTT, Chau NH (2018) Study on silver nanoparticles as a novel explant disinfectant for micropropagation of African violet (*Saintpaulia ionantha* H. Wendl.). Vietnam J Biotech 16(1):87–97

Ozarowski M, Thiem B (2013) Progress in micropropagation of *Passiflora* spp. to produce medicinal plants: a mini-review. Rev Brasil de Farmaco 23(6):937–947

Parthibhan S, Rao MV, Teixeira da Silva JA, Kumar ST (2018) Somatic embryogenesis from stem thin cell layers of *Dendrobium aqueum*. Biol Plant 62(3):439–450

Ramírez-Mosqueda MA, Iglesias-Andreu LG (2016) Direct organogenesis of *Stevia rebaudiana* Bertoni using thin cell layer (TCL) method. Sugar Tech 18:424–428

Ramírez-Mosqueda MA, Iglesias-Andreu LG, Armas-Silva AA, Cruz-Gutiérrez EJ, Torre-Sánchez JF, Leyva-Ovalle OR, Galán-Páez CM (2019) Effect of the thin cell layer technique in the induction of somatic embryos in *Pinus patula* Schl. Et Cham. J For Res 30:1535–1539

Roostika IT, Mariska I (2003) In vitro culture of pineapple by organogenesis and somatic embryogenesis: its utilization and prospect. Buletin Agro Bio 6:34–40

Sabooni N, Shekafandeh A (2017) Somatic embryogenesis and plant regeneration of blackberry using the thin cell layer technique. Plant Cell Tissue Organ Cult 130:313–321

Syu YY, Hung JH, Chen JC, Chuang HW (2014) Impacts of size and shape of silver nanoparticles on *Arabidopsis* plant growth and gene expression. Plant Physiol Biochem 83:57–64

Teixeira da Silva JA (2013) The role of thin cell layers in regeneration and transformation in orchids. Plant Cell Tissue Organ Cult 113:149–161

Teixeira da Silva JA (2015) The in vitro culture of *Ceropegia* species, important medicinal and ornamental plants: a review. Envir Exper Biol 13(1):1–13

Teixeira da Silva JA, Dobránszki J (2013) Plant thin cell layers: a 40-year celebration. Plant Growth Regul 32:922–943

Teixeira da Silva JA, Dobránszki J (2015) Plant thin cell layers: update and perspectives. Folia Hortic 27(2):183–190

Tran Thanh Van K (1973) In vitro control of de novo flower, bud, root and callus differentiation from excised epidermal tissues. Nature 246:44–45

Tripathia D, Raia KK, Raib SK, Rai SP (2018) An improved thin cell layer culture systemfor efficient clonal propagation and in vitro withanolide production in a medicinal plant *Withania coagulans* Dunal. Ind Crop Prod 119(1):172–182

Tu HY, Zhang AL, Xiao W, Lin YR, Shi JH, Wu YW, Wu ST, Zhong CH, Mo SH (2018) Induction and identification of tetraploid *Hedychium coronarium* through thin cell layer culture. Plant Cell Tissue Organ Cult 135:395–406

Tung HT, Nam NB, Huy NP, Luan VQ, Hien VT, Phuong TTB, Le TD, Nhut DT (2018) A system for large scale production of chrysanthemum using microponics with the supplement of silver nanoparticles under light-emitting diodes. Sci Hortic 232:153–161

Tung HT, Thuong TT, Cuong DM, Luan VQ, Hien VT, Hieu T, Nam NB, Phuong HTN, Vinh BVT, Khai HD, Nhut DT (2021) Silver nanoparticles improved explant disinfection, in vitro growth, runner formation and limited ethylene accumulation during micropropagation of strawberry (*Fragaria × ananassa*). Plant Cell Tissue Organ Cult 145(2):393–403

Vaidya BN, Jackson CL, Perry ZD, Dhekney SA, Joshee N (2016) Agrobacterium-mediated transformation of thin cell layer explants of *Scutellaria ocmulgee* small: a rare plant with anti-tumor properties. Plant Cell Tissue Organ Cult 127(1):57–69

Vudala SM, Padial AA, Ribas LLF (2019) Micropropagation of *Hadrolaelia grandis* through transverse and longitudinal thin cell layer culture. South African J Bot 121:76–82

Wang X, Proctor JTA, Krishnaraj S, Saxena PK, Sullivan JA (1999) Rapid somatic embryogenesis and plant regeneration in American ginseng: effects of auxins and explants. J Ginseng Res 23: 148–163

Wattanapan N, Nualsri C, Meesawat U (2018) In vitro propagation through transverse thin cell layer (tTCL) culture system of lady's slipper orchid: *Paphiopedilum callosum* var. sublaeve. Songklanakarin J Sci Technol 40(2):306–313

Winarto B, Yuniarto K, Wegadara DM (2019) Axillary shoots derived from thin cell layer and adenine sulphate application in *in vitro* mass propagation of gerbera [*Gerbera jamesonii* (*H. bolus* ex bolus f.)]. Not Sci Biol 11(1):63–76

Zhong SL, Zhong SG (1992) Morphological and unrestructured characteristics of the embryogenic callus of *American ginseng*. Chinese J Bot 4:92–98

In Vitro Flowering of *Torenia fournieri* 13

Duong Tan Nhut, Tran Trong Tuan, Le Van Thuc, Nguyen Van Binh, and Hoang Thanh Tung

Abstract

Plant cell, tissue, and organ culture techniques along with molecular biology techniques have contributed to providing more knowledge about the flowering process in plants. Until now, besides *Arabidopsis thaliana*, a model plant that has been widely used for flowering research, *Torenia fournieri* has also been used as a source material for in vitro flowering studies. In the study of flowering in plants, in vitro culture conditions are ideal for carrying out further studies of this process. In in vitro conditions, factors such as light intensity, photoperiod, temperature, sugar, minerals, plant growth regulators, adjusted according to the purpose of the study. Changes in media composition, Plant growth regulators (PGRs) or changes in culture conditions can accelerate the growth rate, shorten the vegetative period, and lead to early flowering for further investigation of these phenomena in the physiology of flowering. The objective of this study is to determine the role of several factors such as sugar, PGRs, culture medium, and amino acids on the in vitro flowering of *T. fournieri*.

D. T. Nhut (✉) · H. T. Tung
Tay Nguyen Institute for Scientific Research, VAST, Dalat City, Vietnam

T. T. Tuan
Institute of Tropical Biology, VAST, Ho Chi Minh City, Vietnam

Le Van Thuc
Dalat Nuclear Research Institute, Dalat City, Vietnam

N. Van Binh
Dalat University, Dalat City, Vietnam

D. T. Nhut et al. (eds.), *Plant Tissue Culture: New Techniques and Application in Horticultural Species of Tropical Region*,
https://doi.org/10.1007/978-981-16-6498-4_13

Keywords

Torenia fournieri · In vitro flowering · Sugar · PGR · Amino acid · Culture medium

13.1 Introduction

At least four flowering pathways in plants including photoperiod, energy pathway, growth regulator, and auto-flowering pathway have been studied from the model plant *Arabidopsis thaliana* (Blazquez and Weigel 2000). However, flowering plants will have different stimulating factors suitable for reproduction depending on species and environmental conditions. In addition, the flowering process is influenced by other environmental factors, such as nutrients, ambient temperature, exogenous substances, and biotic stress.

13.1.1 Plant Growth Regulators

The role of cytokinins in the induction of flowering may depend on the control of mitotic activity, the early flowering of the lateral shoot meristems, and the acceleration of the production of appendages by the meristems. In addition, cytokinin is considered as a must-have factor in the induction of flowering (Bernier 1988). BA is a cytokinin that has been shown to be essential for rosebud development (Vu et al. 2006). It also has the ability to regulate flower development through genes that control the activity of the apical meristem (Lindsay et al. 2006).

Cytokinins are widely used for in vitro flowering studies in some orchid species such as *Dendrobium primulinum* (Deb and Sungkumlong 2009), *D. candidum* (Wang et al. 1997), *D. nobile* (Wang et al. 2009), *C. niveo-marginatum* Mak (Kostenyuk et al. 1999), and *B. auricomum* (Than et al. 2009). Thidiazuron (TDZ) is considered to be a more effective plant growth regulator than BA, but it causes less plant growth and faster flowering (Kostenyuk et al. 1999). Zeng et al. (2013) found that TDZ has the ability to induce flowering in mini rose plants, but the flower color changed from red to pink, flowers had no male stamens, sepals changed to leaves, or some flowers had few sepals, without stamen and pistil. Moreover, Vu et al. (2006) found that the induction of flowering by TDZ was unsuccessful, and the shoots were mutated and yellowed despite repeated subcultures to prevent these effects from TDZ. However, when the concentration of inorganic and organic salts in Murashige and Skoog (1962, MS) medium was reduced by ½, the induction of flowering was successful although the rate was lower than that of both BA and ZEA. This proves that the low concentration of mineral salts in MS medium will support TDZ more in flower formation. In addition, Lin et al. (2004) also showed that TDZ affects in vitro flowering in bamboo plants.

Appropriate gibberellic acid content (GA_3) is essential if the flowering occurs, and the lack of GA_3 biosynthesis pathway increases the sensitivity to photoperiod.

High concentration of GA_3 stimulates the induction of flowering and blooming (Kostenyuk et al. 1999). Several previous studies have shown the role of abscisic acid (ABA) in inhibiting in vitro flowering (Vaz and Kerbauy 2008). Exposure to ethylene gas can induce flowering in some species. If the protocorm was grown on media supplemented with ABA, then transfer to the medium containing BA would result in higher flowering rates (Wang et al. 1997). Bloom was successfully using $KMnO_4$ to remove ethylene (Vaz and Kerbauy 2000).

13.1.2 Nutrients

In vitro flowering is influenced by the concentrations and ratios of two major components: carbon sources and micronutrients. Sugar is considered as an important carbon source in the culture medium for successful flower induction and development. Vu et al. (2006) reported that sucrose was only required for the induction and early development of flower buds, while other factors were required for more comprehensive flower development at later stages of in vitro flower morphogenesis (Vu et al. 2006). Taylor et al. (2007) reported that sugars including sucrose, glucose, and fructose were able to induce flowering in *Kniphofia leucocephala*; however, at the same concentration, fructose and glucose were more effective than sucrose, even though the flowers formed were variable.

On MS medium, high nitrogen concentrations generally inhibited flowering and promoted vegetative growth, while halving the mineral content in MS medium or reducing nitrogen concentrations enhanced in vitro flowering in *Cymbidium niveo-marginatum* (Kostenyuk et al. 1999). In some cases, increasing the concentration of phosphorus and potassium in the culture medium promotes favorable in vitro flowering. Studies on *Panax ginseng*, *Lycopersicon esculentum* Mill have demonstrated that BA supplementation and increased phosphorus (P) and potassium (K) contents in the culture medium promoted in vitro flowering faster (Dielen et al. 2001). In apple plant, the number of flowers per plant was linearly correlated with the concentrations of P and K in the leaves. Similar to Cymbidium¸ explants cultured on MS medium supplemented with BA combined with increasing P concentration and decreasing N concentration yielded 2.5 times higher than culture on MS medium supplemented with only ½ mineral salts but the P/N ratio is not determined. In addition, the ratio of NH_4^+/NO_3^- also plays an important role in the development of other organs (Teixeira da Silva et al. 2005). An increase in this ratio inhibits flowering, while a decrease in NH_4^+ concentration promotes flowering (Duan and Yazawa 1994).

13.1.3 Some Recent Studies on In Vitro Flowering

To date, many groups of plants such as ornamentals, commercial plants, medicinal plants, food plants have been used as sources of materials for in vitro flowering (Table 13.1).

Table 13.1 Some recent studies on in vitro flowering

Plant groups	Species	References
Ornamental plant	*D. Madame*	Sim et al. (2008)
	D. chao-praya-smile	Hee et al. (2007)
	D. candidum	Wang et al. (1997)
	C. Goeringii and *C. hybridium*	Li-Ming and Ji-Liang (2006)
	Pharbitis nil	Galoch et al. (2002)
	Passiflora suberosa	Scorza and Janick (1980)
	Lilium longiflorum	Wang (1996)
	Rosa	Wang et al. (2002) Vu et al. (2006) Kachanapoom et al. (2009) Zeng et al. (2013)
Commercial plant	*B. vulgaris* and *D. giganteus*	Rout and Das (1995)
	Bambusa vulgaris var. *vittata* and *B. arundinacea*	John and Nadgauda (1997)
	B. edulis	Lin et al. (2004)
	D. hookeri và cây *D. latiflorus*	Ramanayake (2006) Choun-Sea et al. (2007)
Medicinal plant	*Chenopodium rubrum*	Seidlova et al. (1981)
	Panax ginseng	Tang (2000)
	Cichorium intybus	Li et al. (1996)
	Solanum nigrum	Jabeen et al. (2005)
	Centaurium pulchellum	Cvetić et al. (2004)
	Chamomilla recutita	Kintzios and Michailakis (1999)
	Vitex negundo	Vadawale et al. (2006)
Food plant	*Spinacia oleracea*	Al-Khayri et al. (1992)
	Tomato	Pierik et al. (1994)
	Brassica oleracia var. *botrytis*	Vandana et al. (1995)
	M. paniculata	Jumin and Nito (1996)
	Pisum sativum	Franklin et al. (2000)
	Brassica napus và cây	Koh and Loh (2000)
	Pennisetum glaucum	Devi et al. (2000)
	Pyrus communis	Matsuda et al. (2009)
	Citrus nobilis × *C. deliciosa*	Singh et al. (2006)
	Anethum graveolens	Jana and Shekhawat (2011)
	Cucumis sativus	Kielkowska and Havey (2011)

13.2 Morphological and Structural Changes in Apical Meristem and Metabolic Activity During In Vitro Flowering of *T. fournieri*

The 3-month-old purple-white *T. fournieri* plant (Fig. 13.1a) in the greenhouse was used as material in this study. Leaves are surface disinfected with mild soap and placed under running water for 30 min, and then, shaking the explant with 0.1% $HgCl_2$ solution with 1–2 drops of tween 20 for 5 min in a clean bench. Then, the explant was washed three times with sterilized distilled water and cultured on MS medium supplemented with 30 g/L sucrose, 8 g/L agar, and 0.5 mg/L BA for shoot regeneration (Fig. 13.1b, c). These shoots were subcultured on MS medium to improve shoot quality (Fig. 13.1d).

Fig. 13.1 Plant materials were used in the study. (**a**) The 3-month-old purple-white *T. fournieri* plant in the greenhouse; (**b**) shoots regenerated from leaves; (**c**) shoots under a stereo microscope; (**D**) In vitro shoots (3 cm in height)

13.2.1 Flower Morphology and Changes in Apical Meristem During In Vitro Flowering

Shoots (2 cm) were cultured on MS medium supplemented with 30 g/L sucrose, 8 g/L agar, and without PGRs. Physiological changes during in vitro flowering such as photosynthetic intensity, respiration rate, and total sugar content were recorded after 10, 20, 30, and 40 days of culture.

The plants did not appear flower buds at day 10 and 20; flower buds were recorded only on day 30 and flower buds grew larger at day 40, and fully bloomed at day 50 (Fig. 13.2). In this study, the in vitro flowering in *T. fournieri* was divided into four stages including: stage 1 (10 days after culture), stage 2 (20 days after culture), stage 3 (30 days after culture), and stage 4 (40 days after culture).

After 40 days of culture, the in vitro flower buds had a complete structure, with no phenotypic differences compared with those of in vivo plants. In vitro flowers usually form at the shoot tips and leaf axils (Fig. 13.3). In this study, a problem was noted that the flower color changed to light purple or completely white, which was different from the original research material, which was a purple-white flower variety. The color change of *T. fournieri* flowers in vitro is due to the interaction between the mineral components in the culture medium and not due to the influence of genetic variation (Nhut 2013).

Flowering induction involves many steps, including structural changes in the floral meristem. At the vegetative stage, the meristem is small in size, with no clear distinction. During flowering induction, physiological changes in the apical meristem occur that prepare the plant for the transition from the vegetative to the reproductive stage. This is important for successful flowering (Glover 2007).

The shoot apical meristem of *T. fournieri* in vitro is disc-shaped, flat, structurally indistinct and bears two leaf outlines at the vegetative stage (Fig. 13.3a). The meristem is more structured and changes from disc to dome shape at the flowering induction stage. Cells in the central and peripheral regions begin to protrude, first sepal formation, petal primary, stamen primary, and petal elongation, finally bud flowers develop (Fig. 13.3b).

The anatomical morphology of the in vitro floral meristem of *T. fournieri* (Fig. 13.4) is similar to that of the *A. thaliana* (Elena et al. 2010). The inflorescence

Fig. 13.2 In vitro flower development stages of *T. fournieri*. (**a**) Shoot tip; (**b**) In vitro flower development stages; (**c**) Flower buds under a microscope; (**d**) 50-day-old in vitro flower

Fig. 13.3 Location of vegetative buds and in vitro flower buds of *T. fournieri* under a stereomicroscope. (**a**) Vegetative buds sprouting sides axillary; (**b**) vegetative bud and in vitro flower bud sprouting sides axillary; (**c** and **d**) in vitro flower buds sprouting sides axillary; (**e** and **f**) in vitro flower buds sprouting shoot tip

meristem formed at the shoot tip or the flower meristems (Fig. 13.5), flower buds formed at the vegetative meristems on either side of the leaf axils were obtained on *T. fournieri* (Fig. 13.6). In *T. fournieri* in vitro, the flower sepals are also a type of leaf, so there is no difference from the vegetative leaves (Fig. 13.7c). Flower bud

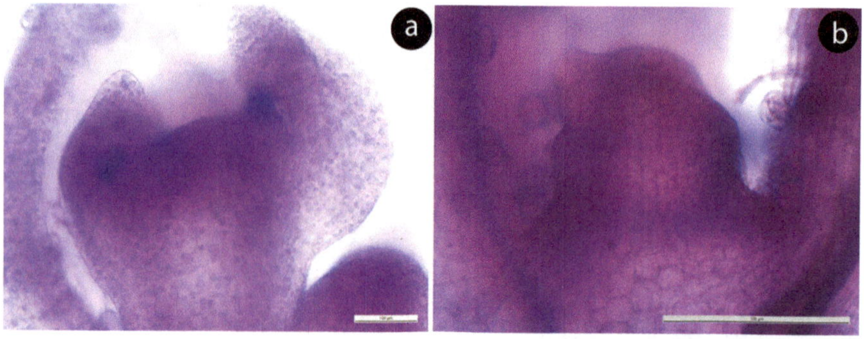

Fig. 13.4 Anatomical morphology of *T. fournieri* shoot meristem. (**a**) Vegetative meristem; (**b**) Flower meristem

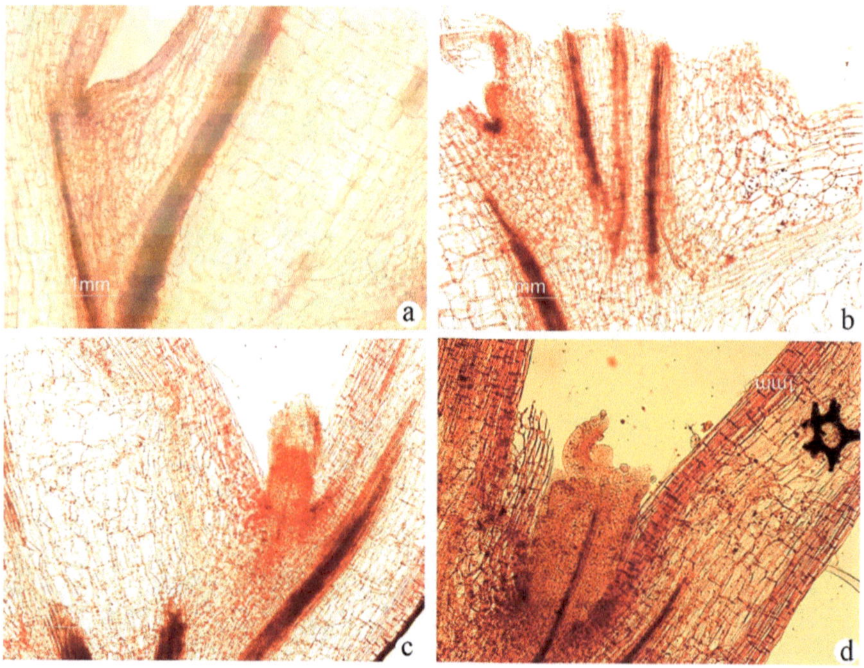

Fig. 13.5 Anatomical morphology of vegetative and floral meristems in lateral shoots of *Torenia fournieri* in vitro under microscope. (**a** and **b**) Cells divide at the vegetative stage; (**c**) vegetative meristem in lateral shoots; (**d**) flower meristem

developed, sepals cracked and petals extended beyond the bud (Fig. 13.7e), and cells outside the central region wrinkled when imaged under the scanning electron microscopy (Fig. 13.7). Summary of *T. fournieri* in vitro flowering derived from ex vitro 3-month old leaf explant is shown in Fig. 13.8.

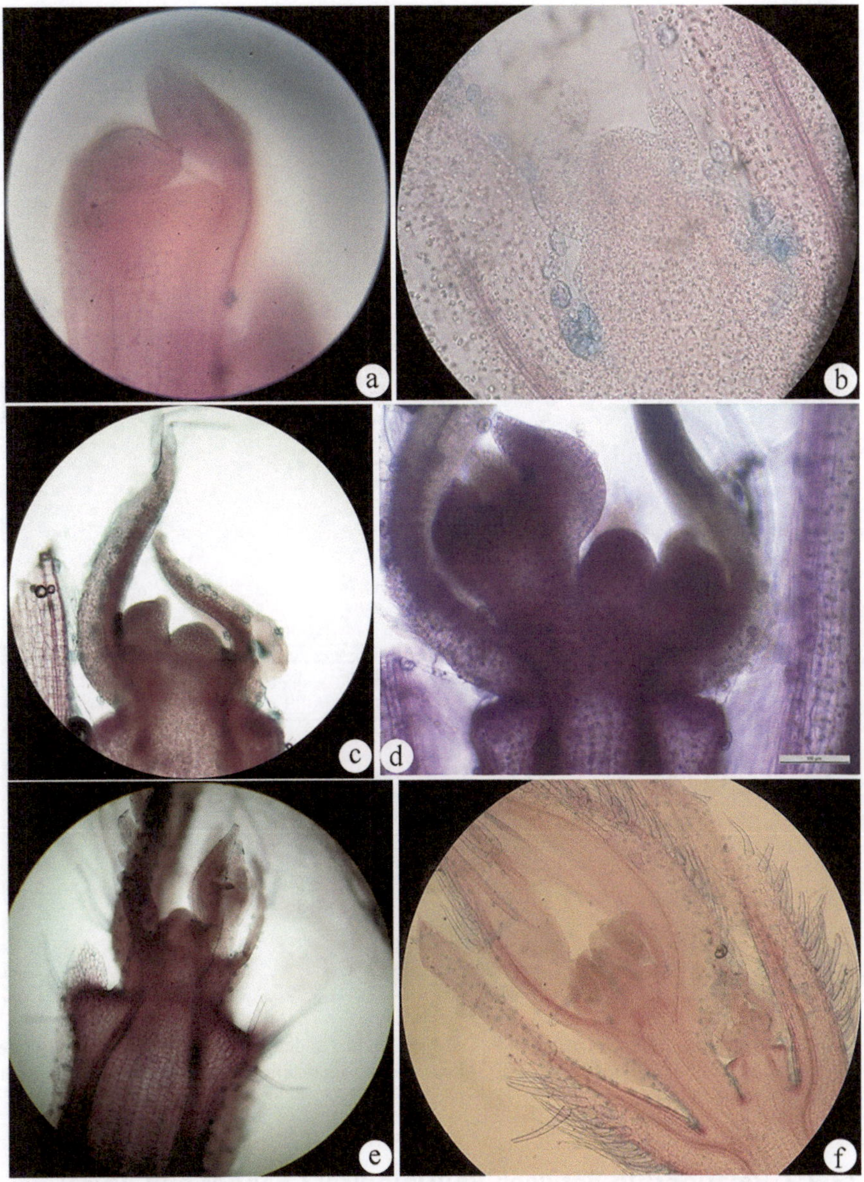

Fig. 13.6 Anatomical morphology of *T. fournieri* in vitro under the microscope. (**a**) Vegetative meristem; (**b**) Flower meristem; (**c, d**) Inflorescence meristem; (**e, f**) flower buds of *T. fournieri* in vitro under microscope

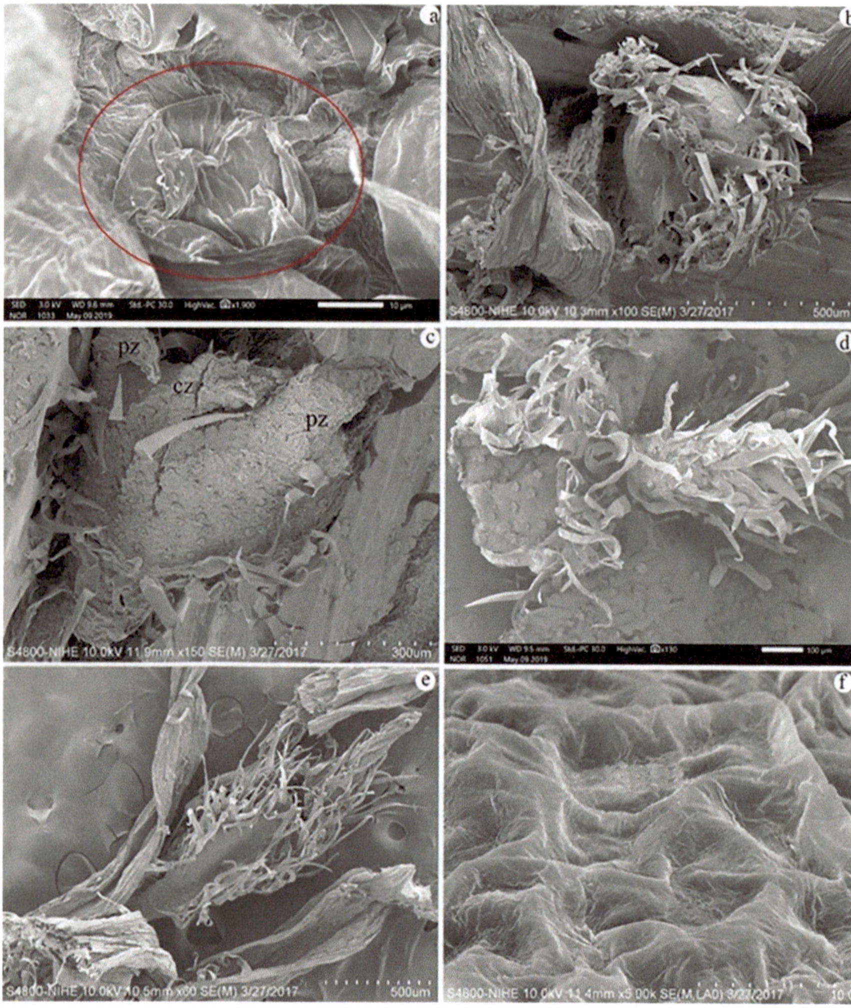

Fig. 13.7 The apical shoot meristem of *T. fournieri* in vitro under a scanning electron microscope (**CZ:** Central zone, **PZ:** Peripheral zone). (**a, b**) Flower meristem; (**c–e**) Flower buds; (**f**) outer surface of the flower bud

13.2.2 Relationship Between Physiological Changes and Apical Meristem Morphology During In Vitro Flowering

During the induction of flowering in plants, plants often undergo physiological changes before morphological changes occur. The respiratory intensity of the explants increased from day 0 to day 10 (0.42–0.55 $\mu mO_2/10\ cm^2/min$), then decreased at day 20 (0.13 $\mu mO_2/10\ cm^2/min$) and increased after day 30 and 40 (Table 13.2). However, shoots of *T. fournieri* from day 10 to 20 had a rapid

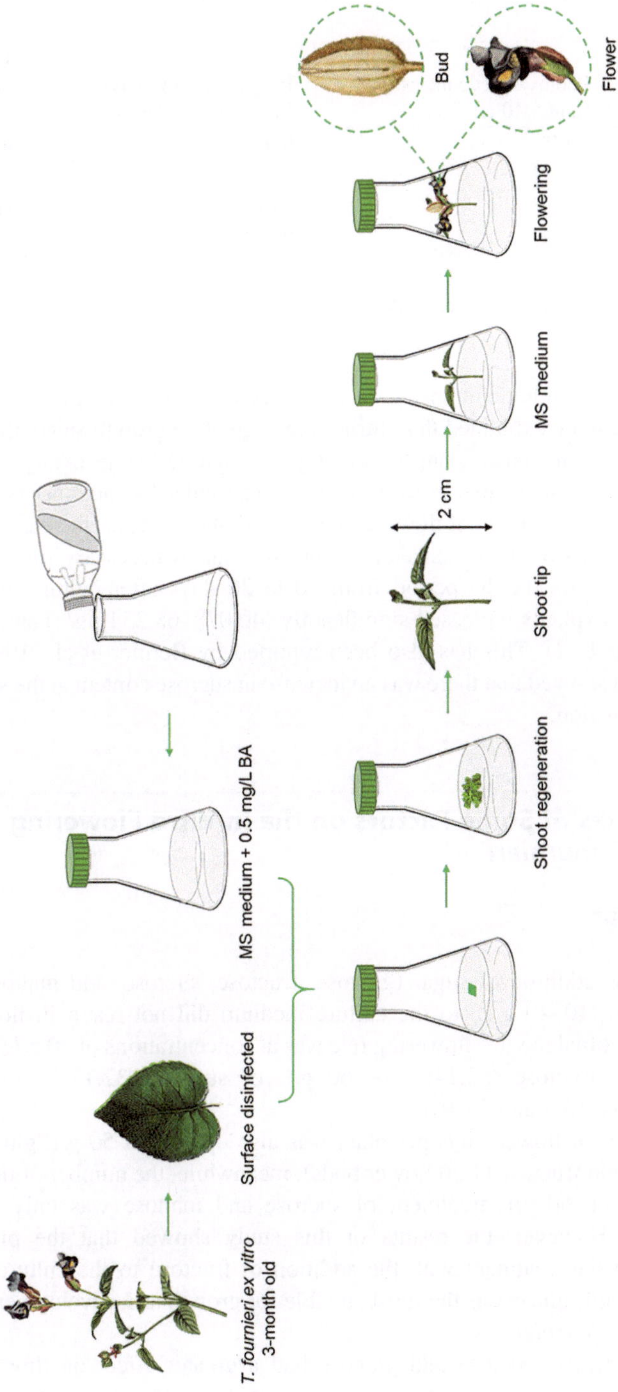

Fig. 13.8 In vitro flowering of *T. fournieri*

Table 13.2 Changes in photosynthetic and respiratory intensity at the in vitro flowering stage of *T. fournieri*

Flowering time (days)	Photosynthetic intensity ($\mu mO_2/10\ cm^2/min$)	Respiratory intensity ($\mu mO_2/10\ cm^2/min$)	Sugar content (μg/g)
0	$0.175^{a,b}$	0.042^{b}	40.08^{c}
10	0.199^{a}	0.055^{a}	52.69^{b}
20	0.229^{a}	0.013^{d}	64.58^{a}
30	0.152^{b}	0.035^{c}	68.25^{a}
40	0.143^{b}	0.041^{b}	48.08^{b}

Note: In the same column, numbers followed by different letters have statistical significance at $p < 5\%$ in Duncan test

increase in photosynthetic intensity, and then decreased at the stage after 30 days of culture. This can be explained that during the vegetative growth stage, the intensity of photosynthesis increases sharply to help plants synthesize necessary nutrients.

In the flowering stage, plants need a lot of energy and substances necessary for the formation and development of flowers, so the respiration process increases rapidly to help break down stored substances to create substances necessary for reactions of biological synthesis. In the period from 10 to 30 days after culture, the sucrose content in the explants increased significantly (40,078–68,254 μg/g) and peaked at day 30 (Table 13.2). This has also been reported by Bernier et al. (1993) and in *Sinapis alba* it showed that there was an increase in sucrose content at the shoot tip of flowering transition.

13.3 Effects of Some Factors on the In Vitro Flowering of *T. fournieri*

13.3.1 Sugar

In general, the addition of sugar (glucose, fructose, sucrose, and maltose) at low concentrations (10–40 g/L) to the culture medium did not result in flowering of shoots. The optimal in vitro flowering rate was at concentrations of 50 g/L of glucose (51.48%) and fructose (53.14%) or 60 g/L of sucrose (32.17%) and maltose (33.13%) (Figs. 13.9 and 13.10).

The number of flower buds per plant was also optimal at 50 g/L glucose (1.20 flower buds) and fructose (1.26 flower buds); meanwhile, the number of flower buds per plant in the 60 g/L treatment of sucrose and maltose was only 0.33–0.46 (Fig. 13.11). However, the results of this study showed that the plants were chlorinated in the treatment with the addition of fructose to the culture medium. Therefore, 50 g/L glucose is the most suitable for growth and development, in vitro flowering in *T. fournieri*.

For *T. fournieri*, fructose and glucose had a greater effect on flowering than sucrose and maltose (Fig. 13.10). The results of this study are similar to those of

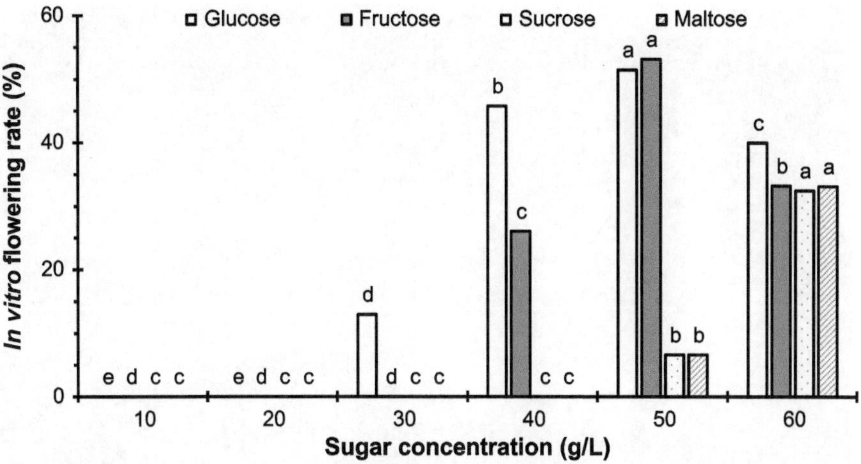

Fig. 13.9 Effects of sugar concentrations on the in vitro flowering rate of *T. fournieri* after 40 days of culture

Taylor et al. (2007) on *K. leucocephala*. This is explained by the effect of this sugar on cell division at the apical shoot meristem, which is important for flowering in plants (Jacqmard et al. 2003). Glucose and fructose do not always induce flowering and may depend on plant species, explant type, and explant responsiveness to different carbon sources.

13.3.2 PGRs

13.3.2.1 GA_3 in Combination with ABA

The results showed that the 1.5 mg/L ABA treatment gave a higher in vitro flowering rate (80%) than the other treatments (ABA combined with GA_3) and control (40%) (Figs. 13.12, 13.13, and 13.14). This suggests that ABA plays an important role in in vitro flowering in *T. fournieri*.

The role of ABA signaling in the flowering transition remains unclear, as both positive and negative roles of ABA on flowering have been previously published (Domagalska et al. 2010; Conti et al. 2014). ABA alters flowering time in some plant species (Conti et al. 2014). Accumulation of ABA during periods of drought leads to early flowering, and ABA signaling may underpin this phenomenon. The mutant *A. thaliana* plants with defects in ABA biosynthesis resulted in delayed flowering due to decreased/reduced expression of the FT and TSF genes. This indicates a positive effect of ABA during flowering (Riboni et al. 2013). Although the addition of ABA had a positive effect on the in vitro flowering of *T. fournieri* in this study, in the case of *A. thaliana*, exogenous ABA delayed the in vitro flowering time (Wang et al. 2013).

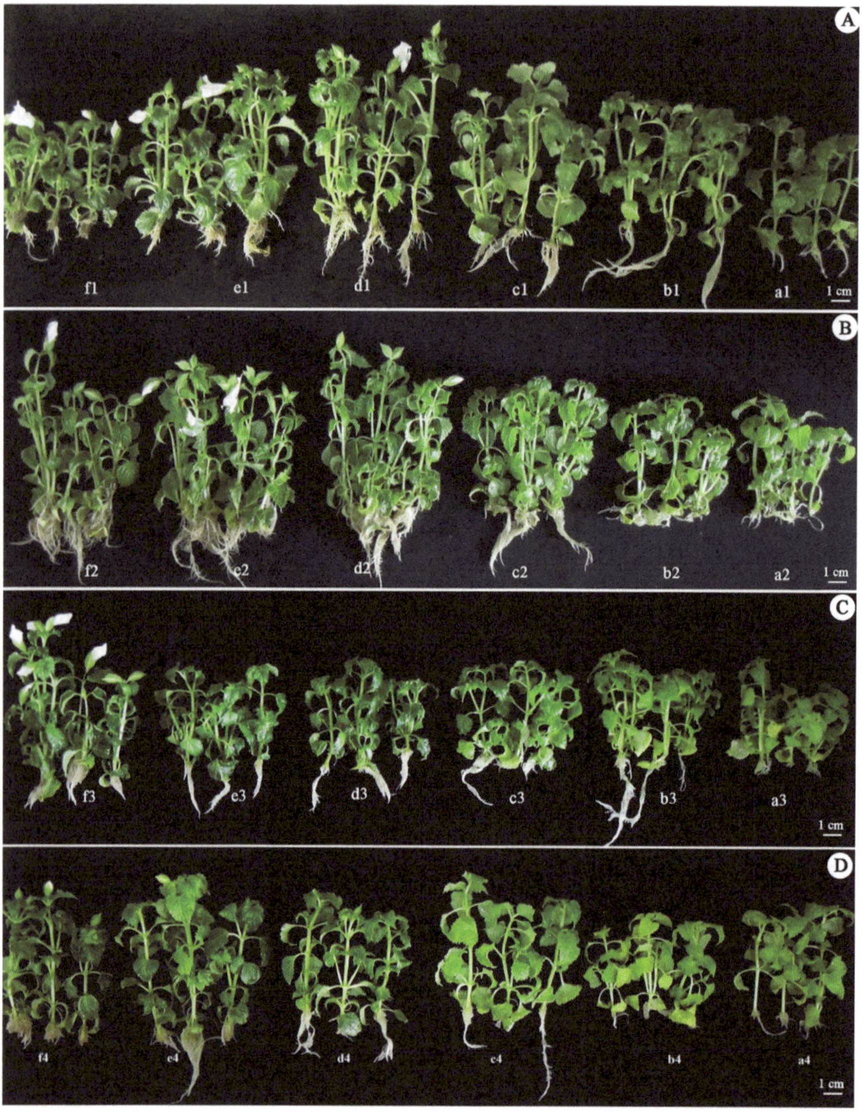

Fig. 13.10 Effects of sugar concentrations on the in vitro flowering of *T. fournieri* after 40 days of culture. (**a**) Glucose; (**b**) fructose; (**c**) sucrose; (**d**) maltose

Several recent studies suggest that the effect of ABA on in vitro flowering can be explained by the expression of genes ABI3, ABI5, and BASIC LEUCINE ZIPPER, which play an important role in in vitro flowering (Finkelstein et al. 2002; Hauser et al. 2011) by regulating *FLC* gene expression (Wang et al. 2013). Another protein involved in the regulation of flowering is HAB1 (HYPERSENSITIVE TO ABA 1) (Saez et al. 2004), which is an inhibitor of the response to ABA signals (Rodriguez

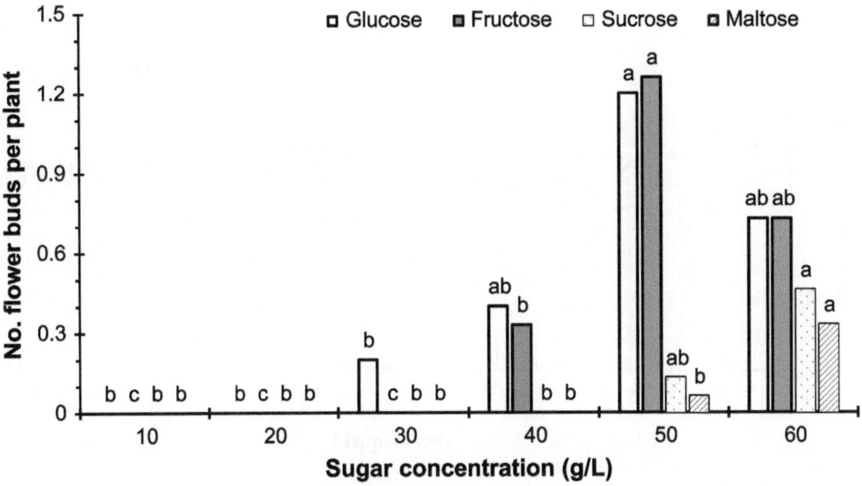

Fig. 13.11 Effects of sugar concentrations on the number of flower buds per plant of *T. fournieri* after 40 days of culture

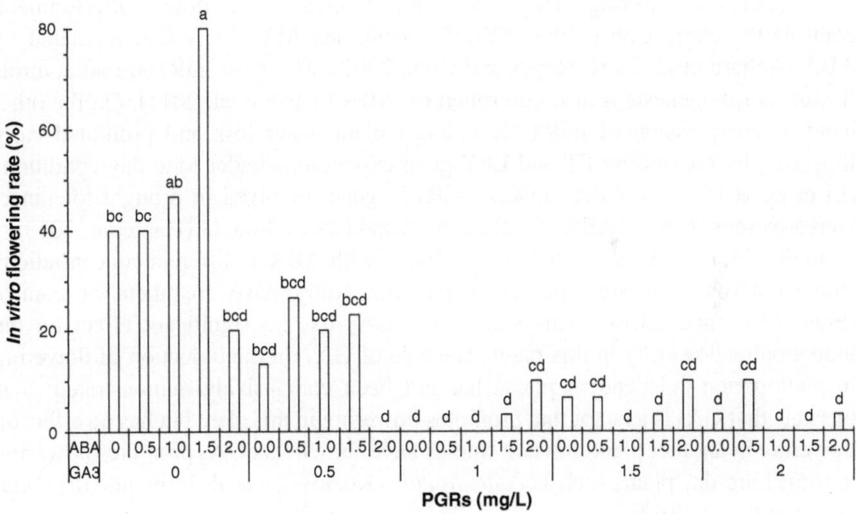

Fig. 13.12 Effects of GA_3 in combination with ABA on the in vitro flowering rate of *T. fournieri* after 40 days of culture

et al. 1998). Expression of HAB1 protein was found to induce flowering in *A. thaliana* (Saez et al. 2004). The HAB1 protein interacts with the SWI/SNF protein complex (Saez et al. 2008). The authors postulated that the interaction of the HAB1 protein and the SWI/SNF complex (chromatin complex) in transcription in response to ABA signals is essential for the regulation of genes involved in the response to ABA signals. Many miRNA genes have also been reported to be involved in

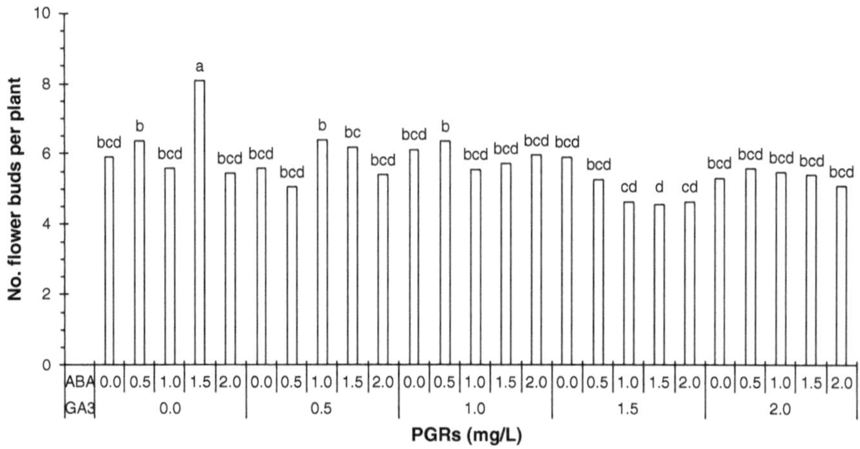

Fig. 13.13 Effects of GA_3 in combination with ABA on the number of flower buds per plant of *T. fournieri* after 40 days of culture

ABA-mediated flowering. The gene miR159, involved in flower development, controls the expression of the MYB101 gene, and MYB33 is also regulated by ABA (Achard et al. 2004; Reyes and Chua 2007). The gene miR160 that controls flower morphogenesis is also controlled by ABA (Yaish et al. 2011). On the other hand, overexpression of miR172c reduced plant water loss and promoted early flowering by controlling FT and LFY gene expressions under long day conditions (Li et al. 2016). The ABA-induced miR172 gene, involved in drought tolerance, increases sensitivity to ABA, accelerating responses to drought (Han et al. 2013).

In this study, GA_3 alone or GA_3 combined with ABA at different concentrations showed a lower in vitro flowering rate than only ABA treatment or control (Figs. 13.12 and 13.14). This shows that GA_3 has no significant effect on the induction of flowering in this plant. The role of GA_3 on the induction of flowering in photoperiod-independent plants has not been conclusively demonstrated; it is possible that GA_3 is a factor that promotes flowering in this plant but has no effect on flowering in another plant. Several studies have shown that GA_3 promotes flowering in some long-day plants such as *Zantedeschia* (Kozłowska et al. 2006) and *Brunonia* (Wahyuni et al. 2011).

Flowering of *A. thaliana* occurs under long day conditions, but flowering can be induced by gibberellin treatment under short day conditions. In general, short-day plants do not respond to gibberellins to induce flowering, while long-day plants do, although there are exceptions to this (Pharis and King 1985). In addition to long-day plants, plants with a long-term requirement, shoot dormancy is controlled by low temperature, and GA_3 can partially replace this cold requirement to help induce flowering (Saez et al. 2008). GA_3 has been used in *Hyacinthus*, *Liatris*, *Muscari*, *Iris*, *Lilium*, *Tulipa*, and *Zantedeschia* species to replace the low temperature conditions required for flowering induction (Achard et al. 2004).

Fig. 13.14 In vitro flowering of *T. fournieri* plants grown on media supplemented with ABA at different concentrations after 40 days of culture. (**a**) 0.0 mg/L ABA; (**b**) 0.5 mg/L ABA; (**c**) 1.0 mg/L ABA; (**d**) 1.5 mg/L ABA; (**e**) 2.0 mg/L ABA; (**f**) Flower morphology and flower position magnified; (**g**) *T. fournieri* plants on medium supplemented with 1.5 mg/L ABA

13.3.2.2 GA_3 in Combination with BA

For *T. fournieri*, the results showed that BA combined with GA_3 did not form flowers, only formed new shoots, long stem segments, and vitreous phenomenon.

Similar results were obtained in *Tylophora indica* (Rani and Rana 2010). GA_3 has been reported to be suitable for shoot regeneration, growth promotion, biomass production, and xylem vascular elongation (Chakraborty et al. 2000; Ericksson et al. 2000). Furthermore, GA_3 can act as a substitute for auxin in shoots.

13.3.2.3 GA_3 in Combination with NAA

Similar to GA_3 combined with BA, the results also showed that GA_3 in combination with NAA did not affect in vitro flowering in *T. fournieri* plants after 40 days of culture. On medium supplemented with 0.5 mg/L NAA, shoot height was higher than that of explants on medium supplemented with high concentration of NAA. In addition, the shoots had root formation and no vitreous phenomena; meanwhile, the explants formed many shoots, weak shoots, and no roots appeared in the medium supplemented with 1–2 mg/L NAA.

13.3.3 Media

In vitro *T. fournieri* shoots (2.0 cm in height) were cultured on a number of different media such as ½MS (MS medium with reduced ½ macronutrient composition), MS½ (MS medium with reduced ½ mineral composition), ¼MS (MS medium with reduced ¼ macronutrient composition), and MS¼ (MS medium with reduced ¼ mineral components) supplemented with 30 g/L sucrose and 8 g/L agar to evaluate the potential in vitro flowering after 40 days of culture. The control was shoots cultured on MS medium. The in vitro flowering rate on ¼MS medium (60.00%) and number flower buds per plant (0.70) was higher than that on the control and the remaining media after 40 days of culture (Figs. 13.15, 13.16, and 13.17).

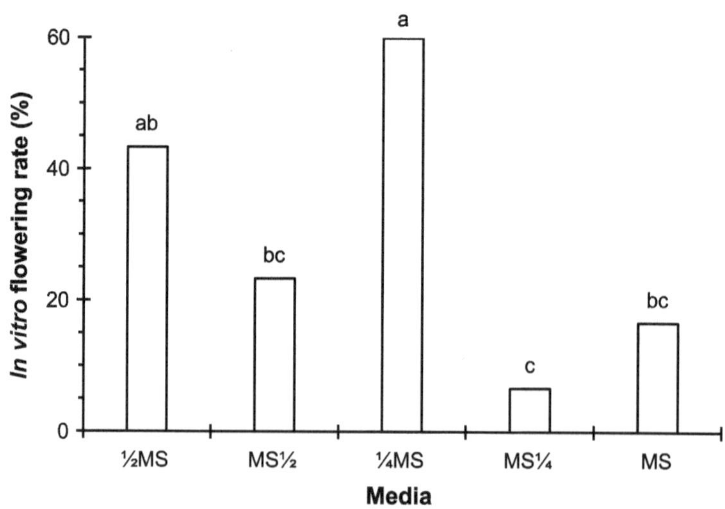

Fig. 13.15 Effects of media on the in vitro flowering rate of *T. fournieri* after 40 days of culture

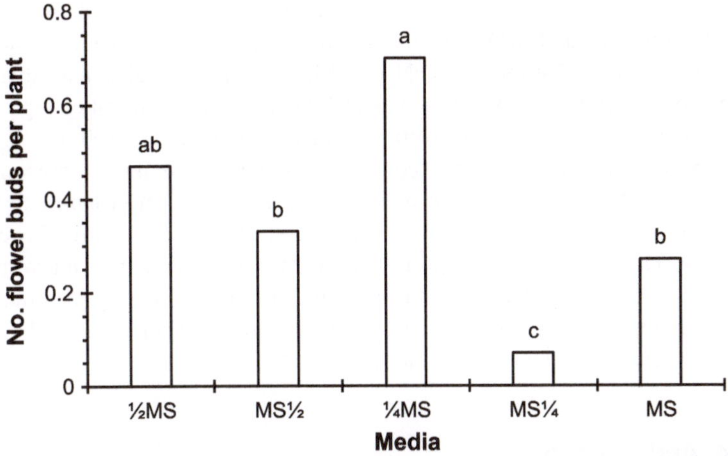

Fig. 13.16 Effects of media on the number of flower buds per plant of *T. fournieri* after 40 days of culture

Fig. 13.17 Effects of media on the in vitro flowering rate of *T. fournieri* after 40 days of culture. (**a**) MS; (**b**) ½MS; (**c**) MS½; (**d**) ¼MS; (**e**) MS¼; (**f**) *T. fournieri* plants cultured on ½MS, MS¼, ¼MS, MS½, and MS media (left to right); (**g**) Flower buds on both sides of leaf axils; (**h**) Flower buds at shoot tip

The macronutrients in ¼MS and ½MS medium were reduced by a ratio of ½:¼; this change resulted in the C/N ratio in the medium being changed. This could be explained when macronutrients, especially N, present in high concentrations partially inhibited the transition from the vegetative phase to the reproductive phase or the formation of flower buds of *T. fournieri*. In the culture medium, the content of macronutrients decreased, so the N content also decreased; while the C content remained constant, leading to a high C/N ratio, which stimulated in vitro flowering, whereas when the C/N ratio was low, vegetative growth prevailed. According to Tanimoto and Harada (1981), the C/N ratio plays an important role in the flowering transition in plants. The ratio of C/N in ¼MS medium is higher than that in MS medium; plants cultured on medium with high C/N ratio will enter reproductive state, and this result is consistent with the study of Wang et al. (2002).

In MS medium, high nitrogen concentration will often inhibit flowering and promote vegetative growth; while a ¼ or ½ reduction in macronutrients of MS medium or a decrease in nitrogen concentration enhanced in vitro flowering in *Cymbidium* (Kostenyuk et al. 1999). Culture media containing ½ macronutrients or reducing N in MS basal medium promoted in vitro flowering in a number of different plants such as rose (Vu et al. 2006), tomato (Dielen et al. 2001), and *Orychophragmus violaceus* (Luo et al. 2000).

13.3.4 Amino Acids

Currently, there have been many research works on amino acids focusing on determining the relationship between amino acids and plant growth, not many studies on the role of amino acids on in vitro flowering. In this study, three amino acids such as arginine, L-tyrosine, or proline were investigated to evaluate the effects of amino acids on the in vitro flowering of *T. fournieri*. The shoots (2.0 cm in height) were cultured on MS medium supplemented with 30 g/L sucrose, 8 g/L agar, and amino acids such as L-tyrosine, arginine, and proline with different concentrations (1.0, 1.5, 2.0, 2.5, 3.0, 3.5, 4.0, 4.5, and 5.0 mg/L). The control was shoots cultured in MS medium without amino acids.

The data in Figs. 13.18 and 13.19 showed that all arginine treatments had in vitro flowering in *T. fournieri* plants after 80 days of culture. In vitro flowering in *T. fournieri* increased with the concentration of arginine added to MS medium and reached the highest at 1.5 mg/L arginine, with an in vitro flowering rate (60.0%) and number of flower buds per plant (0.83), nearly three times higher than the treatment without arginine. At higher concentrations (2–5 mg/L arginine), the in vitro flowering rate and number of flower buds per plant decreased (Fig. 13.19). The results in this study are in contrast to the results of the study on *Plumbago indica*, where 3.1–4.0 M or higher arginine reduced in vitro flowering rate (Nitsch and Nitsch 1967).

To date, there are no published data on the effect of L-tyrosine on in vitro flowering. The results of this study showed that 1–5 mg/L L-tyrosine had negative effects on the in vitro flowering of *T. fournieri* but lower than that without L-tyrosine and 1.5 mg/L arginine (Figs. 13.18, 13.20, and 13.21). In micropropagation, L-tyrosine has a role in growth and development in several plant species such as shoot regeneration from callus in tobacco (Dougall and Shimbayashi 1960) and callus induction from *Gloriosa superba* leaves (Kijwijan et al. 2008).

MS medium supplemented with 2.5 mg/L proline recorded higher in vitro flowering rate (53.33%) and number of flower buds per plant (0.6) compared with other proline concentrations (Fig. 13.19). Increasing proline concentration from 3.0

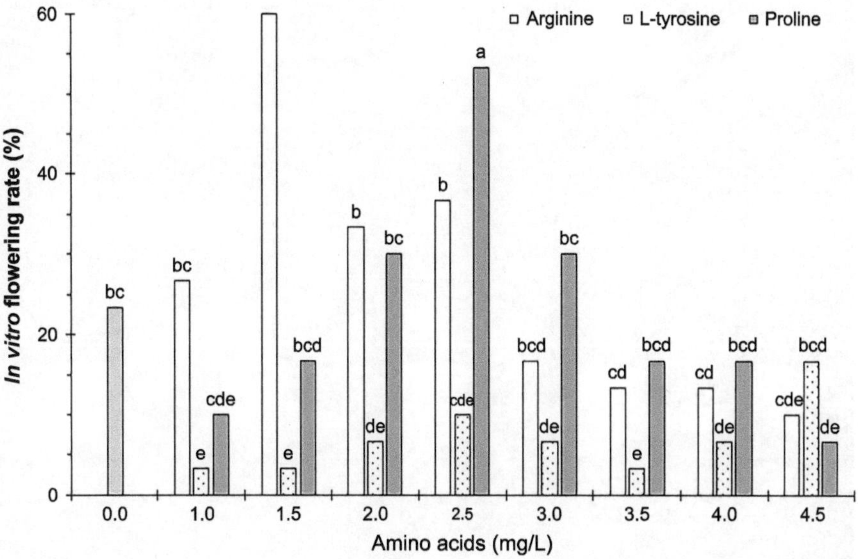

Fig. 13.18 Effects of amino acids on the in vitro flowering rate of *T. fournieri* after 40 days of culture

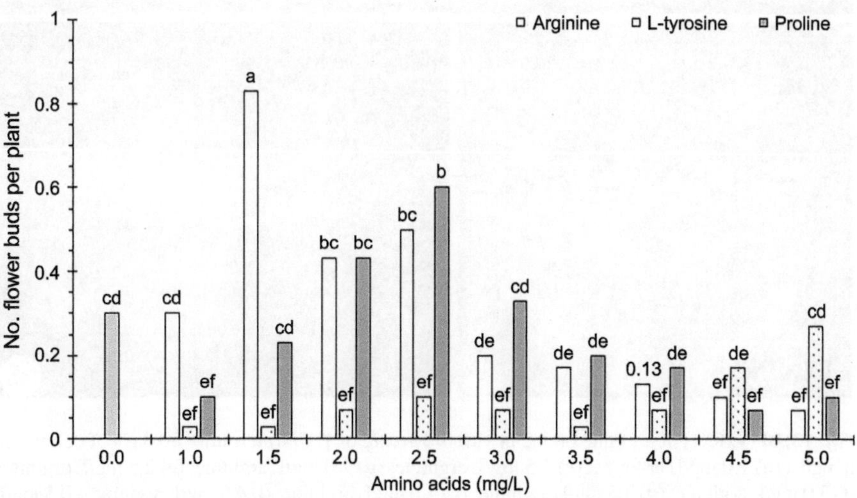

Fig. 13.19 Effects of amino acids on the number of flower buds of *T. fournieri* after 40 days of culture

to 5.0 mg/L resulted in in vitro flowering rate, and the number of flower buds/plant gradually decreased (Fig. 13.22). This shows that proline affects the in vitro flowering of *T. fournieri*. Proline also has an effect on in vitro flowering in *Saccharum officinarum* (Virupakshi et al. 2002) and *Vigna aconitifolia* (Saxena

Fig. 13.20 Effects of arginine on the in vitro flowering of *T. fournieri* after 80 days of culture (**a**) 0 mg/L; (**b**) 1.0 mg/L arginine; (**c**) 1.5 mg/L arginine; (**d**) 2.0 mg/L arginine; (**e**) 2.5 mg/L arginine; (**f**) 3.0 mg/L arginine; (**g**) 3.5 mg/L arginine; (**h**) 4.0 mg/L arginine; (**i**) 4.5 mg/L arginine; (**j**) 5 mg/L arginine; (**k**) flower buds at shoot tip

et al. 2008). Flowering time occurred on day 40 compared to day 80 in this treatment, which indicates that proline did not affect in vitro flowering but positively affected flower development.

Of the three amino acids investigated, the results showed that in vitro flowering varied depending on the type and concentration of amino acids added to the culture

Fig. 13.21 Effects of L-tyrosine on the in vitro flowering of *T. fournieri* after 80 days of culture. (**a**) 0 mg/L L-tyrosine; (**b**) 1.0 mg/L L-tyrosine; (**c**) 1.5 mg/L L-tyrosine; (**d**) 2.0 mg/L L-tyrosine; (**e**) 2.5 mg/L L-tyrosine; (**f**) 3.0 mg/L L-tyrosine; (**g**) 3.5 mg/L L-tyrosine; (**h**) 4.0 mg/L L-tyrosine; (**i**) 4.5 mg/L L-tyrosine; (**j**) 5 mg/L L-tyrosine

medium. For *T. fournieri*, 1.5 mg/L arginine was most suitable for in vitro flowering (Figs. 13.18, 13.19, and 13.20). The role of amino acids in *T. fournieri* in vitro flowering is also unclear. These results obtained that the in vitro flowering only appeared on day 80 in amino acid treatments as compared to those on day 40 in other factors.

13.4 Conclusion

The results in this study showed that all the factors investigated affected the in vitro flowering of *T. fournieri* but these effects were different depending on the type of factor, concentration, and time cultures. In vitro flowering rate (80%) was highest in the treatment using 1.5 mg/L ABA after 40 days of culture. Besides ABA, glucose and fructose influence in vitro flowering better than sucrose and maltose. Most of the treatments are done using amino acids and the induction of in vitro flowering after 80 days of culture.

Fig. 13.22 Effects of proline on the in vitro flowering of *T. fournieri* after 80 days of culture. (**a**) 0 mg/L proline; (**b**) 1.0 mg/L proline; (**c**) 1.5 mg/L proline; (**d**) 2.0 mg/L proline; (**e**) 2.5 mg/L proline; (**f**) 3.0 mg/L proline; (**g**) 3.5 mg/L proline; (**h**) 4.0 mg/L proline; (**i**) 4.5 mg/L proline; (**j**) 5 mg/L proline

References

Achard P, Herr A, Baulcombe DC, Harberd NP (2004) Modulation of floral development by a gibberellin-regulated *microRNA*. Development 131(14):3357–3365

Al-Khayri JM, Huang FH, Morelock TE, Busharar TA (1992) In vitro seed production from sex-modified male spinach plants regenerated from callus cultures. Sci Hort 52(4):277–282

Bernier G (1988) The control of floral evocation and morphogenesis. Annu Rev Plant Physiol Plant Mol Biol 39(1):175–219

Bernier G, Havelange A, Houssa C, Petitjean A, Lejeune P (1993) Physiological signals that induce flowering. Plant Cell 5(10):1147–1155

Blazquez MA, Weigel D (2000) Integration of floral inductive signals in *Arabidopsis*. Nature 404(6780):889–892

Chakraborty D, Mandal AKA, Datta SK (2000) Retrieval of new coloured *chrysanthemum* through organogenesis from sectorial chimera. Curr Sci 78(9):1060–1061

Choun-Sea L, Liang C, Hsaio H, Lin M, Chang W (2007) In vitro flowering of green and albino *Dendrocalamus latiflorus*. New For 34(2):177–186

Conti L, Galbiati M, Tonell C (2014) ABA and the floral transition. In: Zhang DP (ed) Abscisic acid: metabolism, transport and signaling. Springer, Berlin, Germany, pp 365–384

Cvetić T, Budimir SM, Grubišić DV (2004) In vitro flowering of dark-grown Centaurium pulchellum. Arch Biol Sci 56(3–4):21–22

Deb CR, Sungkumlong T (2009) Rapid multiplication and induction of early in vitro flowering in *Dendrobium primulinum* Lindl. J Plant Biochem Biotech 18(2):241–244

Devi P, Zhong H, Sticklen MB (2000) In vitro morphogenesis of pearl millet *Pennisetum glaucum* (L.) R. Br. Efficient production of multiple shoots and inflorescences from shoot apices. Plant Cell Rep 19(6):546–550

Dielen V, Lecouvet V, Dupont S, Kinet JM (2001) In vitro control of floral transition in tomato (*Lycopersicon esculentum* mill.), the model for autonomously flowering plants, using the late flowering unfloral mutant. J Exp Bot 52(357):715–723

Domagalska MA, Sarnowska E, Nagy F, Davis SJ (2010) Genetic analyses of interactions among gibberellin, abscisic acid, and brassinosteroids in the control of flowering time in *Arabidopsis thaliana*. PLoS One 5(11):1–8

Dougall DK, Shimbayashi K (1960) Factors affecting growth of tobacco callus tissue and its incorporation of tyrosine. Plant Physiol 35(3):396–404

Duan JX, Yazawa S (1994) Induction of floral development *Doriella tiny* (*Doritis pulcherrima x Kingiella philippinensis*). Sci Hortic 59(3–4):253–264

Elena RAB, Benítez M, Corvera-Poiré A, Cador ALC, Folter SD, Buen AGD, Garay-Arroyo A, García-Ponce B, Jaimes-Miranda F, Pérez-Ruiz RV, Piñeyro-Nelson A, Sánchez-Corralesa YE (2010) Flower development. Arabidopsis Book 2010(8):1–57

Ericksson ME, Israelsson M, Olssono O, Moritz T (2000) Increased gibberellin biosynthesis in transgenic tree promotes growth, biomass production and xylem fiber length. Nat Biotech 18(7):784–788

Finkelstein RR, Gampala SS, Rock CD (2002) Abscisic acid signaling in seeds and seedlings. Plant Cell 14:15–45

Franklin G, Pius PK, Ignacimuthu S (2000) Factors affecting in vitro flowering and fruting of green pea (*Pisum sativum* L.). Euphytica 115(1):65–73

Galoch E, Czaplewska J, Burkacka-Laukajtys E, Kopcewicz J (2002) Induction and stimulation of in vitro flowering of *Pharbitis nil* by cytokinin and gibberellin. Plant Growth Regul 37(3):199–205

Glover BJ (2007) Understanding flowers and flowering: an integrated approach. Oxford University Press, UK, p 277p

Han Y, Zhang X, Wang Y, Ming F (2013) The suppression of *wrky44* by *gigantea-miR172* pathway is involved in drought response of *Arabidopsis thaliana*. PLoS One 8(11):1–16

Hauser F, Waadt R, Schroeder JI (2011) Evolution of abscisic acid synthesis and signaling mechanisms. Curr Biol 21(9):346–355

Hee KH, Loh C, Yeoh HH (2007) Early in vitro flowering and seed production in culture in *dendrobium* Chao Praya smile (*Orchidaceae*). Plant Cell Rep 26(12):2055–2062

Jabeen FTZ, Venugopal RB, Kiran G, Kaviraj CP, Rao S (2005) Plant regeneration and in vitro flowering from leaf and nodal explants of *Solanum nigrum* (L.): an important medicinal plant. Plant Cell Biotech Mol Biol 6(1–2):17–22

Jacqmard A, Gadisseur I, Bernier G (2003) Cell division and morphological changes in the shoot apex of *Arabidopsis thaliana* during floral transition. Ann Bot 91(5):571–576

Jana S, Shekhawat GS (2011) Plant growth regulators, adenine sulfate and carbohydrates regulate organogenesis and in vitro flowering of *Anethum graveolens*. Acta Physiol Plant 33(2):305–311

John CK, Nadgauda RS (1997) Flowering in *Bambusa vulgaris* var. *vittata*. Curr Sci 73(8):641–643

Jumin HB, Nito N (1996) In vitro flowering of orange jessamine (*Murraya paniculata* L. Jack). Experientia 52(3):268–272

Kachanapoom K, Posayapisit N, Kachanapoom K (2009) In vitro flowering from cultured nodal explants of rose (*Rosa hybrida* L.). Not Bot Hort Agr Cluj-Nap 37(2):261–263

Kielkowska A, Havey MJ (2011) In vitro flowering and production of viable pollen of cucumber. Plant Cell Tissue Organ Cult 109(1):73–82

Kijwijan B, Nokmai J, Muangsan N (2008) Effects of tyrosine and plant growth regulators on growth and development of *Gloriosa superba* Linn. In Vitro Khon Kaen Agr J 36:144–152

Kintzios S, Michailakis A (1999) Induction of somatic embryogenesis and in vitro flowering from inflorescences of camomile (*Chamomilla recutita* L.). Plant Cell Rep 18(7–8):684–690

Koh KL, Loh CS (2000) Direct somatic embryogenesis, plant regeneration and in vitro flowering in rapid-cycling *Brassica napus*. Plant Cell Rep 19(12):1177–1183

Kostenyuk I, Oh BJ, So IS (1999) Induction of early flowering in *cymbidium niveo-marginatum* Mak in vitro. Plant Cell Rep 19(1):1–5

Kozłowska M, Rybus-Zajac M, Stachowiak J, Janowska B (2006) Changes in carbohydrate contents of *Zantedeschia* leaves under gibberellin-stimulated flowering. Acta Physiol Plant 29(1):27–32

Li Y, Zhang H, Bi H (1996) Changes of endogenous hormone contents during floral bud and vegetative bud differentiation in thin cell layer culture of *Cichorium intybus* L. explant. Acta Bot Sin 38(2):131–135

Li M, An F, Li W, Ma M, Feng Y, Zhang X, Guo H (2016) Della proteins interact with *flc* to repress flowering transition. J Int Plant Biol 58(7):642–655

Li-Ming Z, Ji-Liang P (2006) In vitro flowering of cultures from a hybrid of *cymbidium goeringii* and *C. hybridium*. J Plant Physiol Mol Biol 32(3):320–324

Lin CS, Lin CC, Chang WC (2004) Effect of thidiazuron on vegetative tissue-derived somatic embryogenesis and flowering of bamboo *Bambusa edulis*. Plant Cell Tissue Organ Cult 76(1):75–82

Lindsay DL, Sawhney VK, Bonham-Smith PC (2006) Cytokinin-induced changes in clavata1 and wuschel expression temporally coincide with altered floral development in *Arabidopsis*. Plant Sci 170(6):1111–1117

Luo P, Ye Q, Lan Z (2000) A study on floral biology of seedlings in vitro in *Orychophragmus violaceus*: induction of flowers in seedling of *O. violaceus* cultured in vitro. Plant Cell Tissue Organ Cult 63(1):73–75

Matsuda N, Ikeda K, Kurosaka M, Takashina T, Isuzugawa K, Endo T, Omura M (2009) Early flowering phenotype in transgenic pears (*Pyrus communis* L.) expressing the CiFT gene. J Jap Soc Hort Sci 78(4):410–416

Murashige T, Skoog F (1962) A revised medium for rapid growth and bio assays with tobacco tissue cultures. Physiol Plant 15(3):473–497

Nhut DT (2013) Affects of some factors on in vitro flowering of Torenia (*Torenia fournieri* L.). Vietnam. J Sci Technol 51(6):689

Nitsch C, Nitsch JP (1967) The induction of flowering in vitro in stem segments of *Plumbago indica* L. I. The production of vegetative buds. Planta 72(4):355–370

Pharis RP, King RW (1985) Gibberellins and reproductive development in seed plants. Annual Rev Plant Physiol 36(1):517–568

Pierik RLM, Dieleman JA, Heuvelink E (1994) Flowering of tomato in vivo and in vitro in relation to the original position of the axillary bud on the main axis. Sci Hort 59(1):55–60

Ramanayake SMSD (2006) Flowering in bamboo: an enigma. Ceylon J Bio-Sci 35(2):95–105

Rani S, Rana J (2010) In vitro propagation of *Tylophora indica i*nfluence of explanting season, growth regulator synergy, culture passage and planting substrate. J Am Sci 6(12):386–392

Reyes JL, Chua NH (2007) ABA induction of *miR159* controls transcript levels of two *myb* factors during *Arabidopsis* seed germination. Plant J 49(4):592–606

Riboni M, Galbiati M, Tonelli C, Conti L (2013) GIGANTEA enables drought escape response via abscisic acid-dependent activation of the florigens and suppressor of overexpression of *constans1*. Plant Physiol 162(3):1706–1719

Rodriguez PL, Leube MP, Grill E (1998) Molecular cloning in *Arabidopsis thaliana* of a new protein phosphatase 2C (PP2C) with homology to *abi1* and *abi2*. Plant Mol Biol 38(5):879–883

Rout GR, Das P (1995) Isozyme in vitro flowering of *Bambusa vulgaris*. J Plant Biochem Biotech 4(1):43–49

Saez A, Apostolova N, Gonzalez-Guzman M, Gonzalez-Garcia MP, Nicolas C, Lorenzo O, Rodriguez PL (2004) Gain of function and loss of function phenotypes of the protein phosphatase *2C hab1* reveal its role as a negative regulator of abscisic acid signalling. Plant J 37(3):354–369

Saez A, Rodrigues A, Santiago J, Rubio S, Rodriguez PL (2008) *Hab1–swi3b* interaction reveals a link between abscisic acid signaling and putative *swi/snf* chromatin-remodeling complexes in *Arabidopsis*. Plant Cell 20(11): 2972–2988

Saxena SN, Kaushik N, Sharma R (2008) Effect of abscicis acid and proline on in vitro flowering in *Vigna aconitifolia*. Biol Plant 52(1):1831–1833

Scorza R, Janick J (1980) In vitro flowering of *Passiflora suberosa* L. J Am Soc Hortic Sci 105(6):892–897

Seidlova F, Kohli KK, Pavlova L (1981) Effect of abscisic acid on the growth pattern of the shoot apical meristem and on flowering in *Chenopodium rubrum* L. Ann Bot 48(6):777–785

Sim GE, Goh CJ, Loh CS (2008) Induction of in vitro flowering in *dendrobium madame* thong-in (Orchidaceae) seedlings is associated with increase in endogenous N6-(Δ2-isopentynyle)-adenine (iP) and N6-(Δ2-isopentynyle)-adenosine (iPa) levels. Plant Cell Rep 27(8):1281–1289

Singh B, Sharma S, Rani G, Virk GS, Zaidi AA, Nagpa A (2006) In vitro flowering in embryogenic cultures of kinnow mandarin (*Citrus nobilis Lour* x *C. deliciosa* Tenora). Afr J Biotech 5(16):1470–1474

Tang W (2000) High frequency plant regeneration via somatic embryogenesis and organogenesis and in vitro flowering of regenerated plantlets in *Panax ginseng*. Plant Cell Rep 19(7):727–732

Tanimoto S, Harada H (1981) Effects of IAA, zeatin, ammonium nitrate and sucrose on the initiation and development of floral buds in *Torenia stem* segments cultured in vitro. Plant Cell Physiol 22(8):1553–1560

Taylor NJ, Light ME, van Staden J (2007) Monosaccharides promote flowering in *Kniphofia leucocephala* in vitro. Plant Growth Regul 52(1):73–79

Teixeira da Silva JA, Yam T, Fukai S, Nayak N, Tanaka M (2005) Establishment of optimum nutrient media for in vitro propagation of *cymbidium* Sw. (Orchidaceae) using protocorm-like body segments. Prop Ornam Plants 5(3):129–136

Than MMM, Pal A, Jha S (2009) In vitro flowering and propagation of *Bulbophyllum auricomum* Lindl, the royal flower of Myanmar. Acta Hortic 829(14):105–111

Vadawale AV, Barve DM, Dave AM (2006) In vitro flowering and rapid propagation of *Vitex negundo* L.: a medicinal plant. Indian J Biotech 5(1):112–116

Vandana AK, Atul K, Jitendra K (1995) In vitro flowering and pod formation in cauliflower (*brassica oleracia* var. *botrytis)*. Curr Sci 69(6):543–545

Vaz APA, Kerbauy GB (2000) Effects of mineral nutrients on in vitro growth and flower formation of *Psygmorchis pusilla* (Orchidaceae). Acta Hortic 520(15):149–156

Vaz APA, Kerbauy GB (2008) In vitro flowering studies in Psygmorchis pusilla. In: Teixeira da Silva JA (ed) Floriculture, ornamental and plant biotechnology: Advances and topical issue. Global Science Books Ltd., UK, pp 421–426

Virupakshi S, Manjunatha BR, Naik GR (2002) In vitro flower induction in callus from a juvenile explant of sugarcane, *Saccharum officinarum* L. var. CoC 671. Curr Sci 83(10):1195–1197

Vu NH, Anh PH, Nhut DT (2006) The role of sucrose and different cytokinins in the in vitro floral morphogenesis of rose (hybrid tea) cv. "First Prize". Plant Cell Tiss Org Cult 87(3):315–320

Wahyuni S, Krisantini S, Johnston ME (2011) Plant growth regulators and flowering of *Brunonia* and *Calandrinia* sp. Sci Hortic 128(2):141–145

Wang TY (1996) Cytokinins and light intensity regulate flowering of Easter Lily. Hortic Sci 31(6):976–977

Wang G, Xu Z, Chia TF, Chua NH (1997) In vitro flowering of *Dendrobium candidum*. Sci China Life Sci 40(1):35–42

Wang GY, Yuan MF, Hong Y (2002) In vitro flower induction in roses. In Vitro Cell Devel Biol – Plant 38(5):513–518

Wang ZH, Wang L, Ye QS (2009) High frequency early flowering from in vitro seedlings of *Dendrobium nobile*. Sci Hortic 122(2):328–331

Wang Y, Li L, Ye T, Lu Y, Chen X, Wu Y (2013) The inhibitory effect of ABA on floral transition is mediated by *abi5* in *Arabidopsis*. J Exp Bot 64(2):675–684

Yaish MW, Colasanti J, Rothstein SJ (2011) The role of epigenetic processes in controlling flowering time in plants exposed to stress. J Exp Bot 62(11):3727–3735

Zeng SJ, Liang S, Zhang YY, Wu KL, Teixeira da Silva JA, Duan J (2013) In vitro flowering red miniature rose. Biol Plant 57(3):401–409

The Use of Silver Nanoparticles as a Disinfectant and Media Additive in Plant Micropropagation

14

Hoang Thanh Tung, Huynh Gia Bao, Ngo Quoc Buu, Nguyen Hoai Chau, and Duong Tan Nhut

Abstract

Microbial contamination (fungi, bacteria, etc.) is one of the most problematic situations in micropropagation, which causes a reduction of plant quality and loss of value stocks. Therefore, sterilization of culture media and explant surface disinfection is a critical step in plant micropropagation success. The conventional method of autoclaving of media could have reduced the activities of additives and plant growth substances. Some of the common surface disinfectants, such as mercuric chloride, are not eco-friendly and can affect human health. In recent years, silver nanoparticles have been shown to be an effective disinfectant for explants and the culture media and positively affect plant regeneration and eliciting secondary product formation. This chapter highlights recent applications of silver nanoparticles in plant micropropagation.

Keywords

Autoclaving · Explant surface disinfection · Hormesis · Media sterilization · Silver nanoparticles

H. T. Tung · H. G. Bao · D. T. Nhut (✉)
Tay Nguyen Institute for Scientific Research, VAST, Dalat City, Vietnam

N. Q. Buu · N. H. Chau
Institute of Environmental Technology, VAST, Hanoi City, Vietnam

D. T. Nhut et al. (eds.), *Plant Tissue Culture: New Techniques and Application in Horticultural Species of Tropical Region*,
https://doi.org/10.1007/978-981-16-6498-4_14

14.1 Introduction

For aseptic cultures, proper sterilization techniques are necessary to prevent the unwanted growth of bacteria, fungi, and other microorganisms, ensuring the success of the projects. In plant micropropagation, several major steps need to be carried out in preventing contaminations, that is, (1) sterilization of media and tools, (2) surface disinfection of explants using a variety of chemicals, (3) proper handling of media and explants during the culture process, and (4) ensuring proper working conditions of equipment, such as the clean bench. Detailed treatments on disinfection and sterilization procedures can be found in monographs by Block (2001) and McDonnell (2007). Protocols on the handling of plant materials and related techniques for in vitro culture are readily available in the literature and textbooks, such as Vasil and Thorpe (1994), Gamborg and Phillips (1995), Bhojwani and Razdan (1996), Leelavathy and Sankar (2016).

In recent years, nanotechnology is playing an essential role in advancing different fields of plant biology and agriculture, especially tissue culture and biotechnology (Álvarez et al. 2019; Mahendran et al. 2019). Nanoparticles, especially silver nanoparticles, are helpful in many plant micropropagation applications, acting as a disinfectant for explants and improving the plant regeneration process when added at appropriate concentrations to the culture media. Other applications, such as eliciting secondary products from in vitro cultures and improvements to the microponic systems, are also successful (Anjum et al. 2019; Rivero-Montejo et al. 2021; Tung et al. 2018). Furthermore, their use in gene transformation, advancing genetic engineering is being explored (Cunningham et al. 2018; Wang et al. 2019; Lv et al. 2020).

This chapter serves to document recent achievements and augment information from recent reviews, for example, Sarmast and Salehi (2016); Kim et al. (2017); Sanzan et al. 2019; and Álvarez et al. (2019), involving the use of AgNPs on plant micropropagation.

14.2 Sterilization Methods Used in Plant Micropropagation

14.2.1 Autoclaving

Steam sterilization or autoclaving is the most common method for sterilizing nutrient media, glassware, and small instruments. An autoclave is a steam-sterilizing device. The contents within an autoclave are heated using saturated steam under pressure. The autoclave temperature is maintained at 121 °C for 15–30 min, and under these conditions, microorganisms are quickly killed. General operation guidelines and safety issues of autoclaving are readily available in the literature and the web. For sterilizing a small volume of a medium, home pressure cookers such as the Instant Pot® can be used that operates according to the same principle as an autoclave (Swenson et al. 2018).

Autoclaving can have adverse effects on media components. Prolonged heating under pressure may adversely affect the media's quality (Bhojwani and Dantu 2013). For example, the pH of media and some chemicals' properties can be altered (Bragt et al. 1971). Autoclaving can cause hydrolysis of sucrose resulting in glucose and fructose formation (Ball 1953). Heat-sensitive plant growth regulators such as Gibberellic acid can lose most of their activity. Some vitamins are broken down due to high temperatures. Too long a sterilization time can precipitate salts and alter agar properties, and it can also lead to toxic decomposition products such as 5-(hydroxymethyl)-2-furaldehyde and phenolics (Wang and Hsiao 1995). Thus, it is crucial to determine the optimum time for autoclaving culture media and be aware of the heat sensitivity of the chemicals used. Furthermore, Leifert et al. (1994) reported that *Bacillus* could persist after autoclaving at 110–120 °C. Also, 2–5% of the media were contaminated by hand pouring after autoclaving. Therefore, to ensure the media are sterilized and not contaminated, the media should be incubated at 30–35 °C for 24–48 h before use after autoclaving.

14.2.2 Explant Surface Disinfection

Surface sterilization of the experimental material is the first step to kill all the surface microbes before in vitro culture. Different chemicals and procedures have been developed to eliminate surface contaminations. This is accomplished by treating plant organs or tissues with one or more sterilizing agents for a short period, followed by a sterile water rinse. The widely applied chemicals are sodium and calcium hypochlorite, ethanol, mercuric chloride, hydrogen peroxide, fungicide, and antibiotics. Other chemicals have been introduced as surface disinfectants, such as isothiocyanates (Lazo-Javalera et al. 2016) and chlorine dioxide (Duan et al. 2019).

Proper knowledge and understanding of various sterilization agents' positive and negative effects will help to select the best sterilization agent(s) for a specific species and explants (Leelavathy and Sankar 2016). Hypochlorites, such as sodium hypochlorite, are a strong oxidizing agent and are very effective in killing bacteria. Ethanol is a strong sterilizing agent and it is also phytotoxic. A 70% solution serves to denature proteins quickly and inactivating lipophilic viruses. To increase its effectiveness, ethanol at 70% concentration can be applied before or after a hypochlorite treatment for a few minutes or just for a few seconds. **Mercuric chloride** is mainly **bacteriostatic** and has a broad spectrum of activity. Mercuric compounds bind to sulfhydryl groups and are postulated to inhibit enzymatic activity (Phillips and Warshowsky 1958). Although mercuric chloride is an effective sterilant, it is a highly toxic chemical and not environmentally friendly. One needs to be careful in handling this compound as mercuric chloride can be absorbed through the skin and is highly toxic when ingested or inhaled. Extra care should be taken in disposing of mercuric chloride solution after use. As a result, it has not regularly used today. Hydrogen peroxide is an oxidizing agent. At low concentrations, hydrogen peroxide possesses fungicidal and bactericidal activities and is effective in sterilizing seeds. The use of Tween® 20 as a surfactant improves the wettability of explants and seeds to increase the sterilization procedure's efficiency. Antibiotics are toxic to explants and are rarely used because of their inadequate and adverse effects on the growth of explants (Teixeira da Silva et al. 2003). Continuous use of antibiotics can lead to the development of antibiotic-resistant strains of bacteria.

The success of removing surface contaminants also requires an understanding of the plant materials chosen for the studies. This includes consideration of the source of explants, that is, field-grown vs. greenhouse-grown plants, the age of the explants, and the types of organs and tissues selected. A more vigorous sterilization regime and careful selection of disinfectants will be needed for field-grown materials. There is no universal sterilization protocol that is suitable for all explants. One needs to test various agents and optimize the procedure used. Recommendations and protocols for explant sterilization can be found in published literature.

The current disinfectants used in the sterilization of ex vitro explants in laboratories have advantages and disadvantages. Because of their antimicrobial action, silver compounds, especially silver nanoparticles, have become a popular disinfectant for plant micropropagation. At optimum concentrations, silver compounds are effective and more eco-friendly, and compatible with various tissue types.

14.3 The Use of Silver Nanoparticles (AgNPs) in Plant Micropropagation

Diagram depicting the use of AgNPs in micropropagation was shown in Fig. 14.1.

14.3.1 Silver Nanoparticles as an Explant Disinfectant

In recent years, silver nanoparticles are widely used as an antimicrobial agent and serve as an explant disinfectant. Many reviews, for example, Ahmad et al. (2020), Hamad et al. (2020), and Deshmukh et al. (2019), summarize recent findings and suggest mechanisms of action of AgNPs on microorganisms. The antimicrobial ability of AgNPs depends on factors such as particle shape, size distribution, metallic silver content, synthesis procedure, and the coating agent of the nanoparticles (Andújar et al. 2020). AgNPs exhibit more effective antibacterial properties than silver salts because of their very large surface area, allowing better attachment with microorganisms. AgNPs smaller than 10 nm interact readily with bacteria. Once attached to cell membranes, AgNPs can enter the bacteria, releasing silver ions into bacterial cells, enhancing bactericidal activity (Morones et al. 2005; Hamad et al. 2020; Ahmad et al. 2020).

In plant micropropagation, AgNPs have been successfully used as a surface disinfectant. In strawberries, 200 and 500 mg/L AgNPs effectively reduced contaminations compared to those treated with mercuric chloride (Tung et al. 2021). Furthermore, leaf explanted disinfected with 200 mg/L AgNPs for 20 min yield higher biomass of callus and higher shoot regeneration rates when compared to

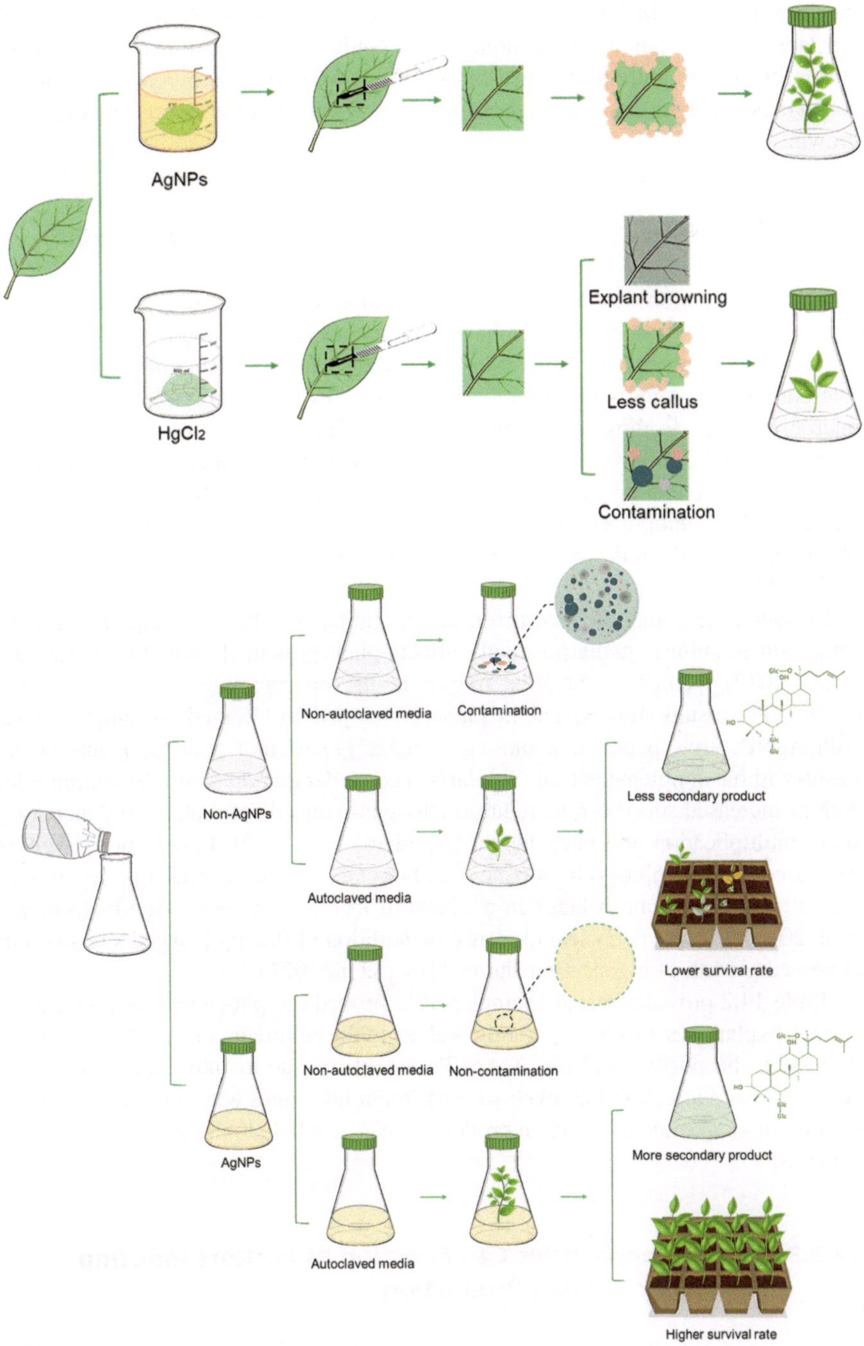

Fig. 14.1 The use of AgNPs in plant micropropagation

the control. Table 14.1 identifies recent publications that use AgNPs as a surface sterilant. AgNPs can be used alone or in conjunction with other disinfectants. According to the authors, the use of AgNPs at an appropriate concentration can achieve positive results and appears to have fewer adverse effects on explants' growth.

14.3.2 The Effects of Silver Nanoparticles in Autoclaved Media

AgNPs can function further as a disinfectant in autoclaved media. Some explants, especially of woody species, often have endogenous bacterial and fungal infections. These microorganisms are difficult to kill by the surface sterilization process alone. The addition of AgNPs to culture media can destroy or minimize the growth of microorganisms leading to "clean" cultures. The addition of AgNPs to the Murashige and Skoog (1962; MS) medium reduced the number of infected barley embryos in vitro and promoted seedling growth (Krupa-Malkiewicz et al. 2019). In Vanilla, using a temporary immersion system to promote shoot growth, the addition of 50 mg/L AgNPs to the medium resulted in zero contamination (Spinoso-Castillo et al. 2017).

Besides functioning as a sterilizer, incorporating AgNPs at an appropriate concentration in culture media positively affects plant growth (Landa 2021). Sarmast et al. (2012) recorded a positive impact in the micropropagation of *Araucaria excelsa*. The results showed that the plants developed on MS medium supplemented with AgNPs grew better, maintaining a darker green color than the plants in the media without supplementation. Similarly, *Tecomella undulata* supplementing with AgNPs increased survival rate, reduced ethylene-induced defoliation, and increased shoot multiplication and plant height (Aghdaei et al. 2012). Furthermore, AgNPs decreased the 1-aminocyclopropane-1-carboxylate synthase transcript levels and may alter the hormone balance in explants in media with added AgNPs (Sarmast et al. 2015). In strawberry leaf explants, an addition of 0.2 mg/L AgNPs resulted in increased growth of regenerated shoots (Tung et al. 2021).

Table 14.2 provides recent examples of improved morphogenetic response from various explant systems incorporating AgNPs in the culture media. As Landa (2021) discussed, the positive effects of AgNPs are likely due to inhibition of ethylene perception and affecting the levels of other plant hormones within tissues. Changes in nutrient uptake and antioxidant production can also help to improve the growth of explants.

14.3.3 Silver Nanoparticles Can Function as Elicitors Inducing Secondary Product Production

Besides improving the performance of explants in culture, the inclusion of AgNPs enhances secondary products' accumulation. Plants react to environmental stress and elicitors, generating physiological responses which often involved in the production

Table 14.1 Selected recent publications on the use of AgNPs as a surface sterilant of explants

Species	Treatment and outcomes	Reference
Capparis decidua fruits	AgNPs were green synthesized from the fruit of *Capparis decidua*. Explants soaking for 20 min. With 100 mg/L AgNPs gave good survival and decontamination percentage. An increased AgNPs concentration gave 100% bacterial and 98.6% for fungal decontamination but a reduced survival percentage (68.5%)	Ahlawat et al. (2020)
Fragaria x ananassa	Chemically synthesized AgNPs were used as a surface sterilant and a culture medium additive. Application of 200 mg/L for 20 min was effective in explant disinfection and shoot regeneration. AgNPs stimulated the growth of shoot and plantlet and shortened the duration of root formation compared to those in the control without AgNPs during micropropagation. There was an overall improvement in performance	Tung et al. (2021)
Hevea brasiliensis	Used as a surface disinfectant and a culture medium additive. Application of 10 ppm AgNPs for 20 min gave the best result as a surface sterilant. When incorporated into culture media, at a concentration of 4 ppm, the highest percentage of explant survival with a low percentage of browning was observed compared to the controls. The effects depended on the stage of leaf development	Moradpour et al. (2016)
Kappaphycus striatus	The treatment with AgNPs at 500 ppm resulted in no contamination. A high explant survival rate (80%) and callus induction rate (54.4%) were obtained	Mo et al. (2020)
Ocimum seeds and tissues	The use of 10–100 mg/L biosynthesized nanosilver was effective resulting in no contamination. AgNPs also have a stimulatory effect on the formation of callus. At a high concentration of 100 mg/L with a longer exposure time (60 min), the nanosilver posed no harmful effect on the seeds, tissues, and callus induction; instead, it stimulated callus formation	Adebomojo and Abdul Rahaman (2020)
Pennisetum alopecuroides	AgNPs were obtained from a commercial source. When explants were soaked in a high concentration of AgNPs (100–250 mg/L) alone or in combination with 70% ethanol and 2% NaOCl, explants were successfully disinfected, especially the nodal explants. Supplementation of the media with a low concentration (4 mg/L) gave 40% contamination-free explants	Parzymies et al. (2019)
Phoenix dactylifera L.	The application of 5 mg/L AgNPs was suitable as a surface disinfectant. AgNP treatment resulted in better survival of explants (88.9%) and 0% mortality	El-Sharabasy et al. (2017)

(continued)

Table 14.1 (continued)

Species	Treatment and outcomes	Reference
Phoenix dactylifera L. cv Sewi and Medjool	AgNPs were used as a surface sterilant and a culture medium additive. The application of 500 ug/L was successful as a surface disinfectant. Lower concentrations resulted in good callus formation and a higher number of somatic embryos and shoot numbers	El-Kosary et al. (2020)
Psidium friedeichsthalianum (O. Berg) Nied	Commercially purchased AgNPs were added onto the surface of the explants on a culture medium during the in vitro establishment phase. The addition of 5 mg/L Argovit® AgNPs as a permanent bilayer reduced shoot contamination rate to 50% compared to 80% for the controls, and this treatment also promoted leaf growth and shoot multiplication rate	Andújar et al. (2020)

of specialized metabolites. These metabolites allow for better adaptation to adverse environmental conditions (Anjum et al. 2019). Since many of these secondary metabolites are of commercial value, the elicitation using nanomaterials has generated many interests. Besides using traditional cultures, that is, callus and organ cultures, the hairy root explant systems are being explored for secondary metabolite production (Kaur et al. 2021). The hairy root system reacts well to elicitation using nanoparticles, for example, hairy root culture of *Cucumis* (Chung et al. 2018a). Current reviews on this topic, for example, Hatami et al. (2019), Kralova and Jampilek (2021), Lala (2021), and Rivero-Montejo et al. (2021) summarize successes on using AgNPs as an elicitor of secondary metabolites through micropropagation. Table 14.3 showcases recent successes using silver nanoparticles as elicitors of secondary metabolites.

The action of AgNPs is likely to induce the formation of reactive oxygen species and signaling messengers, leading to secondary metabolite production (Rivero-Montejo et al. 2021). The positive action of AgNPs on stimulating secondary production is that they serve as an elicitor of hormesis (Spinoso-Castillo et al. 2017). Hormesis is defined as the stimulatory effects caused by low levels of inhibitors (Stebbing 1982). The addition of AgNPs at low concentrations activates plant stress defense mechanisms by inducing antioxidant compounds and reactive oxygen species. This results in generating hormetic effects leading to growth promotion and secondary metabolites' production. Some of the effects of AgNPs function as elicitors of secondary metabolites are shown in examples in Table 14.3.

14.3.4 Silver Nanoparticles as a Media Sterilizing Agent in Non-Autoclaved Media

The direct addition of AgNPs to culture media is especially useful in the culture of plantlets using microponic systems (Tung et al. 2018, also see Chap. 10).

Table 14.2 Selected recent publications on the effects of AgNPs on explant micropropagation and morphogenesis

Species	Treatment and outcomes	Reference
Campomanesia rufa (O. Berg) Nied	AgNPs did not affect the in vitro multiplication of nodal segments at low concentrations. Moreover, at a higher concentration (15.5 mg/L), a reduction in the number of shoots was found	Timoteo et al. (2019)
Dianthus caryophyllus cv Noblesse, cv. Antigua, and cv. Mariposa	AgNPs at 6–8 mg/L increased the number of shoots per explant and the highest regeneration rate. High concentration (12 mg/L) enhanced rooting response, the number of roots/plant, and root length. The total antioxidant activity and reducing power potential of regenerated plants varied depending on the AgNPs in the medium	Zia et al. (2020)
Hostacapitata	AgNPs were obtained from a commercial source. The addition of AgNPs to the regeneration medium stimulated the growth of isolated meristem. The meristem-derived plants were free of the Hosta virus X gene through RAPD analysis. The inclusion of AgNPs helped generate virus-free and genetically stable plants	Pe et al. (2020)
Hordeum vulgare	The addition of 6 and 8 mg/L AgNPs in the MS medium limited the number of infected barley embryos cultured. At 4 and 8 mg/L concentration, the best seedling growth with the longest roots was observed	Krupa-Malkiewicz et al. (2019)
Lavandula angustifolia Mill.	Chemically synthesized AgNPs stimulated shoot formation and increased plant weight. Longer roots were also observed in the regenerated shoots in the presence of AgNPs. The concentration of AgNPs also affected the size of secretory trichomes in leaves. Nanoparticles also increased antioxidant enzyme activities	Jadczak et al. (2019, 2020)
Maerua oblongifolia	Biosynthesized AgNPs from *Ochradenusarabicus* leaf extracts. AgNPs at 20 mg/L increased the shoot number, shoot length, fresh and dry weights, and chlorophyll content of plantlets. AgNPs at 1 mg/L gave the highest protein content. A high concentration (50 mg/L) increased the levels of proline, superoxide dismutase, and catalase activities	Shaikhaldein et al. (2020)
Musa acuminate cv "Grand Nain"	Low concentrations of AgNPs promoted the growth of explants and increased	El-Mahdy et al. (2019)

(continued)

Table 14.2 (continued)

Species	Treatment and outcomes	Reference
	pigment content. There was a notable increase in proline, hydrogen peroxide content, and total phenolic and enzymes associated with phenolic metabolism	
Oryza sativa	Biosynthesized AgNPs from *Partheniumhysterophorus* plant extract promoted callus induction frequency, callus regeneration, and rhizogenesis at concentrations of 5 and 10 mg/L. The primary plant growth regulator responsive genes were downregulated upon exposure to PHAgNPs (5 mg/L and 10 mg/L) compared with the control treatment. However, above 10 mg/L, the expression of all the genes markedly increased	Manickavasagam et al. (2019).
Rosa hybrida L. "Baby Love"	Using chemically synthesized AgNPs shoot explants performed best at 2 mg/L AgNPs. There was a positive effect of AgNPs in overcoming the yellowing of leaf, leaf abscission, and improved quality of explants. AgNPs inhibited ethylene gas biosynthesis and the pectinase and cellulase enzyme activities. Treatment of explants at the micropropagation stage also helped to improve the adaptability and growth of Rose plantlets at the nursery stage	Ngan et al. (2020)
Vanilla planifolia Jack. Ex Andrews	Commercial Agrovit® AgNPs were added to the medium of a temporary immersion system for continual shoot explant culture. No contamination was found at a concentration of 50 mg/L and above. Growth stimulation was also observed	Spinoso-Castillo et al. (2017)

Microponics combines micropropagation and hydroponic techniques to develop high-quality plantlets generated from in vitro cultures. The microponic method is best applied at the rooting stage of plant propagation. With the addition of AgNPs as a media disinfectant and in conjunction with liquid nutrient media without a carbon source such as sucrose, fungal and bacterial contaminations can be kept to a minimum. By combining rooting induction with the acclimatization stage, the overall time of plant propagation is significantly reduced, and the production cost is also minimized (Hahn et al. 1996, 2000a, 2000b). While effective, autoclaving is a major expense for many tissue culture laboratories (Chen 2016). The direct addition of AgNPs to media minimizes the routine use of the autoclaving step. Since a high-temperature treatment of media is not needed, the direct application of AgNPs can reduce the negative effects of the conventional autoclaving process in media

Table 14.3 Selected recent publications on the effects of AgNPs as an elicitor of secondary metabolites in cell cultures

Species	Treatment and outcomes	Reference
Brassica rapa hairy root culture	Biosynthesized AgNPs elicited high concentrations of reactive oxygen species. Glucosinolates and their transcripts were significantly enhanced in elicitated hairy roots	Chung et al. (2018b)
Caralluma tuberculata	Commercially purchased AgNPs at 60 μg/L combined with 2,4-D and BA resulted in the highest fresh and dry biomass accumulation of callus. At 90 μg/L, AgNPs enhanced the production of phenolics, flavonoids, PAL, and antioxidant activities	Ali et al. (2019)
Cucumisanguria hairy root culture	Biosynthesized AgNPs were more efficient in promoting biomass increase in hairy root cultures as compared to silver nitrate. Furthermore, AgNPs also elicited a higher amount of phenolic compounds and flavonoids than silver nitrate	Chung et al. (2018a)
Echinacea purpurea	AgNPs were obtained from a commercial source. The cell suspension was first established. AgNPs were more effective in stimulating cichoric acid production compared to silver nitrate. The application of 2 mg/L of AgNPs increased the production of cichoric acid significantly	Ramezannezhad et al. (2019)
Linum usitatissimum L.	AgNPs were added to cell suspension cultures to elicit the production of lignans and neolignans. At day ten, the beginning of the log growth phase, AgNPs elicited the highest production of lignans, neolignans, total phenolic and flavonoid content, and biomass	Zahir et al. (2019)
Stevia rebaudiana	The lower concentration of AgNPs (12.5–50 mg/L) promoted the greatest shoot production and length per explant. Adding AgNPs to the culture medium significantly affected shoot multiplication and length. AgNPs were located in stem epidermal cells, within vascular bundles, and in intermembrane spaces using fluorescence microscopy. In leaves, they were observed in ribs and stomata	Castro-González et al. (2019)
Stevia rebaudiana	AgNPs were obtained from a commercial source. AgNPs at 45 mg/L promoted callus growth and elicited the highest amount of stevioside from callus culture	Golkar (2019)
Swertiachirata	Using biosynthesized AgNPs from leaf extracts of *S. chirata*, 4 mg/L AgNPs in the presence of BA gave a maximum average number of shoots per explants. Changes to reactive oxygen species content and antioxidant enzyme activities were noted	Saha and Gupta (2018)

preparation, which is better for plant growth and development. This approach can be extended to open culture conditions and other liquid culture systems.

In the chrysanthemum microponic system, plant shoots were cultured in media supplemented with different AgNPs concentrations added directly to the boiled culture medium without autoclaving (Tung et al. 2018). They showed that the addition of silver nanoparticles at a concentration of 7.5 ppm in microponic culture reduced the microbial content of 8 bacterial species, that is, *Corynebacterium* sp., *Enterobacter* sp., *Arthrobacter* sp., *Agrobacterium* sp., *Xanthomonas* sp., *Pseudomonas* sp., *Micrococcus* sp., and *Bacillus* sp. and 3 species of molds consist of *Aspergillus* sp., *Fusarium* sp. and *Alterneria* sp. As shown in their study, the addition of AgNPs to the culture medium effectively serves as a disinfectant for the culture medium, replacing the conventional autoclaving step and promotes rooting at the same time.

14.4 Conclusion

There is little doubt that the use of AgNPs as a surface disinfectant or as an additive to culture media results in positive outcomes and improves morphogenetic processes and plant performance. Higher concentrations (>100 mg/L) are needed when AgNPs are used as a surface disinfectant, and lower concentrations (<10 mg/L) are preferred when used as a medium additive. For best results, the optimum concentration of AgNPs for different processes, such as serving as a surface disinfectant or as an additive to media, needs to be determined separately to achieve the desirable effects.

Different factors could influence the outcome of the results. It is well established that the antimicrobial and morphogenetic abilities of AgNPs depend on factors such as particle shape, size distribution, metallic silver content, synthesis procedure, and the coating agent of the nanoparticles (Andújar et al. 2020). These factors can determine how the nanoparticles interact with the microbes and explants in culture. In recent years, green synthesis of AgNPs is becoming popular as synthesis is rapid, eco-friendly, and easily scaled-up. The nanoparticles prepared from medicinal plants are more pharmacologically active, likely due to the attachment of pharmacologically active residues (Singh et al. 2016). However, as Pryshchepa et al. (2020) indicated, detailed information concerning the synthesis of silver nanoparticles is often incomplete. The results obtained "could be affected by artifacts introduced with exploited equipment, the utilized technique or sample preparation stages" (Pryshchepa et al. 2020). Thus, it is necessary to document the characteristics and evaluate the properties of each batch of AgNPs synthesized before use.

Continual research using different approaches will further widen the use of nanoparticles in plant science. A proper theoretical understanding of the interactions between the applied nanoparticles and endogenous plant factors will ensure success in plant micropropagation.

References

Adebomojo AA, AbdulRahaman AA (2020) Surface sterilization of *Ocimum* seeds and tissues with biosynthesized nanosilver and its effects on callus induction. IOP Conf Ser: Mater Sci Eng 805: 012024

Aghdaei M, Salehi H, Sarmast MK (2012) Effects of silver nanoparticles on *Tecomellaundulata* (Roxb.) Seem. Micropropagation Adv Hort Sci 26:21–24

Ahlawat J, Sehrawat AR, Choudhary R, Yadav SK (2020) Biologically synthesized silver nanoparticles eclipse fungal and bacterial contamination in micropropagation of *Capparis decidua* (FORSK.) Edgew: a substitute to toxic substances. Indian J ExpBiol 58:336–343

Ahmad SA, Das SS, Khatoon A, Ansari MT, Afzal M, Md H, Nayak AK (2020) Bactericidal activity of silver nanoparticles: a mechanistic review. Mater Sci Technol 3:756–769

Ali A, Mohammad S, Khan MA, Raja NI, Arif M, Kamil A, Mashwani ZUR (2019) Silver nanoparticles elicited in vitro callus cultures for accumulation of biomass and secondary metabolites in *Carallumatuberculata*. Artif Cells Nanomed Biotechnol 47(1):715–724

Álvarez SP, Tapia MAM, Vega MEG, Ardisana EFH, Medina JAC, Zamora GLF, Bustamante DV (2019) Nanotechnology and plant tissue culture. In: Prasad R (ed) Plant nanobionics, Advances in the understanding of nanomaterials research and applications, vol 1. Springer, Cham, pp 333–370

Andújar I, González N, García-Ramos JC, Bogdanchikova N, Pestryakov A, Escalona M, Concepción O (2020) Argovit™ silver nanoparticles reduce contamination levels and improve morphological growth in the *in vitro* culture of *Psidiumfriedrichsthalianum* (O. Berg). Nied SN ApplS ci 2:2110

Anjum S, Anjum I, Hano C, Kousar S (2019) Advances in nanomaterials as novel elicitors of pharmacologically active plant specialized metabolites: Current status and future outlooks. RSC Adv 9:40404–40423

Ball E (1953) Hydrolysis of sucrose by autoclaving media, a neglected aspect in the technique of culture of plant tissues. Bull Torrey Bot Club 80:409–411

Bhojwani SS, Dantu PK (2013) Plant tissue culture: An introductory text. Springer, London, UK, pp 39–43

Bhojwani SS, Razdan MK (1996) Plant tissue culture: Theory and practice, a revised edition. Elsevier, Amsterdam, The Netherland, pp 618–766

Block SS (ed) (2001) Disinfection, sterilization and preservation, 5th edn. Lippincott Williams & Wilkins, New York, pp 881–917

Bragt JV, Mossel DAA, Pierik RML, Veldstra H (1971) Effects of sterilization on components in nutrient media. In: Veenman H, Zonon NV (eds) Misc. Papers 9. Wageningen, the Netherlands, pp 38–104

Castro-González CG, Sánchez-Segura L, Gómez-Merino FC, Bello-Bello JJ (2019) Exposure of stevia (*Stevia rebaudiana* B) to silver nanoparticles in vitro: transport and accumulation. Sci Rep 9(1):1–10

Chen C (2016) Cost analysis of plant micropropagation of *Phalaenopsis*. Plant Cell Tissue Organ Cult 126:167–175

Chung IM, Rajakumar G, Thiruvengadam M (2018a) Effect of silver nanoparticles on phenolic compounds production and biological activities in hairy root cultures of *Cucumisanguria*. Biol Futura 69:97–109

Chung IM, Rekha K, Rajakumar G, Thiruvengadam M (2018b) Influence of silver nanoparticles on the enhancement and transcriptional changes of glucosinolates and phenolic compounds in genetically transformed root cultures of *Brassica rapa* ssp. *rapa*. Bioprocess Biosyst Eng 41: 1665–1677

Cunningham FJ, Goh NS, Demirer GS, Matos JL, Landry MP (2018) Nanoparticle-mediated delivery towards advancing plant genetic engineering. Trends Biotechnol 36:882–897

Deshmukh SP, Patil SM, Mullani SB, Delekar SD (2019) Silver nanoparticles as an effective disinfectant: a review. Mater Sci Eng C Mater Biol Appl 97:954–965

Duan Y, Zhang H, Sun M, Zhao F, Xue T, Xue J (2019) Use of chlorine dioxide to sterilize medium for tissue culture of potato. Sci Rep 9:10232

El-Kosary S, Abd Allatif AM, Stino RG, Hassan MM, Kinawy AA (2020) Effect of silver nanoparticles on micropropagation of Date palm (*Phoenix dactylifera* L. cv. Sewi and Medjool). Plant Arch 20:9701–9706

El-Mahdy MT, Radi AA, Shaaban MM (2019) Impacts of exposure of banana to silver nanoparticles and silver ions in vitro. Middle East J Appl Sci 9(3):727–740

El-Sharabasy SF, Ghazzawy HS, Munir M (2017) In vitro application of silver nanoparticles as explant disinfectant for date palm cultivar Barhee. J Appl Hort 19(2):106–112

Gamborg OL, Phillips GC (1995) Plant cell, tissue and organ culture. Springer, Heidelberg, Berlin, pp 1–35

Golkar P, Moradi M, Garousi GA (2019) Elicitation of Stevia glycosides using salicylic acid and silver nanoparticles under callus culture. Sugar Tech 21:569–577

Hahn EJ, Lee YB, Ahn CH (1996) A new method on mass-production of micropropagated chrysanthemum plants using microponic system in plant factory. Acta Hortic 440:527–532

Hahn EJ, Bea JH, Lee YB (2000a) Growth and photosynthetic characteristics ofChrysanthemumplantlets as affected by pH and EC of nutrient solution in microponic culture. J Korean Soc Hort Sci 41(1):12–15

Hahn EJ, Kozai T, Paek KY (2000b) Blue and red light-emitting diodes with or without sucrose and ventilation affect in vitro growth of *Rehmanniaglutinose*plantlets. J Plant Biol 43:247–250

Hamad A, Khashan KS, Hadi A (2020) Silver nanoparticles and silver ions as potential antibacterial agents. J Inorg Organomet Polym 30:4811–4828

Hatami M, NaghdiBadi H, Ghorbanpour M (2019) Nano-elicitation of secondary pharmaceutical metabolites in plant cells: a review. J Med Plants 18:6–36

Jadczak P, Kulpa D, Bihun M, Przewodowski W (2019) Positive effect of AgNPs and AuNPs in *in vitro* cultures of *Lavandulaangustifolia* Mill. Plant Cell Tissue Organ Cult 139:191–197

Jadczak P, Kulpa D, Drozd R, Przewodowski W, Przewodowska A (2020) Effect of AuNPs and AgNPs on the antioxidant system and antioxidant activity of Lavender (*Lavandulaangustifolia* Mill.) from *in vitro* cultures. Molecules 25:5511

Kaur G, Prakash P, Srivastava R, Verma PC (2021) Enhanced secondary metabolite production in hairy root cultures through biotic and abiotic elicitors. In: Ramawat KG, Ekiert HM, Goyal S (eds) Plant cell and tissue differentiation and secondary metabolites. Springer, pp 625–660

Kim DH, Gopal J, Sivanesan L (2017) Nanomaterials in plant tissue culture: the disclosed and undisclosed. RSC Adv 7:36392–36505

Kralova K, Jampilek J (2021) Response of medicinal and aromatic plants engineered nanoparticles. Appl Sci 11:1813

Krupa-Małkiewicz M, Oszmiański J, Lachowicz S, Szczepanek M, Jaśkiewicz B, Pachnowska K, Ochmian I (2019) Effect of nanosilver (nAg) on disinfection, growth, and chemical composition of young barley leaves under *in vitro* conditions. J Integr Agric 18(8):1871–1881

Lala S (2021) Nanoparticles as elicitors and harvesters of economically important secondary metabolites in higher plants: a review. IET Nanobiotechnol 15:28–57

Landa P (2021) Positive effects of metallic nanoparticles on plants: overview of involved mechanisms. Plant Physiol Biochem 161:12–24

Lazo-Javalera MF, TroncosoRojas R, Tiznado Hernandez ME, Martinez-Tellez MA, Vargas Arispuro I, Islas-Osuna MA, Rivera-Dominguez M (2016) Surface disinfection procedure and *in vitro* regeneration of grapevine (*Vitisvinifera* L) axillary buds. SpringerPlus 5:Article number: 453

Leelavathy S, DeepaSankar P (2016) Curbing the menace of contamination in plant tissue culture. J Pure Appl Microbiol 10(3):2145–2152

Leifert C, Morris CE, Waites WM (1994) Ecology of microbial saprophytes and pathogens in tissue culture and field-grown plants: reasons for contamination problems in vitro. Crit Rev Plant Sci 13(2):139–183

Lv Z, Jiang R, Chen J, Chen W (2020) Nanoparticle-mediated gene transformation strategies for plant genetic engineering. Plant J 104:880–891

Mahendran D, Geetha N, Venkatachalam P (2019) Role of silver nitrate and silver nanoparticles on tissue culture medium and enhanced the plant growth and development. In: Kumar M, Muthusamy A, Kumar V, Bhalla-Sarin N (eds) In vitro plant breeding towards novel agronomic traits. Springer, Singapore, pp 59–74

Manickavasagam M, Pavan G, Vasudevan V (2019) A comprehensive study of the hormetic influence of biosynthesized AgNPs on regenerating rice calli of Indica cv. IR64. Sci Rep 9:8821

McDonnell GE (2007) Antisepsis, disinfection, and sterilization: types, action, and resistance. Wiley, ASM Press, Washington, pp 304–345

Mo VT, Cuong LK, Tung HT, Huynh TV, Nghia LT, Khanh CM, Lam NN, Nhut DT (2020) Somatic embryogenesis and plantlet regeneration from the seaweed *Kappaphycus striatus*. Acta Physiol Plant 42:104

Moradpour M, Aziz MA, Abdullah SNA (2016) Establishment of *in vitro* culture of rubber (*Heveabrasiliensis*) from field-derived explants: effective role of silver nanoparticles in reducing contamination and browning. J Nanomed Nanotechnol 7:375

Morones JR, Elechiguerra JL, Camacho A, Holt K, Kouri JB, Ramírez JT, Yacaman MJ (2005) The bactericidal effect of silver nanoparticles. Nanotechnology 16(10):2346–2353

Murashige T, Skoog F (1962) A revised medium for rapid growth and bioassays with tobacco tissue cultures. Physiol Plant 15(1):473–497

Ngan HTM, Cuong DM, Tung HT, Nghiep ND, Le BV, Nhut DT (2020) The effect of cobalt and silver nanoparticles on overcoming leaf abscission and enhanced growth of Rose (*Rosa hybrida* L. 'Baby Love') plantlets cultured *in vitro*. Plant Cell Tissue Org Cult 141(2):393–405

Parzymies M, Pudelska K, Poniewozik M (2019) The use of nano-silver for disinfection of *Pennisetumalopecuroides* plant material for tissue culture. Acta Sci Pol Hortorum Cultus 18(3):127–135

Pe PPW, Naing AH, Soe MT, Kang HH, Park KI, Kim CK (2020) Establishment of meristem culture for virus-free and genetically stable production of the endangered plant *Hosta capitata*. Sci Hortic 272:109591

Phillips CR, Warshowsky B (1958) Chemical disinfectants. Ann Rev Microbiol 12:525–550

Pryshchepa O, Pomastowski P, Buszewski B (2020) Silver nanoparticles: Synthesis, investigation techniques, and properties. Adv Colloid Interf Sci 284:102246

Ramezannezhad R, Aghdasi M, Fatemi M (2019) Enhanced production of cichoric acid in cell suspension culture of *Echinacea purpurea* by silver nanoparticle elicitation. Plant Cell Tissue Organ Cult 139:261–273

Rivero-Montejo SDJ, Vargas-Hernandez M, Torres-Pacheco I (2021) Nanoparticles as novel elicitors to improve bioactive compounds in plants. Agriculture 11:134

Saha N, Gupta SD (2018) Promotion of shoot regeneration of *Swertiachirata* by biosynthesized silver nanoparticles and their involvement in ethylene interceptions and activation of antioxidant activity. Plant Cell Tissue Org Cult 134(2):289–300

Sanzan I, Leone A, Ambrosome A (2019) Nanotechnology in plant science: to make a long story short. Front Bioeng Biotechnol:Article 120

Sarmast MK, Salehi H (2016) Silver nanoparticles: an influential element in plant nanobiotechnology. Mol Biotechnol 58(7):441–449

Sarmast MK, Salehi H, Khosh-Khui M (2012) Micropropagation of *Araucaria excelsa* R. Br. var. *glauca* Carrière from orthotropic stem explants. Physiol Mol Biol Plants 18(3):265–271

Sarmast MK, Niazi A, Salehi H, Abolimoghadam A (2015) Silver nanoparticles affect *ACS* expression in *Tecomellaundulatain vitro* culture. Plant Cell Tissue Organ Cult 121:227–236

Shaikhaldein HO, Al-Qurainy F, Nadeem M, Khan S, Tarroum M, Salih AM (2020) Biosynthesis and characterization of silver nanoparticles using *Ochradenusarabicus* and their physiological effect on *Maeruaoblongifolia* raised in vitro. Sci Rep 10:17569

Singh P, Kim Y-J, Zhang D, Yang D-C (2016) Biological synthesis of nanoparticles from plants and microorganisms. Trends Biotechnol 34:588–600

Spinoso-Castillo J, Chavez-Santoscoy R, Bogdanchikova N, Pérez-Sato J, Morales-Ramos V, Bello-Bello J (2017) Antimicrobial and hormetic effects of silver nanoparticles on *in vitro* regeneration of vanilla (*Vanilla planifolia* Jacks. ex Andrews) using a temporary immersion system. Plant Cell Tissue Org Cult 129(2):195–207

Stebbing ARD (1982) Hormesis: the stimulation of growth by low levels of inhibitors. Sci Total Environ 22:213–234

Swenson VA, Stacy AD, Gaylor MO, Ushijima B, Philmus B, Cozy LM, Videau NM, Videau P (2018) Assessment and verification of commercially available pressure cookers for laboratory sterilization. PLoS One 13(12):e0208769

Teixeira da Silva JA, Duong NT, Michi T, Seiichi F (2003) The effect of antibiotics on the *in vitro* growth response of chrysanthemum and tobacco stem transverse thin cell layers (tTcLs). SciHortic 97:397–410

Timoteo C, Paiva R, dos Reis MV, Cunha Claro PI, Corrêa da Silva DP, Marconcini JM, de Oliveira JE (2019) Silver nanoparticles in the micropropagation of *Campomanesiarufa* (O. Berg) Nied. Plant Cell Tissue Organ Cult 137:359–368

Tung HT, Nam NB, Huy NP, Luan VQ, Hien VT, Phuong TTB, Le TD, Nhut DT (2018) A system for large scale production of Chrysanthemum using microponics with the supplement of silver nanoparticles under light-emitting diodes. SciHortic 232:153–161

Tung HT, Bao HG, Cuong DM, Ngan HTM, Hien VT, Luan VQ, Vinh BVT, Phuong HTN, Nam NB, Trieu LN, Truong NK, Hoang PND, Nhut DT (2021) Silver nanoparticles as the sterilant in large-scale micropropagation of chrysanthemum. In Vitro Cell Dev Biol Plant:1–10

Vasil IK, Thorpe TA (eds) (1994) Plant cell and tissue culture. Kluwer Academic Publishers, Dordrecht, pp 293–312

Wang XJ, Hsiao KC (1995) Sugar degradation during autoclaving: Effects of duration and solution volume on breakdown of glucose. Physiol Plant 94:415–418

Wang JW, Grandio EG, Newkirk GM, Demirer GS, Butrus S, Giraldo JP, Landry MP (2019) Nanoparticle-mediated genetic engineering of plants. Mol Plant 12:1037–1040

Zahir A, Nadee M, Ahmad W, Giglioli-Guivarc'h N, Hano C, Abbasi BH (2019) Chemogenic silver nanoparticles enhance lignans and neolignans in cell suspension cultures of *Linumusitatissimum* L. Plant Cell Tissue Organ Cult 136:589–596

Zia M, Yaqoob K, Mannan A, Nisa S, Raza G, Rehman RU (2020) Regeneration response of carnation cultivars in response of silver nanoparticles under in vitro conditions. Vegetos 33:11–20

Enhanced Growth and Overcoming Abnormal Phenomena in Micropropagation by Nanoparticles

15

Duong Tan Nhut, Ha Thi My Ngan, Nguyen Thi Nhu Mai, Phan Le Ha Nguyen, Bui Van Le, and Hoang Thanh Tung

Abstract

This chapter provides background technical information and demonstrates some positive effects of different metal nanoparticles on the growth and development of in vitro cultured plantlets. Metal nanoparticles were obtained by the aqueous solution method with the core composing of silver, iron, or cobalt (Ag^o, Fe^o, Co^o). The shell of silver nanoparticles was made of chitosan, and the shell component for iron and cobalt nanoparticles was carboxymethyl cellulose. The abscission phenomenon in Rose plantlets (*Rosa hybrida* L. "Baby Love") was suppressed when cobalt nanoparticles (CoNPs) were incorporated in the medium. During the micropropagation of Gerbera (*Gerbera jamesonii* "Revolution yellow"), silver nanoparticles (AgNPs) promoted plantlet growth and overcame the vitrification phenomenon. Iron nanoparticles (FeNPs) improved the quality of carnation (*Dianthus caryophyllus* "Express golem") plantlets, increasing rooting efficiency and tolerance to culture conditions through increased activity of antioxidant enzymes. AgNPs and CoNPs helped to limit ethylene gas biosynthesis and action. The plantlets derived from in vitro cultures supplemented with AgNPs, CoNPs, and FeNPs gave the highest survival percentage and improved development at the nursery stage compared to the control without metal nanoparticle treatments.

D. T. Nhut (✉) · H. T. M. Ngan · N. T. N. Mai · P. L. H. Nguyen · H. T. Tung
Tay Nguyen Institute for Scientific Research, VAST, Dalat City, Vietnam

B. Van Le
University of Science, VNU—HCMC, Hochiminh City, Vietnam

D. T. Nhut et al. (eds.), *Plant Tissue Culture: New Techniques and Application in Horticultural Species of Tropical Region*,
https://doi.org/10.1007/978-981-16-6498-4_15

Keywords

Antioxidant enzymes · Core-shell · *Dianthus caryophyllus* · Ethylene · *Gerbera jamesonii* · Hydrolysis enzymes · Metal nanoparticles · *Rosa hybrida*

15.1 Introduction

Plant biotechnology, especially plant cell, tissue, and organ culture, has become a standard propagation method for many horticultural plants. Micropropagation methods allow the generation of a large number of genetically homogenous and disease-free plants. The in vitro approach also provides research on plant physiology, overcoming some of the difficulties when using traditional methods (Mahmood et al. 2017). However, micropropagation still has unresolved limitations. This includes abnormal morphologies and anatomical features, microbial contaminations, vitrification of cells and tissues, leaf abscission, and explant browning. These limitations result in low quality of in vitro shoots, decreased propagation coefficient, and plantlet survival at the nursery stage (Kharrazi et al. 2011; Bhojwani and Dantu 2013). Therefore, improving and optimizing culture systems have always been one of the main goals in commercial plant production.

Different species react to the in vitro environment differently. The frequency of vitrification, leaf abscission, tissue browning, oxidative stresses, and other abnormal symptoms differ among species cultured. The causes of these phenomena are likely due to the high concentration of carbon dioxide and humidity inside the small culture vessels ($\geq$95%), improper temperature optimization, the suboptimal light intensity on photosynthetic efficiency and pigment biosynthesis, and the types and concentrations of plant growth regulators and gelling agents used (Gaspar et al. 1987). Also, the production and excessive accumulation of gases, especially ethylene by plant tissues, the activities of reactive oxygen species (ROS), and the formation of hydrolytic enzymes, can result in poor growth of plantlets reducing their quality (Lai et al. 2005). Changing media components and their concentrations, altering culture conditions such as aeration, and adding ethylene biosynthesis inhibitors can improve plantlet regeneration and development. Moreover, some of these treatments could result in lower propagation rates, increase the risk of toxicity to the explants, and are not suitable for all cultured plantlets (Bahmani et al. 2009; MacDonald et al. 2011; Merelo et al. 2017).

Nanotechnology is considered the industrial revolution, promoting development in all fields, especially biomedicine, energy, environment, agriculture, and information technology. Recently, nanomaterials affecting plant growth have attracted researchers around the world. Studies on the effects of metal nanoparticles on plants have been shown (Kole et al. 2016; Sanzari et al. 2019). Both the positive and negative effects of nanomaterials on plant growth depend on many factors such as composition, structure, concentration of nanoparticles, and plant species (Siddiqui et al. 2015; Kim et al. 2017).

This chapter aims to provide some background and technical information on metal nanoparticles, that is, silver (AgNPs), iron (FeNPs), and cobalt (CoNPs) and their effects on cultured plantlets. In addition to *Gerbera*, we summarized our recent findings on the effects of metal nanoparticles on Rose (Ngan et al. 2020a) and Carnation (Ngan et al. 2020b) plantlets below. These plants have beautiful flowers with high economic values and are most popular today because of the varieties of colors and sizes, flower shapes, and commercial applications.

15.2 Metal Nanoparticles

The term nanotechnology was introduced in 1959 when the American physicist Richard Feynman mentioned the ability to manufacture materials at microscopic sizes by assembling atoms and molecules in his talk at an annual meeting of the American Physical Society. In the 1980s, with the invention of analytical devices, including the scanning tunneling microscope, allowing the visualization and manipulation of atoms or molecules, nanotechnology began to be of great invested interest. Tremendous advances continue to be made in recent years. A brief history of nanoscience and nanotechnology has been summarized in an article by Bayda et al. (2019).

15.2.1 Nanoparticles and Plant Growth and Development

Interaction of metal nanoparticles with plants causes changes in the physiology and biochemistry of plants. A proper understanding of their action mechanisms is critical for applying nanotechnology in plant micropropagation (Sanzari et al. 2019). Recent studies show that metal nanoparticles can positively or negatively affect plants depending on particle composition, size, physical and chemical properties, nanoparticle concentration, and plant species (Ma et al. 2010). The positive effects of metal nanoparticles on plants were demonstrated through their ability to increase germination rates (Gopinath et al. 2014) and to enhance growth and physiological activities (Salama 2012). Other positive effects such as increasing the efficiency of fertilizer and water usage (Vakhrouchev and Golubchikov 2007), limiting abscission of plant organs (Seif et al. 2011), increasing crop yield (Kole et al. 2013), stimulating nodules formation of leguminous plants (Taran et al. 2014), and enhancing the production of secondary compounds (Fouad and Hafez 2018) have also been noted. Even though the exact mechanisms of actions are not clear, scientists have recognized the importance of metal nanoparticles in plant biotechnology and agriculture.

15.2.2 General Information on How to Prepare Metal Nanoparticles

Nanomaterials are natural or artificial materials with external, internal, or surface structures at nano-sized (1–100 nm). Based on the origin, three types of nanoparticles have been identified, that is, natural, random or incidental, and designer or engineered nanoparticles. Natural nanoparticles have existed since the earth's formation and found in dirt, moon dust, minerals, photochemical reactions, forest fires, and erosion. Random nanoparticles are generated by artificial industrial processes such as gasoline/diesel exhaust, coal burning, and industrial emissions. Designer nanoparticles can be divided into five categories: carbon-based, metal-based, magneto, dendrimers, and mixed nanoparticles (Banerjee and Kole 2016).

The metal nanoparticles are the focus of this chapter. They have a core-shell structure. The core is composed of zero-valence metal atoms such as Au, Ag, Cu, Fe, Ni, Co, Zn. Depending on the method of synthesis and application, the shell can be chitosan, thin nanolayer (silica nano, gold nano, etc.), metal oxides/hydroxides (Fe_2O_3, $Fe(OH)_2$), carboxymethyl cellulose (CMC), polymers (polyvinyl alcohol, PVA), polyethylene glycol (PEG), and polyvinyl pyrrolidone (PVP). The shell functions to protect and stabilize the core and to disperse the nanoparticles in the environment (Fedlheim and Foss 2001). Different methods of nanoparticle synthesis are available in the literature (Kulkarni and Muddapur 2014), and ready-made metal nanoparticles are also available commercially, for example, Ted Pella Inc., Redding, CA, USA.

AgNPs (smaller than 20 nm in size) have a core-shell structure where the core is silver atoms (Ag^o), and the coating is chitosan or nanolayer silica, nano-gold, carboxymethyl cellulose (CMC). The aqueous solution method is the most frequently used in preparing spherical AgNPs. Silver nitrate ($AgNO_3$) is used as the silver ion source and sodium borohydride ($NaBH_4$) as the reducing agent. AgNPs are produced in a homogenous solution of $AgNO_3$ with β-chitosan as stabilizer at ambient temperature, according to the ratio of 750–1000 ppm $AgNO_3$, 250–300 ppm β-chitozan, 200 ppm $NaBH_4$, with a molar ratio of $NaBH_4$ and $AgNO_3$(1:4), and a dripping rate of $NaBH_4$ from 10 to 12 drops per min.

FeNPs (20–60 nm in size) also have a core-shell nanostructure. The core is composed of iron atoms (Fe^o). The shell can be iron oxide/hydroxides or other common coating ingredients such as CMC and polymers. The aqueous solution with $NaBH_4$ as a reducing agent and CMC as a stabilizer is the most commonly used method to manufacture iron nanoparticles. For preparing 1 g FeNPs: dissolve $FeSO_4 \cdot 7H_2O$ (18 mmol) in deionized water (160 mL) with 0.5% CMC (40 mL) and stir vigorously (1500 rpm) for 15 min. Add two g $NaBH_4$ to the mixture and stir for 10 min. Vacuum-dry the residue at 70 °C to obtain the final product.

CoNPs (20–60 nm in size) are synthesized by reducing the precursor $CoCl_2.6H_2O$ with $NaBH_4$ with CMC as a stabilizer (Fig. 15.1). To prepare 1 g CoNPs, dissolve 18 mmol $CoCl_2 \cdot 6H_2O$ in 160 mL deionized water, mix with 40 mL of CMC solution (5%), and stir vigorously (1500 rpm) for 15 min. Add 1 g $NaBH_4$ into the mixture with continuous stirring for another 10 min. Collect the precipitate and vacuum-dried at 70 °C. The size of CoNPs is determined by transmission electron microscopy and X-ray diffraction analysis. CoNPs are dispersed in solution by ultrasound with a particle size of about 20–60 nm and a purity >95%.

Fig. 15.1 Diagram for preparation of CoNPs

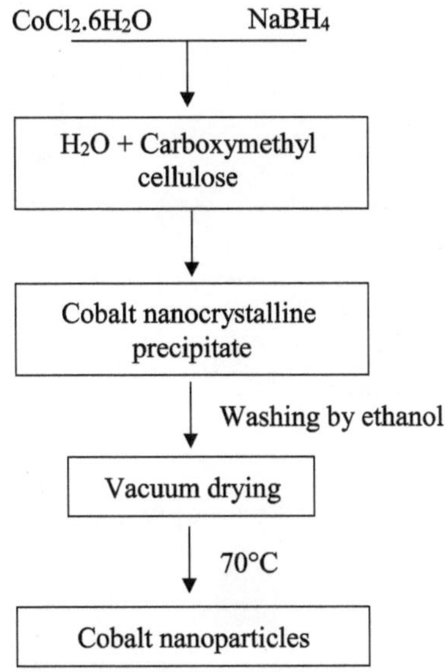

The shelf-lives of metal nanoparticles produced by the aqueous solution method stored at room temperature have been estimated to be at least 7 months with no difference in particle size observed under an electron microscope.

15.3 Materials

15.3.1 Metal Nanoparticles Solutions

The solutions of AgNPs (500 ppm), FeNPs (500 ppm), CoNPs (1000 ppm) were produced in the Institute of Environmental Technology (Chau et al. 2008). Different concentrations (0; 1; 2; 3; 5; 7 mg/L) of AgNPs were added directly to culture media. However, different concentrations of FeNPs (0; 1.40;2.80; 4.20; 5.60;11.20 mg/L) and CoNPs (0;1.55; 3.10; 4.65; 6.20; 12.40 μg/L) were supplemented as a substitute for the metal salts (5.60 mg/L Fe-EDTA or 6.20 μg/L$CoCl_2$) in the original medium formulation (Fig. 15.2).

15.3.2 Determination of Growth Parameters

The ethylene gas content in culture vessel was determined by gas chromatography (GC) with flame ionization probe (FID). One mL of a head gas sample was obtained directly from a culture vessel and injected into GC (GC-CP 3380) manually with a syringe (BD Tuberculin syringe 1 mL). A stainless steel column (3 m × 1.5 mm in size) filled with an adsorbent (Porapack R—particles 80–100 Mesh in size) served as

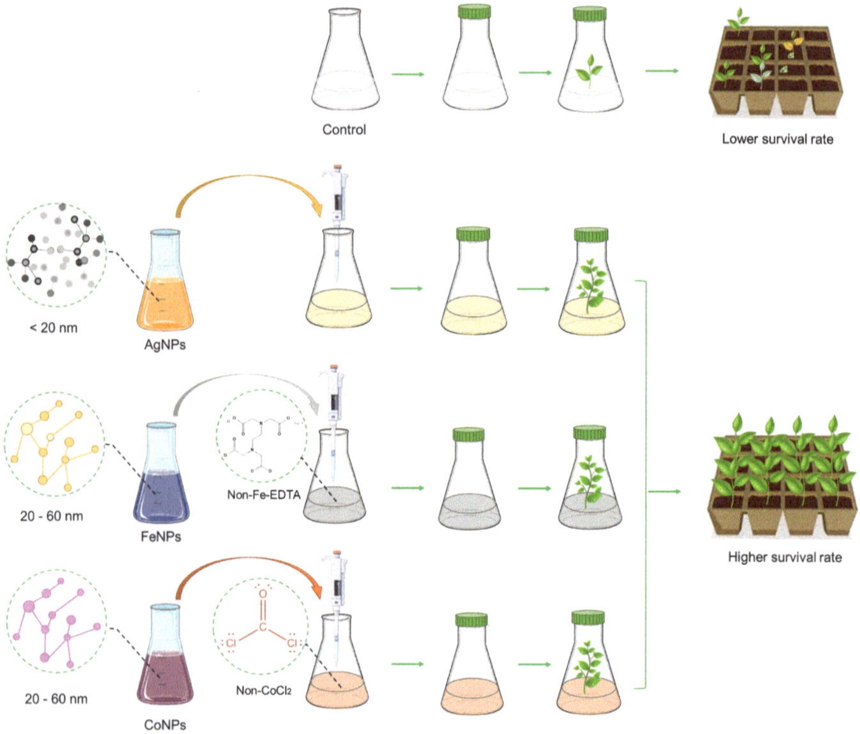

Fig. 15.2 The procedure of adding nano into the culture medium

the stationary phase component. The galvanometer sensitivity was 10^{-12} Am/V and nitrogen gas (N_2) was used as the carrier gas (Cristescu et al. 2012).

Cellulase activity was determined based on the hydrolysis of carboxymethyl cellulose (CMC) by cellulase enzyme (pH = 5, at 40 °C). The reduced sugar produced gave a color reaction with the 3,5-dinitrosalicylic acid (DNS) as the reagent. The color product was determined by a colorimetric method measured at 540 nm using a spectrophotometer. A unit of cellulase enzyme activity was calculated as the number of glucose (mg) produced by 1 mL (or 1 g of the preparation) at 40 °C for 15 min (Zhang et al. 2009).

The enzyme extract was incubated with soluble pectin (1%) at 37 °C for 60 min to determine pectinase activity. The amount of reducing sugar produced (D-galacturonic acid) gave a color reaction with DNS. The change in the optical density was determined using a spectrophotometer at 575 nm. One pectinase activity unit is the number of μmol of D-galacturonic acid produced by 1 mL of enzyme solution. The concentration of reducing sugar produced was calculated based on the D-galacturonic acid standard curve (Vatanparast et al. 2014).

The chlorophyll a and chlorophyll b content were assessed by absorption spectrometric analysis of leaf extracts (1 g in 50 mL of acetone solution) using an ultraviolet spectrophotometer (Hitachi, UV-2900, Tokyo, Japan). The absorbance

(OD) was measured at 662 and 645 nm (Lichtentaler and Wellburn 1985). The chlorophyll a or b contents were checked on the formula: Chlorophyll a = (11.75 A662 − 2.35 A645); Chlorophyll b = (18.61 A645 − 3.96 A662).

The SPAD index (chlorophyll a + b content in leaves) was measured with a SPAD-502 instrument (Minolta Co., Ltd., Osaka, Japan) in the third leaf (from the shoot-tip) of the plantlet. Meanwhile, the dry mass ratio was determined by formula: Dry matter ratio (%) = Dry weight (mg) × 100 per fresh weight (mg).

The activities of antioxidant enzymes, that is, superoxide dismutase (SOD), ascorbate peroxidase (APX), and catalase (CAT) (U/g) were determined by absorption spectroscopy of enzyme extracts from leaf tissues (300 mg/sample) (Ngan et al. 2020b). SOD activity was determined according to the method of Marklund (1974). The SOD in the explant catalyzes the decomposition of the peroxide (-O–O-) radicals, inhibiting pyrogallol's self-oxidation. The inhibition ratio reflects the SOD activity in the sample at 320 nm. The APX (Nakano and Asada 1981) present in the test explant oxidizes and reduces the maximum absorption of the ascorbate at 290 nm in the presence of H_2O_2 within a short time. The CAT activity was checked by reacting the explant with 100 μL H_2O_2 (65 mM) in 2 min. The remaining H_2O_2 after the reaction was combined with 100 μL ammonium molybdate ($NH_4Mo_7O_{24}$) to give a yellow complex, and the SOD was measured at 405 nm. One unit of catalase activity (U/g) is equivalent to 1 μmol H_2O_2 hydrolyzed for 1 min (Goth 1991).

15.3.3 The Plant Material

The following tissue culture protocols allow the generation of explants to illustrate the nanoparticles' efficacies in micropropagation. Axillary buds of *Rosa hybrida* and *Dianthus caryophyllus* plants were washed under 30 min running water; then, 30 s soaked in 70% ethanol, and sterilized further with $HgCl_2$ (0.1%) for 7 min. The buds were then rinsed three times with sterilized distilled water and cultured on MS (Murashige and Skoog 1962) medium supplemented with 1.5 mg/L6-benzyladenine (BA), 30 g/L sucrose, and 8 g/L agar to induce adventitious shoots. The newly formed shoot tips (1.5 cm in length) were cultured on half-strength MS medium (½ MS) supplemented with 0.2 mg/L 3-indolebutyric acid (IBA), 30 g/L sucrose, and 8 g/L agar (Senapati and Rout 2008). The different concentrations of CoNPs (at 1.55 μg/L; 3.10 μg/L; 4.65 μg/L; 6.20 μg/L; 12.40 μg/L corresponding to ¼, ½, ¾, equal and twofold cobalt content in ½ MS medium) were added to the modified MS medium with the removal of 6.20 μg/L $CoCl_2$ from the original MS formulation. The MS medium without the added CoNPs was used as control.

The young receptacles of *Gerbera jamesonii* flowers were 30-min washed under running water, then, 10-min sterilized with 0.1% $HgCl_2$, and 3-time rinsed with sterile distilled water. These receptacles were cut into 0.5 mm-thick sections and cultured on MS medium supplemented with 0.02 mg/L thidiazuron (TDZ), 0.8 mg/Ladenine, 10% coconut water, 30 g/L sucrose, and 8 g/L agar for inducing adventitious shoots. The newly formed shoots were then cultured on MS medium supplemented with 0.9 mg/L IBA, 30 g/L sucrose, and 8 g/L agar (Yen et al. 2013). The different concentrations AgNPs (1 mg/L; 2 mg/L; 3 mg/L; 5 mg/L; 7 mg/L) were added to the culture medium. For the control, the explants were cultured in the medium without AgNPs.

The *Dianthus caryophyllus* shoot tips (1.5 cm) were cultured on ½ MS medium supplemented with 1.0 mg/L α-Naphthaleneacetic acid (NAA), 30 g/L sucrose, and 8 g/L agar (Khatun et al. 2013). The different concentrations of FeNPs (at 1.40 mg/L; 2.80 mg/L; 4.20 mg/L; 5.60 mg/L; 11.20 mg/L corresponding to ¼, ½, ¾, equal, and twofold iron content in ½ MS medium) were added to the modified MS medium with the removal of 5.60 mg/L Fe-EDTA from the original MS formulation. The MS medium without FeNPs was used as the control. All media were adjusted to pH 5.8 before autoclaving at 121 °C for 30 min.

15.4 Enhanced Growth and Overcoming Vitrification of Gerbera In Vitro Culture Using Silver Nanoparticles

Many studies have demonstrated that the addition of AgNPs to the culture medium improves plant growth and overcoming abnormal phenomena (Fig. 15.3). After 4 weeks of culture, the effects of different AgNPs concentrations on the in vitro growth and vitrification of Gerbera plantlets are shown in Fig. 15.4. The results indicated that the development of Gerbera plantlets on medium supplemented with

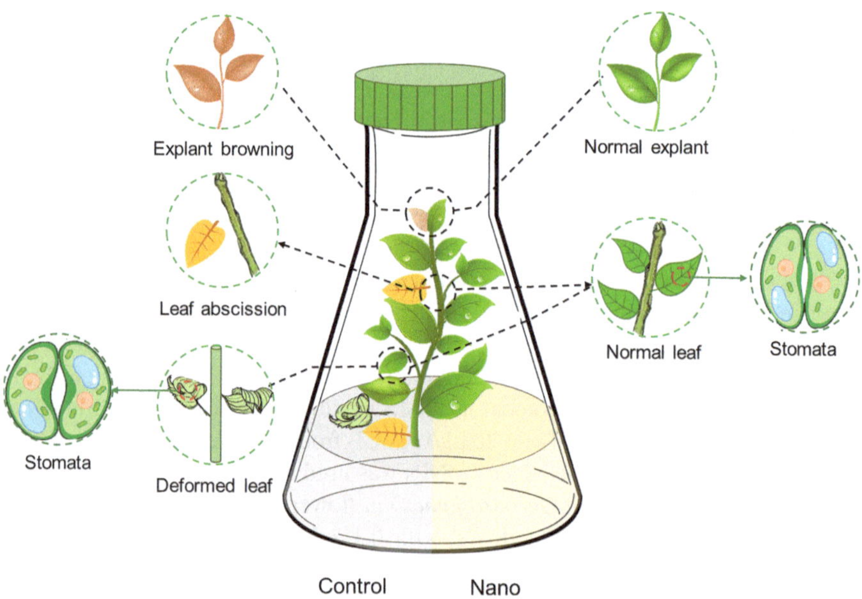

Fig. 15.3 Enhanced growth and overcoming abnormal phenomena in micropropagation by nanoparticles

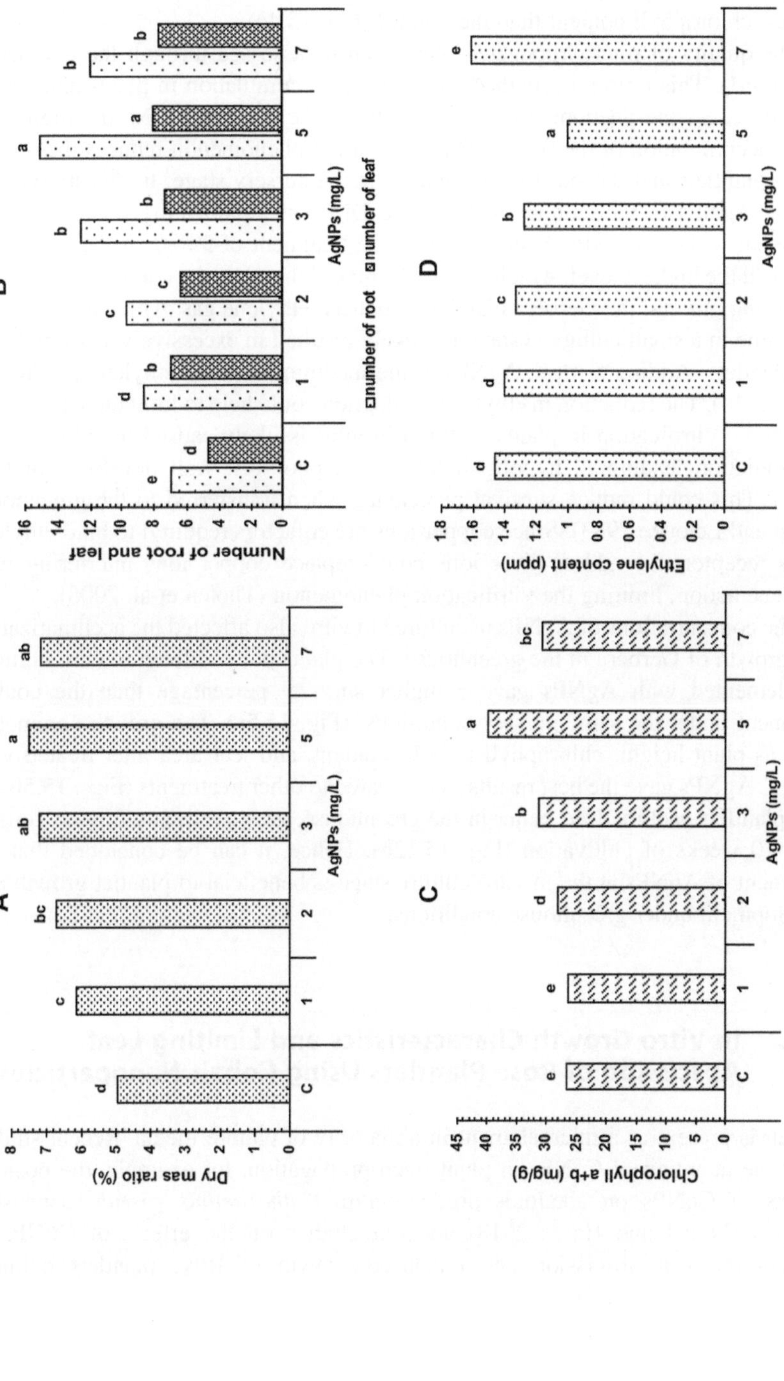

Fig. 15.4 Effects of different AgNPs concentrations on in vitro Gerbera growth and development. *C* Control

5 mg/L AgNPs was significantly better than other treatments. The number of roots and leaves produced was higher at 5 mg/L AgNPs than the control treatment (Fig. 15.4b). The chlorophyll (a + b) content reflects the photosynthesis efficiency and pigment biosynthesis in leaves. The addition of 5 mg/L AgNPs resulted in the highest chlorophyll content than the control (Fig. 15.4c).

The quality of micropropagated plants can be assessed through the dry matter ratio (DM). This ratio reflects the level of water accumulation in plants and can be used to assess the phenomenon of vitrification. The lower the DM, the higher the water accumulation of the plants. The vitrification phenomenon reduces the quality of the plantlets and the survival percentage at the nursery stage. In this study, after adding AgNPs to the culture medium, the DM was significantly higher than the treatment without AgNPs. Specifically, in the treatment of 5 mg/LAgNPs, the DM achieved the highest level, which was 1.51 times higher than the control (Fig. 15.4a).

During growth, the Gerbera plantlets produced ethylene gas, which accumulated over time in a small culture system and likely resulted in excessive water uptake by the plantlets. Adding 5 mg/L AgNPs to the medium inhibited ethylene production (Fig. 15.4d). The reduction in ethylene production could help to limit the vitrification symptom. Vitrification in plants cultured in vitro is likely caused by ethylene gas accumulation leading to hypolignification and poor cell wall development (Ziv 1991). This could reduce survival percentage when transferred to the greenhouse (Phan and Letouze 1983). Since copper ions are cofactors required to bind ethylene to its receptor, the added silver ions could replace copper ions interfering with ethylene action, limiting the vitrification phenomenon (Tholen et al. 2006).

The concentrations of AgNPs in cultured in vitro also affected the acclimatization and growth of Gerbera in the greenhouse. The plants derived from in vitro cultures supplemented with AgNPs gave a higher survival percentage than the control treatment under the same nursery conditions. (Fig. 15.5a). The growth parameters such as plant height, chlorophyll (a + b) content, and leaf area after treated with 5 mg/L AgNPs gave the best results as compared to other treatments (Figs. 15.5b–d). The plantlets acclimatized better in the greenhouse, grew well, and flowers formed after 10 weeks of cultivation (Fig. 15.12b). Hence, it can be concluded that the treatment of AgNPs at the in vitro culture stage is beneficial to plantlet growth and development under greenhouse conditions.

15.5 In Vitro Growth Characteristics and Limiting Leaf Abscission of Rose Plantlets Using Cobalt Nanoparticles

Cobalt is present as a microelement in a majority of culture media. Recent studies draw the attention of CoNPs in plant micropropagation, for example, the positive effects of CoNPs on alkaloids production in *Catharanthus roseus* suspension cultures (Fouad and Hafez 2018) and our studies on the effects of CoNPs in overcoming leaf abscission and enhanced growth of Rose plantlets cultured in vitro (Ngan et al. 2020a).

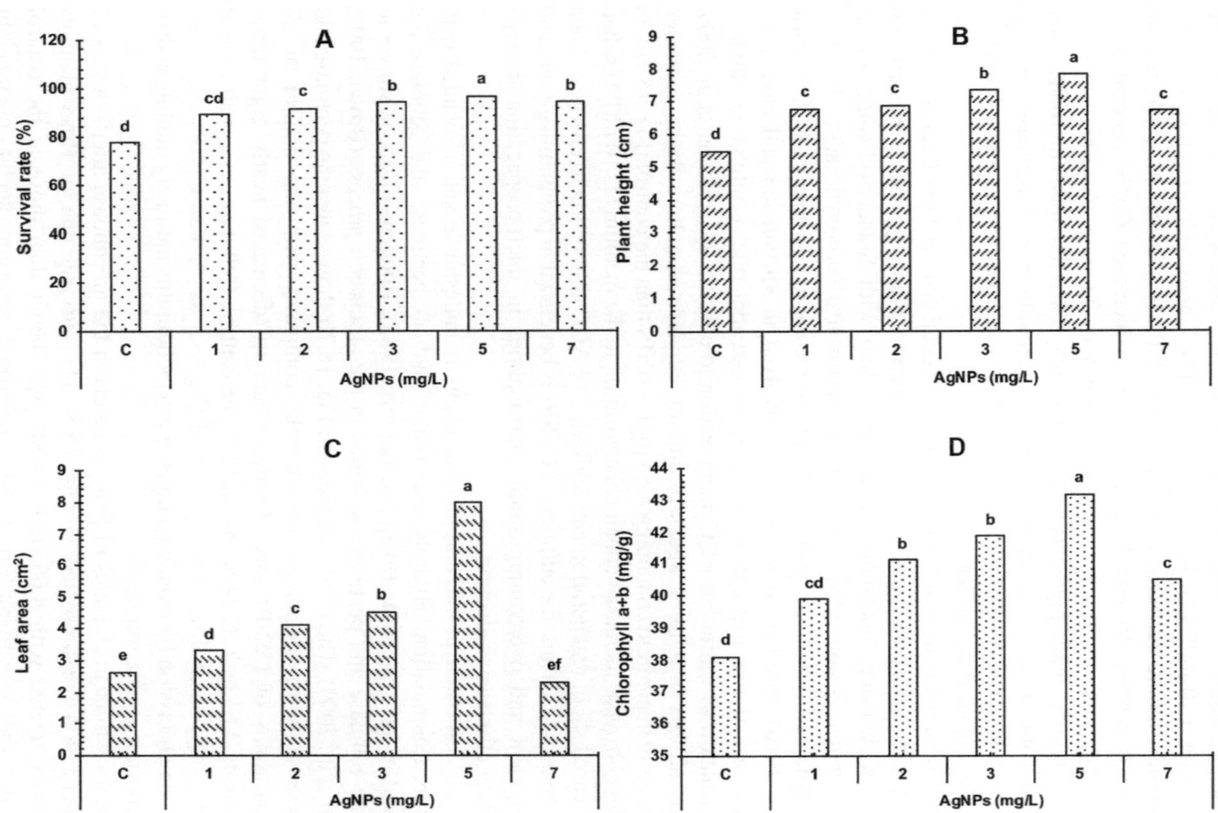

Fig. 15.5 Adaptability and growth of Gerbera plantlets derived from medium supplemented with AgNPs after 4 weeks in the greenhouse. *C* Control

CoNPs effect on the growth and development of Rose plantlets after 4 weeks of culture were shown in Fig. 15.6. The results showed that shoots cultured in medium supplemented with 4.65 μg/LCoNPs performed better than those in the control. The plant height, number of roots, root length, and leaf area were all higher than the control with 6.20 μg/L $CoCl_2$ (Fig. 15.6a–c). The highest chlorophyll content was also recorded in the treatment with 4.65 μg/LCoNPs. It is important to note that the chlorophyll content decreased significantly with increased CoNP concentrations (Fig. 15.6d). Our results are similar to the study by Jaleel et al. (2009) on *Vigna radiate*; photosynthetic pigments such as chlorophyll a and b, total chlorophyll, and carotene content increased at low cobalt concentrations and decreased at high concentrations. The results indicate that CoNPs can improve metabolic efficiency leading to enhance the growth of plantlets.

Ethylene is a unique gaseous plant hormone, and it can exert its effect at very low concentrations (0.01–1.0 ppm) in micropropagation. Plants synthesize ethylene gas through a two-step biochemical reaction starting with S-adenosyl-L-methionine (SAM), which is converted to l-aminocyclopropane-1-carboxylic acid (ACC) by the enzyme ACC synthase; then, ACC was converted to ethylene by ACC oxidase (Chang 2016). Ethylene is a key factor involved in vitrification and abscission. Studies have shown that cobalt ions play an essential role in inhibiting ethylene gas formation by inhibiting ACC's conversion to ethylene gas (Kumar et al. 2009; Thao et al. 2015). After 4 weeks of culture, ethylene in culture flasks decreased significantly in medium containing 4.65 μg/L CoNPs than the control (Fig. 15.7a). In *Swertia chirayita*, increased cobalt concentration in the medium (1–10 μM) resulted in increased shoot regeneration and inhibition of ACC conversion to ethylene (Saha and Gupta 2018). Thus, the addition of CoNPs is beneficial in promoting growth and development and overcoming some abnormalities in micropropagation through inhibiting ethylene production.

Organ abscission in micropropagated plants is undesirable and associated with ethylene accumulation. Hydrolytic enzymes such as pectinase and cellulase are involved in leaf abscission (Phan and Letouze 1983). Ethylene has been shown to regulate cellulase and pectinase activities in the abscission process (Brown 1997; Agustí et al. 2009). The results recorded in Fig. 15.7b showed that the pectinase and cellulase activities decreased significantly compared with the control at all concentrations of CoNPs used. Pectinase activity decreased nearly eight times when using 4.65 μg/L CoNPs compared to the control. Cellulase activity decreased approximately three times than the control (Ngan et al. 2020a). CoNPs exert their effects on abscission by reducing ethylene production; this indirectly inhibits hydrolytic enzyme production.

The growth characteristics of Rose plantlets in the greenhouse after 4 weeks of acclimatization are presented in Fig. 15.8. The survival rate of Rose plantlets previously treated with 4.65 μg/L CoNPs was better than those in the control (without CoNPs) treatment (Fig. 15.8a). The growth indicators, such as plant height and leaf area, reached the highest at the treatment supplement with 4.65 μg/L (Fig. 15.8b, c). Plantlets derived from cultured in vitro (4.65 μg/L CoNPs) have the higher quality and rooted-shoot, thick leaves with a high chlorophyll content

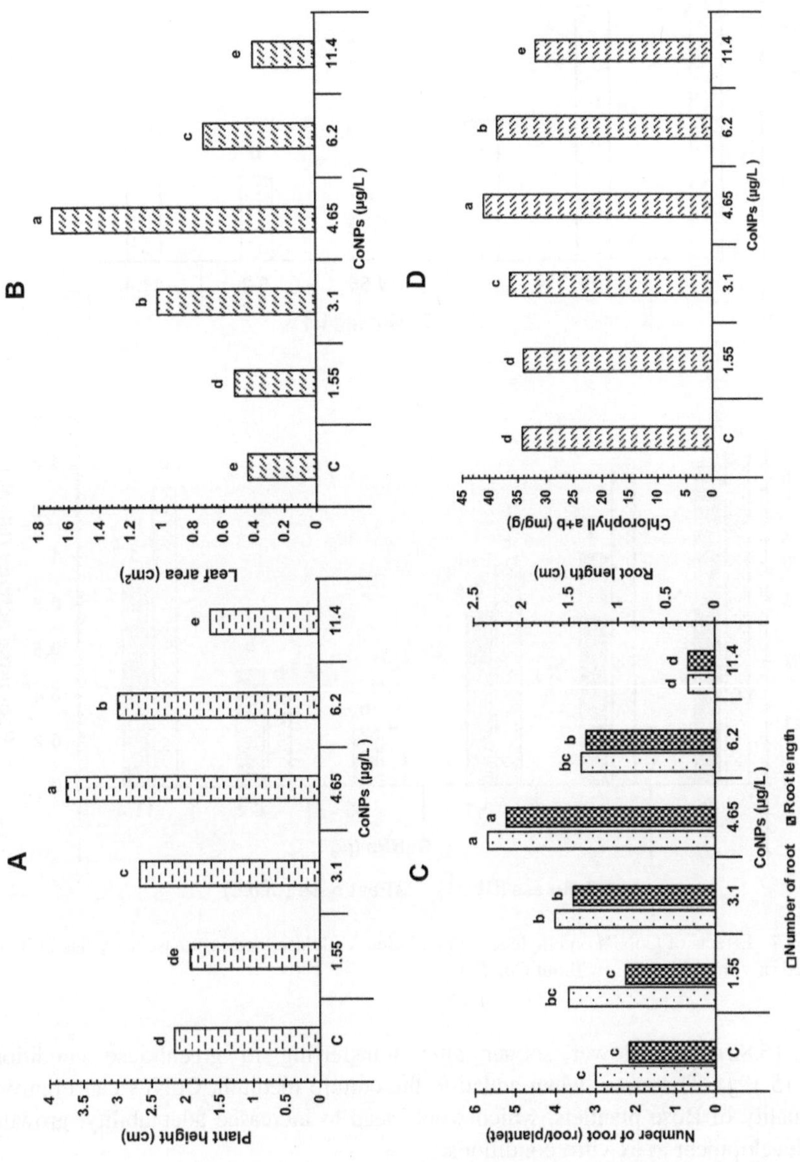

Fig. 15.6 Effects of different CoNPs concentrations on in vitro Rose plantlet growth and development. *C* control (without CoNPs)

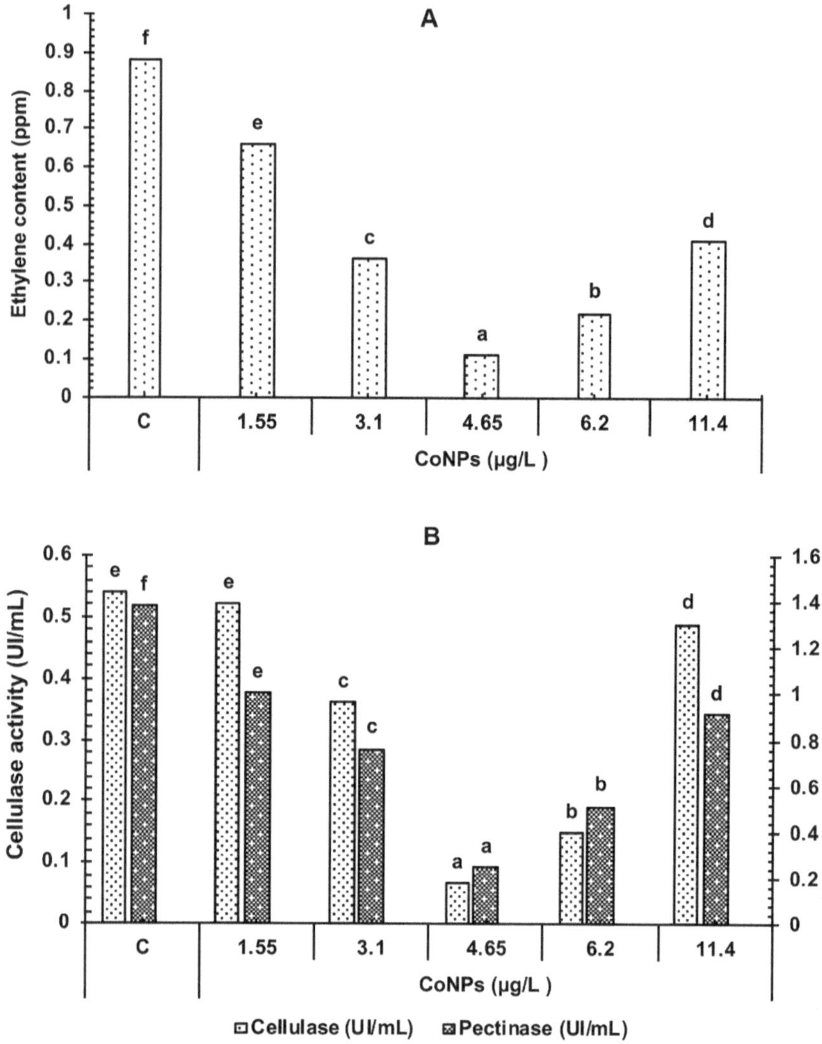

Fig. 15.7 Effects of CoNPs on ethylene accumulation, cellulase, and pectinase activities in Rose cultured in vitro. *C* control (without CoNPs)

(Fig. 15.8d), and flower sooner after transferring to greenhouse condition (Fig. 15.12a). Therefore, when added to the culture medium, CoNPs can improve the quality of Rose plantlets, which would lead to increased adaptability, growth, and development at ex vitro conditions.

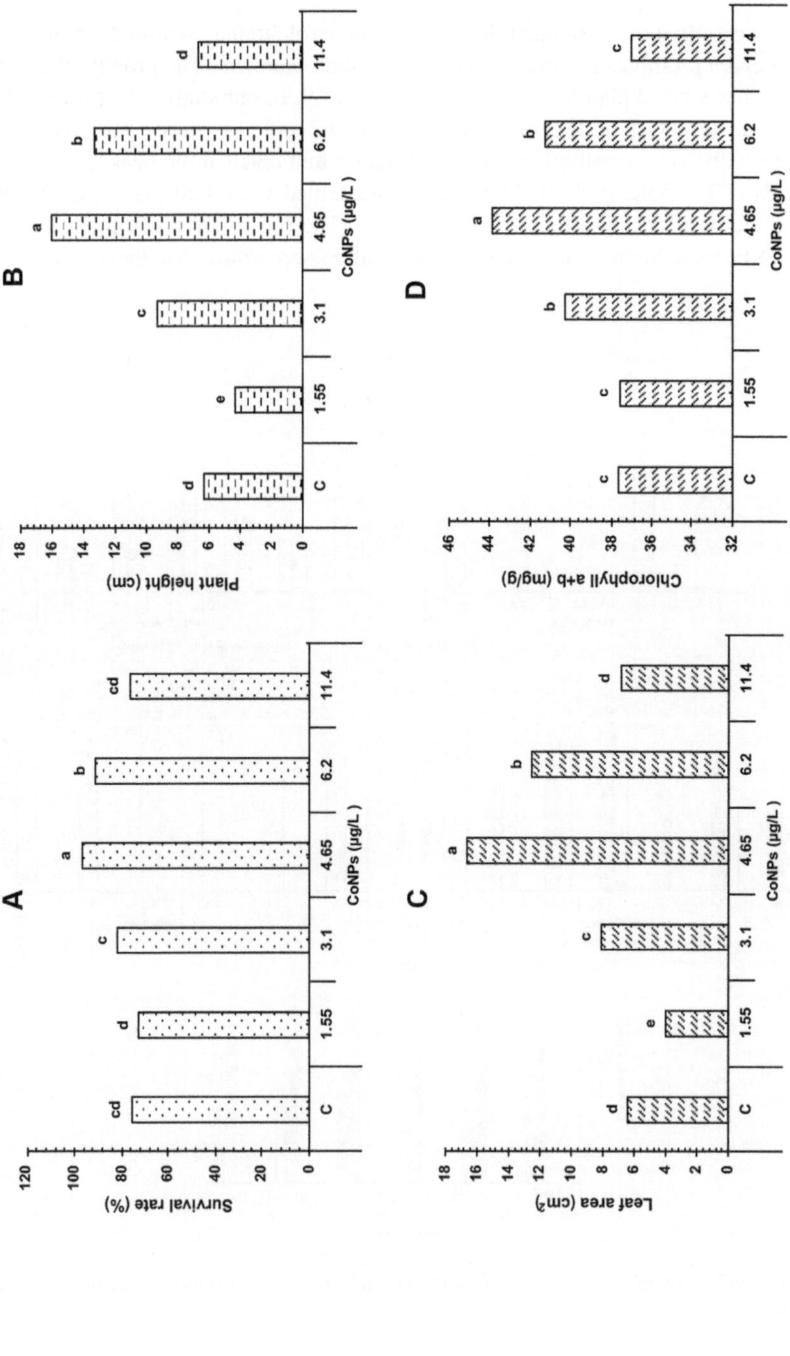

Fig. 15.8 The acclimatization and growth of CoNPs derived- Rose plantlets after 4 weeks in the greenhouse. *C* control (without CoNPs)

15.6 In Vitro Root Growth and Antioxidant Enzyme Activity of Carnation Plantlets Cultured In Vitro Using Iron Nanoparticles

Iron is an essential microelement that has a diverse role to play within the plant. The use of FeNPs promotes in vitro growth of carnation plantlets. The growth of FeNPs derived- 4-week-old plantlets are shown in Fig. 15.9. In our study, plant height, the number of leaves, leaf length, chlorophyll a, b, and dry mass ratio increased proportionally to increasing FeNPs concentration and reached the peak at 4.20 mg/L FeNPs (Fig. 15.9a, c, d, e). Medium supplemented with 4.20 mg/L FeNPs also yielded the best rooting efficiency (Fig. 15.9b). The rooted-shoots in the treatment with FeNPs were higher, more lateral roots, and less browning than those in control.

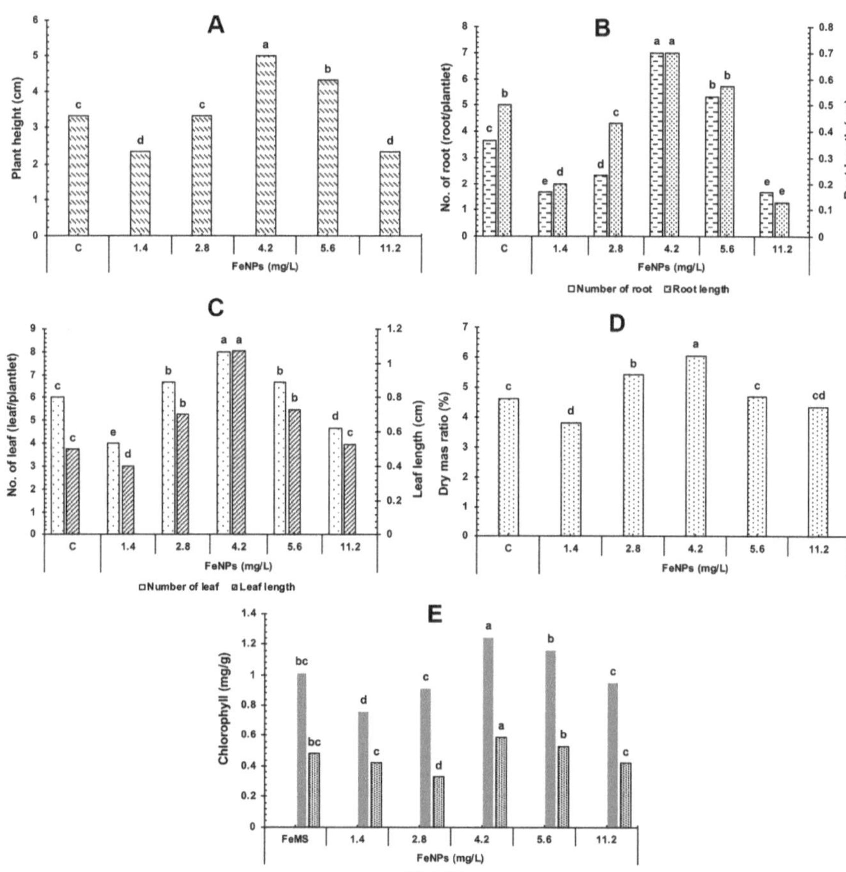

Fig. 15.9 Effects of different FeNPs concentrations on in vitro carnation growth and development. *C* control

Ma et al. (2013) also indicated that the medium added to higher 0.9 mM FeNPs increased the growth and rooted-shoot efficiency in *Populous deltoids* × *Populous nigra* plants. Kim et al. (2014) also obtained that iron stimulated root formation and development in *Arabidopsis thaliana*. FeNPs at 0.1 mM concentration increased chlorophyll content in *Glycine max* leaves (Sheykhbaglou et al. 2018). Moreover, 0.3 mMFeNPs were toxic and decreased the peanut root system's growth and development (Li et al. 2015). In carnation plantlets, the culture medium supplemented with 4.20 mg/L FeNPs improved root growth, overcame the vitrification, improved photosynthetic efficiency and chlorophyll content, and the overall plantlet quality.

Biotic or abiotic stresses can be increased reactive oxygen species (ROS) and induced detoxification responses in plants. Ngan et al. (2020b) showed that superoxide dismutase (SOD), catalase (CAT), and ascorbate peroxidase (APX) play protective roles as scavengers. In this study, the activities of SOD, CAT, and APX of carnation plantlets cultured in a medium supplemented with FeNPs were determined. All three enzymes reached the highest activity within plantlets in the medium containing 4.20 mg/L FeNPs (Fig. 15.10a–c). Our results are the same with the study of Li et al. (2013), where the addition of Fe_2O_3NPs to medium leads to an increase in antioxidant activities in watermelon. Li et al. (2013) noted increasing antioxidant enzyme activities enhanced cell elongation, photosynthesis, respiration, and transpiration. One of the functional roles of FeNPs in the culture of carnation plantlets is to increase antioxidant enzymes' activities. This will undoubtedly aid in combating biotic and abiotic stresses, enhancing the growth of plantlets.

The treatment of plantlets with FeNPs during micropropagation also influenced plantlets' growth at the subsequent nursery stage in the absence of FeNPs. After 4 weeks in the greenhouse, the results showed that carnation plantlets had significant differences in growth characteristics. Ex vitro growth parameters were the highest (Fig. 15.11b–d) in plants previously treated with 4.20 mg/L FeNPs. Furthermore, the survival percentage (91.67%) was also higher than the control (71.67%) (Fig. 15.11a). The flowering of carnation was also recorded after 10 weeks at nursery condition. The flowers were larger in size, and longevity was improved than those in the control treatment (Fig. 15.12c).

In basil, FeNPs positively affected rooting, plant height, chlorophyll content, and dry weight (Peyvandi et al. 2011). The effects of FeNPs on stem elongation, flowering, and harvest stage of *Calendula officinalis* were the highest after adding 0.1 mMFeNPs (Amuamuha et al. 2012). Applying a 0.04% Fe_2O_3NPs solution by spraying on leaves of wheat resulted in higher yield and protein content than those in the control (without Fe_2O_3NPs) (Mitra et al. 2015). The addition of FeNPs at an appropriate concentration is beneficial to plant propagation. Our studies showed that in carnation plantlets, FeNPs promote plantlet growth, limiting vitrification, enhancing acclimatization, and the subsequent growth of carnation plants under greenhouse conditions.

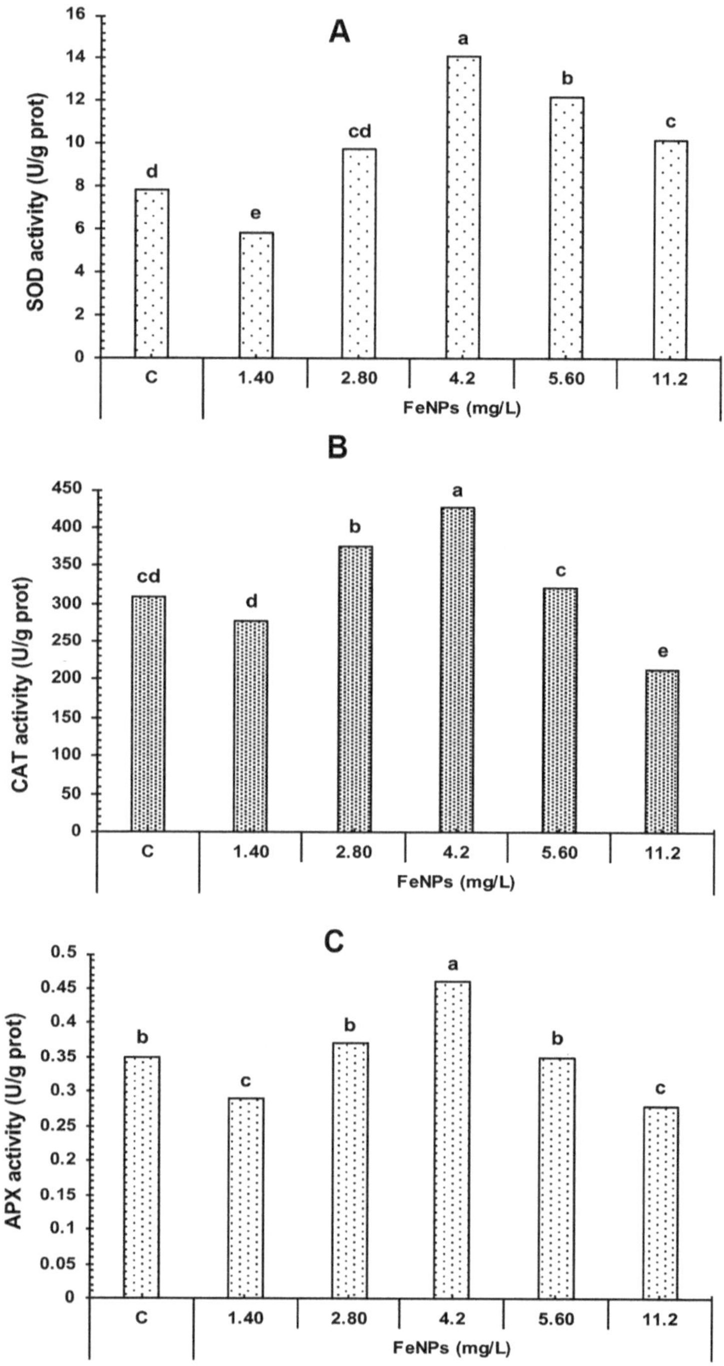

Fig. 15.10 Effect of FeNPs on activity of SOD, CAT, and APX in carnation cultured in vitro. *C* Control, *U* unit, *Prot.* protein

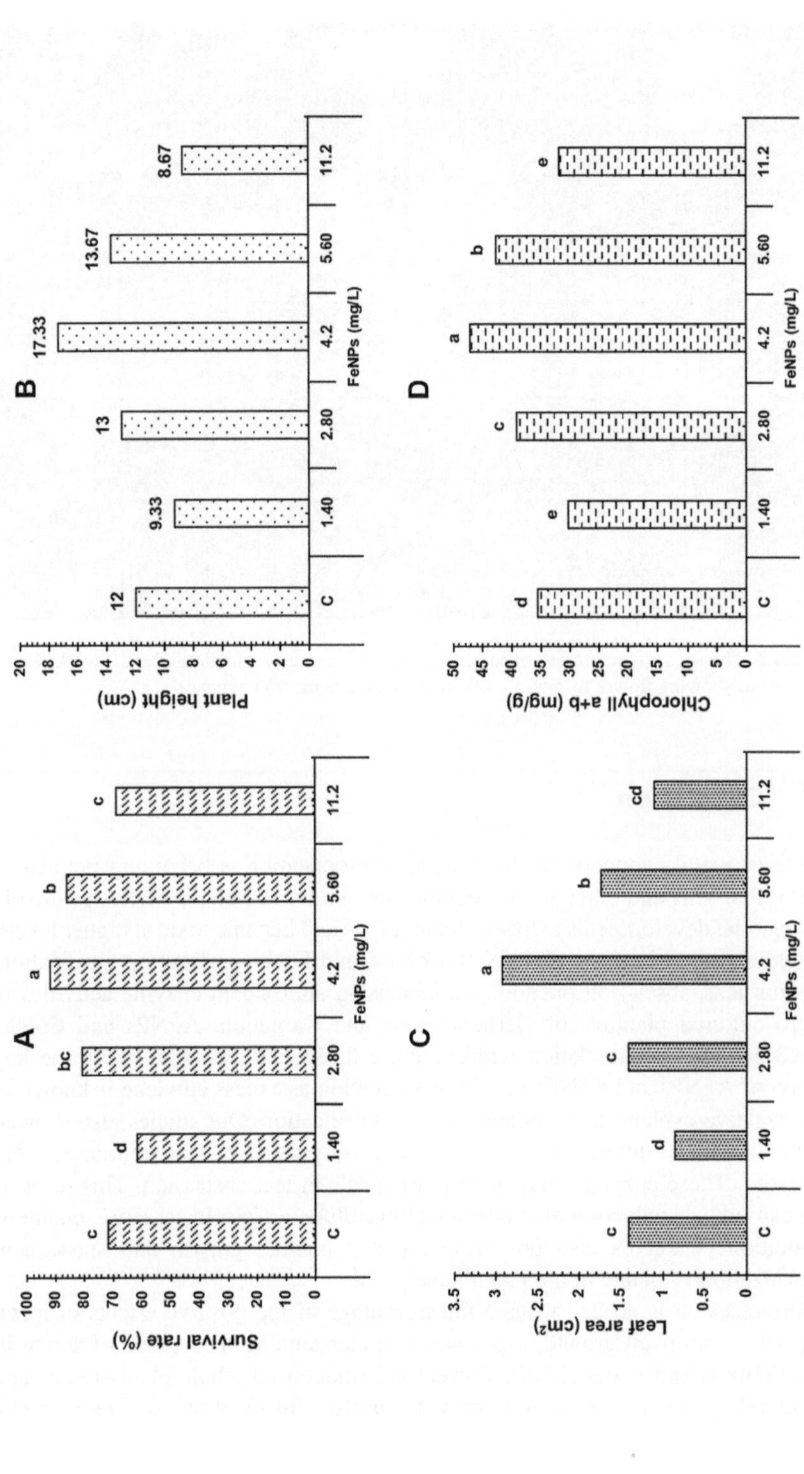

Fig. 15.11 Adaptability and growth of carnation plantlets derived from in vitro medium culture supplemented with FeNPs after 10 weeks in the greenhouse. *C* Control (iron in ½ MS medium)

Fig. 15.12 The subsequent growth of rose, Gerbera, and carnation plantlets after 10 weeks in the nursery stage showing flower formation. (**a**) Rose; (**b**) Gerbera; (**c**) Carnation

15.7 Conclusion

The use of metal nanoparticles in plant micropropagation is becoming popular in recent years. This and other studies demonstrate that metal nanoparticles positively affect plantlet developments at low concentrations and become toxic at higher levels. Our studies showed effects of AgNPs, CoNPs, and FeNPs in limiting vitrification, yellowing leaf, abscission phenomena, increasing antioxidant enzyme activities in in vitro cultured plantlets of Gerbera, Rose and Carnation. AgNPs and CoNPs reduced ethylene accumulation within culture flasks. This is most likely the key function of AgNPs and CoNPs in micropropagation as excess ethylene is known to cause negative explant development, such as vitrification. Our studies also showed that the levels of hydrolytic enzymes, that is, pectinase, and cellulase, were suppressed. These enzymes play an important role in leaf abscission. This result is likely caused by a reduction of ethylene within culture vessels. In carnation plantlets, the addition of FeNPs also positively affected plantlet growth and subsequent acclimatization of plants in the greenhouse.

More research is needed to take full advantage of the positive effects of metal nanoparticles on plant growth, especially in understanding their mode of action in plants (Sanzari and Leone 2019). Current information on whole plant studies and agricultural practices does not provide useful information on how metal nanoparticles function in plant micropropagation. To fully appreciate their functions, we need to focus on specific culture systems and address particular issues and problems in in vitro systems. For example, do metal nanoparticles cause oxidative stress and induce the production of antioxidants through gene activation at the same time? Do they have a direct effect on organogenesis, that is, root formation? Information on the molecular genetics of plant responses after metal nanoparticle application is urgently needed. Many challenges await our attention. At present, one can undoubtedly conclude that metal nanoparticles are useful additives that can promote explant development in vitro.

References

Agustí J, Merelo P, Cercás M, Tadeo F (2009) Comparative transcriptional survey between laser-micro dissected cells from laminar abscission zone and petiolar cortical tissue during ethylene-promoted abscission in citrus leaves. BMC Plant Biol 9:1–20

Amuamuha L, Pirzad A, Hadi H (2012) Effect of varying concentrations and time of nanoiron foliar application on the yield and essential oil of Pot marigold. IRJABS 3(10):2085–2090

Bahmani R, Karami O, Gholami M (2009) Influence of carbon sources and their concentrations on rooting and hyperhydricity of apple rootstock MM.106. Appl Sci 6(11):1513–1517

Banerjee J, Kole C (2016) Plant nanotechnology: an overview on concepts, strategies, and tools. In: Kole C, Kumar DS, Khodakovskaya MV (eds) Plant nanotechnology: principles and practices. Springer, Switzerland, pp 1–14

Bayda S, Adeel M, Tuccinardi T, Cordani M, Rizzolio F (2019) The history of nanoscience and nanotechnology: from chemical-physical applications to nanomedicine. Molecules 25(1):112

Bhojwani SS, Dantu PK (2013) Plant tissue culture: an introductory text. Springer, India, pp 260–263

Brown KM (1997) Ethylene and abscission. Physiol Plant 100:567–576

Chang C (2016) Q&A: How do plants respond to ethylene and what is its importance? BMC Biol 14(1):1–7

Chau H, Bang L, Buu N, Dung T, Ha H, Quang D (2008) Some results in manufacturing of nanosilver and investigation of its application for disinfection. Adv Nat Appl Sci 9(2):241–248

Cristescu SM, Mandon J, Arslanov D, De Pessemier J, Hermans C, Harren FJ (2012) Current methods for detecting ethylene in plants. Ann Bot 111(3):347–360

Fedlheim DL, Foss CA (2001) Metal nanoparticles: synthesis, characterization, and applications. CRC Press, Boca Raton, USA, pp 289–312

Fouad AS, Hafez RM (2018) Effect of cobalt nanoparticles and cobalt ions on alkaloids production and expression of CrMPK3 gene in Catharanthusroseus suspension cultures. Cell MolBiol (Noisy-le-Grand, France) 64(12):62–69

Gaspar T, Kevers C, Debergh P, Maene L, Paques M, Boxus P (1987) Morphological, physiological and ecological aspects. In: Bonga JM, Durzan DJ (eds) Cell and tissue culture in forestry I. Minnesota Publishers, Dordrech, The Netherlands, pp 152–166

Gopinath K, Gowri S, Karthika V, Arumugam A (2014) Green synthesis of gold nanoparticles from fruit extract of *Terminaliaarjuna*, for the enhanced seed germination activity of *Gloriosasuperba*. J Nanostructure Chem 4(3):1–11

Goth L (1991) A simple method for determination of serum catalase activity and revision of reference range. Clin Chim Acta 196(2-3):143–151

Jaleel CA, Changxing Z, Jayakumar K, Iqbal M (2009) Low concentration of cobalt growth, biochemical constituents, mineral status and yield in *Zea mays*. J Sci Res 1(1):128–137

Kharrazi M, Tehranifar A, Nemati S, Bagheri A, Sharifi A (2011) In vitro culture of carnation (*Dianthus caryophyllus* L.) focusing on the problem of vitrification. J Biol Environ Sci 5(13):1–6

Khatun M, Rahman MM, Roy PK (2013) In vitro regeneration and field evaluation of carnation (*Dianthus caryophyllus* L.) through shoot tip and node culture. J Appl Res Technol 9(1):93–99

Kim JH, Lee Y, Kim EJ, Gu S, Sohn EJ, Seo YS, Chang YS (2014) Exposure of iron nanoparticles to *Arabidopsis thaliana* enhances root elongation by triggering cell wall loosening. Environ Sci Technol 48(6):3477–3485

Kim DH, Gopal J, Sivanesan I (2017) Nanomaterials in plant tissue culture: the disclosed and undisclosed. RSC Adv 7:36492–36505

Kole C, Kole P, Randunu KM, Choudhary P, Podila R, Ke PC (2013) Nanobiotechnology can boost crop production and quality: first evidence from increased plant biomass, fruit yield and phytomedicine content in bitter melon (*Momordicacharantia*). BMC Biotechnol 3(1):37–47

Kole C, Kumar DS, Khodakovskaya MV (eds) (2016) Plant nanotechnology: principles and practices. Springer, Switzerland, pp 263–303

Kulkarni N, Muddapur U (2014) Biosynthesis of metal nanparticles: a review. J Nanotechnol 510246:1–8

Kumar V, Parvatam G, Ravishankar GA (2009) $AgNO_3$—A potential regulator of ethylene activity and plant growth modulator. Electron J Biotechnol 12(2):1–15

Lai CC, Lin HM, Nalawade SM (2005) Hyperhydricity in shoot cultures of *Scrophulariayoshimurae* can be effectively reduced by ventilation of culture vessels. J Plant Physiol 162(2):355–361

Li J, Chang PR, Huang J, Wang Y, Yuan H, Ren H (2013) Physiological effects of magnetic iron oxide nanoparticles towards watermelon. J Nanosci Nanotechnol 13(8):5561–5567

Li X, Yang Y, Gao B, Zhang M (2015) Stimulation of peanut seedling development and growth by zero-valent iron nanoparticles at low concentrations. PLoS One 10(4):1–12

Lichtentaler HK, Wellburn AR (1985) Determination of total carotenoids, chlorophyll aand bof leaf in different solvents. Biochem Soc Trans 11(5):591–592

Ma X, Geiser-Lee J, Deng Y, Kolmakov A (2010) Interactions between engineered nanoparticles (ENPs) and plants: phytotoxicity, uptake and accumulation. Sci Total Environ 408(16):3053–3061

Ma X, Gurung A, Deng Y (2013) Phytotoxicity and uptake of nanoscale zerovalent iron (nFe) by two plant species. Sci Total Environ 443(1):844–849

MacDonald MT, Lada RR, Dorais M, Pepin S (2011) Endogenous and exogenous ethylene induces needle abscission and cellulase activity in post-harvest balsam fir (*Abiesbalsamea* L.). Trees 25(5):940–947

Mahmood S, Reza R, Hossain G, Hauser B (2017) Response of cytokinins on in vitro shoot multiplication of Rose cv. Frisco. Res Rev: J AgricSciTechnol 5(2):6–11

Marklund S, Marklund G (1974) Involvement of the superoxide anion radical in the autooxidation of pyrogallol and a convenient assay for superoxide dismutase. Eur J Biochem 47(3):469–474

Merelo P, Agustí J, Arbona V (2017) Corrigendum: cell wall remodeling in abscission zone cells during ethylene-promoted fruit abscission in citrus. Front Plant Sci 8(3):23–29

Mitra B, Payam M, Behzad S (2015) The effect of iron nanoparticles spraying time, concentration on wheat. Biol Forum: Int J 7(1):679–683

Murashige T, Skoog F (1962) A revised medium for rapid growth and bio-assays with tobacco tissue cultures. Plant Physiol 1(3):473–497

Nakano Y, Asada K (1981) Hydrogen peroxide is scavenged by ascorbate specific peroxidase in spinach chloroplasts. Plant Cell Physiol 22(5):867–880

Ngan HTM, Cuong DM, Tung HT, Nghiep ND, Le BV, Nhut DT (2020a) The effect of cobalt and silver nanoparticles on overcoming leaf abscission and enhanced growth of Rose (*Rosahybrida*L. 'Baby Love')plantlets cultured in vitro. Plant Cell Tissue Organ Cult 141(2):393–405

Ngan HTM, Tung HT, Le BV, Nhut DT (2020b) Evaluation of root growth, antioxidant enzyme activity and mineral absorbability of carnation (*Dianthus caryophyllus* "Express golem") plantlets cultured in two culture systems supplemented with iron nanoparticles. Sci Hortic 272:109612

Peyvandi M, Parandeh H, Mirza M (2011) Comparing the effect of nano iron chelate and iron chelate on growth parameters, antioxidant enzymes activity of basil (*Ocimumbasilicum* L). Iran J Mod Cell Mol Biotechnol 1:89–99

Phan C, Letouze R (1983) A comparative study of chlorophyll, phenolic and protein contents, and of hydroxycinnamate: CoA ligase activity of normal and 'vitreous' plants (*Prunusavium* L.) obtained in vitro. Plant Sci Lett 31(2):323–327

Saha N, Gupta SD (2018) Promotion of shoot regeneration of *Swertiachirata* by biosynthesized silver nanoparticles and their involvement in ethylene interceptions and activation of antioxidant activity. Plant Cell Tissue Organ Cult 134(2):289–300

Salama HMH (2012) Effects of silver nanoparticles in some crop plants, common bean (*Phaseolus vulgaris* L.) and corn (*Zea mays* L.). Int Res J Biotechnol 3(10):190–197

Sanzari I, Leone A, Ambrosone (2019) Nanotechnology in plant science: to make a long story short Front Bioeng Biotech 7: Article 120

Seif SM, Sorooshzadeh A, Rezazadehs H, Naghdibadi HA (2011) Effect of nanosilver and silver nitrate on seed yield of borage. J Med Plant Res 5(2):171–175

Senapati S, Rout G (2008) Study of culture conditions for improved micropropagation of hybrid rose. Hortic Sci 35(1):27–34

Sheykhbaglou R, Sedghi M, Fathi-Achachlouie B (2018) The effect of ferrous nano-oxide particles on physiological traits and nutritional compounds of soybean (*Glycine max* L.) seed. An Acad Bras Cienc 90(1):485–494

Siddiqui MH, Al-Whaibi MH, Mohammad F (2015) Nanotechnology and plant sciences. Springer International Publishing Switzerland, London, 289p

Taran NY, Gonchar OM, Lopatko KG, Batsmanova LM, Patyka MV, Volkogon MV (2014) The effect of colloidal solution of molybdenum nanoparticles on the microbial composition in rhizosphere of *Cicer arietinum* L. Nanoscale Res Lett 9(1):289–297

Thao NP, Khan MIR, Thu NBA, Hoang XLT, Asgher M, Naffees AK, Son PTL (2015) Role of ethylene and its cross talk with other signaling molecules in plant responses to heavy metal stress. Plant Physiol 169(1):73–84

Tholen D, Poorter H, Voesenek L (2006) Ethylene and plant growth. In: Naffes AK (ed) Ethylene action in plants. Springer-Verlag, Berlin, Heidelberg, pp 35–46

Vakhrouchev AV, Golubchikov VB (2007) Numerical investigation of the dynamics of nanoparticle systems in biological processes of plant nutrition. J Phys Conf Ser 61(1):31–35

Vatanparast M, Hosseininaveh V, Ghadamyari M, Sajjadian SM (2014) Plant cell wall degrading enzymes, pectinase and cellulase, in the digestive system of the red palm weevil *Rhynchophorusferrugineus*. Plant Prod Sci 50(4):190–198

Yen NTK, Huy NP, Cuong HV, Linh NTN, Nam NB, Luan VQ, Nhut DT (2013) Effect of activated charcoal and aerated culture on growth and development growth of in vitro and ex vitro *Gerbera jamesonii*. Vietnam J Sci Tech 51(4):435–446

Zhang YP, Hong J, Ye X (2009) Cellulase assays. In: Mielenz JR (ed) Biofuels: Methods in molecular biology (Methods and Protocols), vol 581. Humana Press, Totowa, pp 213–231

Ziv M (1991) Quality of micropropagated plants-vitrification. In Vitro Cell Dev Biol 27(2):64–69

A Protocol of Shoot Regeneration and Polyploid Plantlet Production in *Paphiopedilum villosum*

16

Duong Tan Nhut, Do Thi Thuy Tam, Vu Quoc Luan, Nguyen Thi Thanh Hien, and Hoang Thanh Tung

Abstract

This chapter reports a protocol of polyploid induction of *Paphiopedilum villosum* using colchicine as a mutagen agent. Adventitious shoots of *P. villosum* were induced from nodal explants (from elongated in vitro shoots) by using cytokinins (BA, KIN, and TDZ). The shoot (1.5 cm in high) was exposure to different concentrations and durations of colchicine solutions and transferred into the in vitro rooting medium. The results showed that the treatment of 50 μM colchicine in 6 days gave a polyploidy induction rate of 19.88%. The flow cytometric analysis and chromosome counts of root tip squash were used for the analysis of tetraploids and mixoploids. The shoots derived from stem nodes treated with colchicine solution were effective for the production of polyploid plantlets for subsequent propagation of *P. villosum* and other endangered orchids.

Keywords

Chromosome count · Colchicine solution · Flow cytometric · Stem node · *Paphiopedilum villosum* · Polyploidy

D. T. Nhut (✉) · D. T. T. Tam · V. Q. Luan · H. T. Tung
Tay Nguyen Institute for Scientific Research, VAST, Dalat City, Vietnam

N. T. T. Hien
Ton Duc Thang University, Hochiminh City, Vietnam

D. T. Nhut et al. (eds.), *Plant Tissue Culture: New Techniques and Application in Horticultural Species of Tropical Region*,
https://doi.org/10.1007/978-981-16-6498-4_16

16.1 Introduction

Polyploidization plays an important role in the production of new, premium plant varieties, especially in the orchid industry. Artificial polyploid induction is of particular importance because polyploid orchids can be created from their diploid varieties (Nakasone 1960). Although new orchids produced from cross-lines and between species, a high rate of infertility is common in derived plants of these crosses (Watrous and Wimber 1988). In contrast, autopolyploidy orchids could maintain fertility, especially in heterozygous hybrid plants. This is because of the replication of chromosome in heterozygous plants, which can restore fertility (Ranney 2006).

16.2 Polyploidy in Plant Breeding

Ramsey and Schemske (1998) pointed out that polyploidy has a prominent role in the evolution of new species of higher plants. Several studies have shown that polyploid plants have many advantages such as faster metabolism and more secondary metabolites (Dhawan and Lavania 1996), larger nutrient storage organs (Adaniya and Shirai 2001), disease resistance and better tolerance to adverse environmental conditions (Song et al. 2012). Polyploid plants are often used for breeding new cultivars than diploid plants because of their higher yields and better quality. So far, many polyploid varieties have been produced successfully worldwide (Song et al. 2012).

There are different methods to induce polyploidy in plants. The most common way is using chemicals such as colchicine, oryzalin, or amiprophos-methyl (Niazian and Nalousi 2020). Among them, colchicine is the most commonly used compound as it can give high polyploid induction when plants are treated in vitro (Ganga and Chezhiyan 2002; Sun et al. 2009; Abu-Qaoud and Shtaya 2014). Colchicine can inhibit kinetochore fibers' formation to stop mitosis in the metaphase stage when chromosomes have duplicated, but the cell division has not yet occurred. The disruption in cell division leads to the formation of polyploid cells. The subsequent development of these cells with double the number of chromosomes can affect the size and number of stomata, pollen diameter, and other morphological characteristics of derived plants (Blakeslee and Avery 1937).

Many food crops and fruit trees have been successfully produced polyploid cultivars (Udall and Wendel 2006; Sattler et al. 2016). For example, colchicine treatment from somatic embryo culture resulted in tetraploid individuals of grapevine (*Vitis vinifera* L.) (Yang et al. 2006). The effect of simultaneous treatment of 0.05% colchicine and 2% DMSO at 24, 48, and 72 h under different temperatures was tested in the induction of autotetraploid pea. The highest survival rate was obtained after 48 h of treatment and under 15–16 °C, and polyploid induction was also the highest under these conditions (9.5%) (Wang et al. 2009). Treatment with colchicine can also produce doubled haploids (diploids derived from haploid pollen) in wheat microspore culture, where the average frequency was 81.73% (Islam 2010).

Studies on producing polyploidy of ornamental plants have also been conducted. Important polyploid lines of Chrysanthemum were selected via in vitro colchicine treatment (Liu and Gao 2007). Tetraploids of *Alocasia* × *Amazonica* were also successfully induced through colchicine and oryzalin treatments (Thao et al. 2003). In orchids, tetraploid plants were produced from their diploid varieties in *Cymbidium* (Wimber and van Cott 1966), *Vanda* (Sanguthai et al. 1973), *Dendrobium* (Sanguthai et al. 1973; Jiang et al. 2014), *Phalaenopsis* (Chen et al. 2009; Azmi et al. 2016), and *Paphiopedilum* (Watrous and Wimber 1988). Polyploid *Dendrobium officinale* was induced by colchicine (Jiang et al. 2014). Results of polysaccharide accumulation analysis indicated that the polysaccharide contents in leaf, stem, and protocorm-like body (PLB) of autotetraploid were 1.83, 1.31, and 1.95-fold to diploid, respectively. However, a comparison of the growth rate showed mutant plantlets were significantly lower than the control (Jiang et al. 2014).

Various methods have been used to inducing autopolyploidy in plants. Efficacy can be improved by micropropagation with different in vitro plants material sources, such as axillary bud, shoot, tissue, cell, and other organs. These methods are more suitable for polyploidy as compared with in vivo shoot treatments (Predieri 2001). However, polyploidy induction in in vitro orchid has not been widely used.

16.3 Polyploidy Induction of *Paphiopedilum villosum*: A Case History

Paphiopedilum, one of the most charming and unique flowers, has shoe-shaped labellum (lip), so, the commercial name is Venus slipper orchid. Due to their high economic value, they are over-exploited in the wild for commercial purposes. Therefore, most *Paphiopedilum* species and in particular *P. villosum* are considered by CITES to be rare and endangered orchid species. *P. villosum* is found in China, India, Myanmar, and Vietnam. It has a long life cycle, very low fruiting rate, and seeds germinate in difficult natural conditions. Therefore, researching and propagating this species in vitro are essential solutions to produce a large number of plantlets in a relatively short time (Long et al. 2010).

In vitro propagation of *Paphiopedilum* orchids was recently summarized by Zeng et al. (2016). This chapter presents information on polyploidy induction of *P. villosum* using a colchicine treatment of adventitious shoots derived from stem node cultures. Various methods such as stomatal size measurement, flow cytometric analysis, and chromosomal counting for ploidy determination are also outlined. The purpose of this chapter is to provide a basic method for generating *P. villosum* polyploid plantlets.

16.3.1 Plant Material and Methodology

The *P. villosum* plants were maintained at the greenhouse of the Tay Nguyen Institute for Scientific Research (Dalat city, Lam Dong province, Vietnam). Flowers

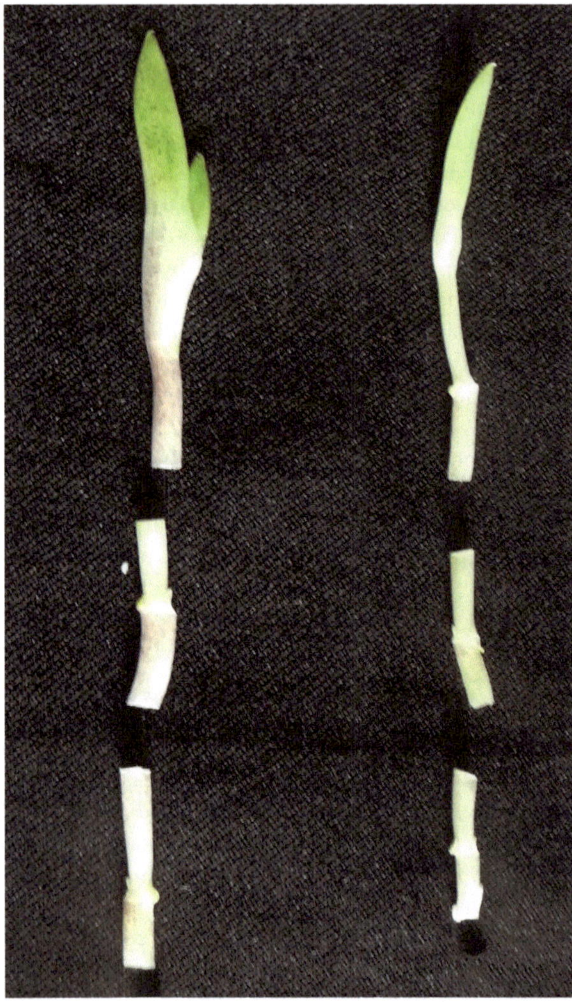

Fig. 16.1 Elongation of *Paphiopedilum villosum* shoots after 3 months of in vitro culture

were hand pollinated and mature capsules were harvested. Mature capsules were first surface-sterilized with 20% (v/v) Sunlight solution (Vietnam) and placed under running tap water for 2–3 h followed by soaking in 1 mg/mL streptomycin solution for 30 min. Seeds of mature capsules were cultured on Knudson C medium (Knudson 1946). The 3-month young shoot derived from seedling was excised and cultured on MS medium (Murashige and Skoog 1962) added to 2.0 mg/L BA (Nhut et al. 2005) to generate 2-month shoot with uniform-sized.

Since the newly formed shoots were highly condensed and individual nodes were difficult to separate, the shoots were induced to elongate using in vitro stem node elongation method (Luan et al. 2015) (Fig. 16.1). Young shoots with 5 leaves were cultured on SH medium (Schenk and Hildebrandt 1972) added of 0.5 mg/L 6-Benzylaminopurine (BA), 0.5 mg/L α-Naphthalene acetic acid (NAA), 30 g/L

Fig. 16.2 Procedure for explant sterilization and shoot elongation

sucrose, 9.0 g/L agar and 1.0 g/L activated charcoal (AC). The explants were placed under darkness condition without subculture in 30, 60, 90, and 120 days to investigate the effect of light conditions on shoot elongation. For control, the explants were cultured under fluorescent lights condition (Fig. 16.2).

The elongated shoots were used as materials for the production of additional in vitro shoots for polyploid induction. This step was essential to generate a large number of shoots for polyploid induction study.

16.3.1.1 Production of In Vitro Shoots for Polyploid Induction

Elongated shoots were cut into separate nodes and cultured on the SH medium added to 1.0 mg/L Kinetin (KIN) or 1.5 mg/L Thidiazuron (TDZ) or 0.5 mg/L BA. For the control, stem nodes were cultured on a plant growth regulator-free SH medium. The treatments were put in culture room at 25 ± 2 °C under the fluorescent light (40–45 μmol m^{-2} s^{-1}) with a 16 h photoperiod and humidity of 55–60% (Huy et al. 2019).

Data on shoot induction and the number of shoots per stem node were recorded after 2 months of culture. Shoot multiplication of nodes in different positions, shoot height, and the number of shoots were also compared. The shoot multiplication efficiencies were compared by shoot multiplication coefficients (SMCs) defined as:

$$SMCs = [(\%\text{viable explants}) \times (\text{Average number of shoots per explant})]/100$$

16.3.1.2 Colchicine Treatment

The colchicine solution was prepared by dissolving colchicine (Sigma-AldrichTM, INC., MO 63178 USA) in a volume of not less than 90% alcohol, add sterile water to adjust to desired concentration; and then filter-sterilize by Millipore® Sterivex™ with a diameter of 0.22 μm (Sigma-Aldrich, Germany). Colchicine solutions of different concentrations were added to autoclaved liquid medium vessels.

The *P. villosum* shoots (1.5 cm in length) were culture in liquid SH medium added to colchicine solutions (2, 10, or 50 μM). Treatments were placed under darkness condition in 3, 6, or 9 days. At the end of the treatment period, shoots were 3-time washed with sterile distilled water in a laminar flow cabinet and cultured on the SH medium supplemented with 0.5 mg/L BA, 0.5 mg/L NAA, 9.0 g/L agar, 30 g/L sucrose, and 1.0 g/L AC (Luan et al. 2015). For the control, colchicine-untreated shoots were used (Fig. 16.3). The survival percentage of shoots was determined after 2 months of culture.

16.3.1.3 Stomatal Guard Cell Observation

Polyploid plantlet was recorded by examining the guard cells length of new leaves from each plantlet (each treatment and the control) (Russell 2004). If guard cell length was ≥1.25× the control, a plant would be determined that it could be polyploid. To determine the size of guard cells, epidermal peels from new leaves of plantlets were taken from the leaves' abaxial surface using forceps, then placed on a slide with a drop of water, a coverslip was placed on it and observed under a microscope (OLCH30/R, Olympus, Tokyo, Japan) at 400× magnification. The guard cells length (twenty stomata of each leaf and two leaves from each plantlet)

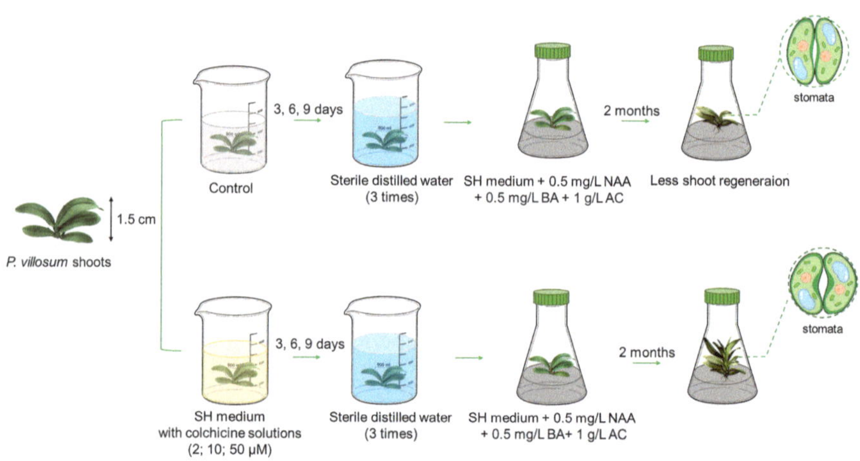

Fig. 16.3 Explant treatment procedure with colchicine solutions

were checked using an eye-piece micrometer. The polyploid induction rate was determined as follows:

$$\text{Polyploid induction rate } (\%) = \frac{\text{No.of Polyploid plantlets}}{\text{No.of shoots used}} \times 100$$

16.3.1.4 Flow Cytometric Analysis

The ploidy levels of plantlets were determined by flow cytometric analysis using Ploidy Analyzer PA I (Partec, Germany) based on the modified method of Dpooležel et al. (1989). The nuclear DNA contents plantlets were expressed relative to the control value (diploid *P. villosum*). Fresh weight (50 mg) of leaf tissues from each polyploid plantlet were excised (5 × 5 mm) and placed on a petri dish on ice. Solution A (1 mL) was consisting of cold 14.3 mL $MgSO_4$ solution, 15 mg dithiothreitol, 300 μL Propidium Iodide stock solution, and 375 μL Triton X-100 stock solution was added. Explants were cut into tiny pieces by a sharp blade, the extracts were filtered through a 33 μm nylon filter and centrifuged at 15,000 rpm for 15–20 s. The unused supernatant, and the residue was dissolved in 200 μL of solution B including 3 mL solution A, 7.5 μL RNase A solution, and 3.0 CRBC (chicken red blood cell) and incubated at 37 °C in 15 min before analysis.

16.3.1.5 Chromosome Counting

For chromosome counting, root tips from in vitro plantlets were excised and collected at approximately 10:00 h. The explants were exposed in 2% colchicine (w/v) in 3 h. The root tips were pretreated in an aqueous solution containing 0.075 M potassium chloride (KCl) and fixed in Carnoy solution for 40 min (Miething et al. 2006). The root tips were then washed with 90% ethanol (v/v), and stored in 70% ethanol (v/v). Just prior to staining, the root tips were hydrolyzed in 1 N HCl in 20 min before staining, then rinsed with 45% acetic acid (v/v). The root tip meristems (1 mm-long) were stained with acetocarmine, then they were squashed on a slide and covered with a coverlip. The spreadout cells were examined under a microscope to count the chromosomes (Fig. 16.4).

16.3.1.6 Polyploid Cloning and Acclimatization

Individual polyploid lines were obtained from proliferating polyploid plantlets (diploid plantlets removed). The plantlets with vigorous roots of each polyploid line were planted in pots containing sphagnum moss for acclimatization and growth.

16.3.1.7 Statistical Analysis

The experiment was repeated three times with 5 vessel culture (3 explants each). The mean of each treatment method was followed one-way ANOVA and compared by Duncan's multiple range test at $p < 0.05$ (Duncan 1955).

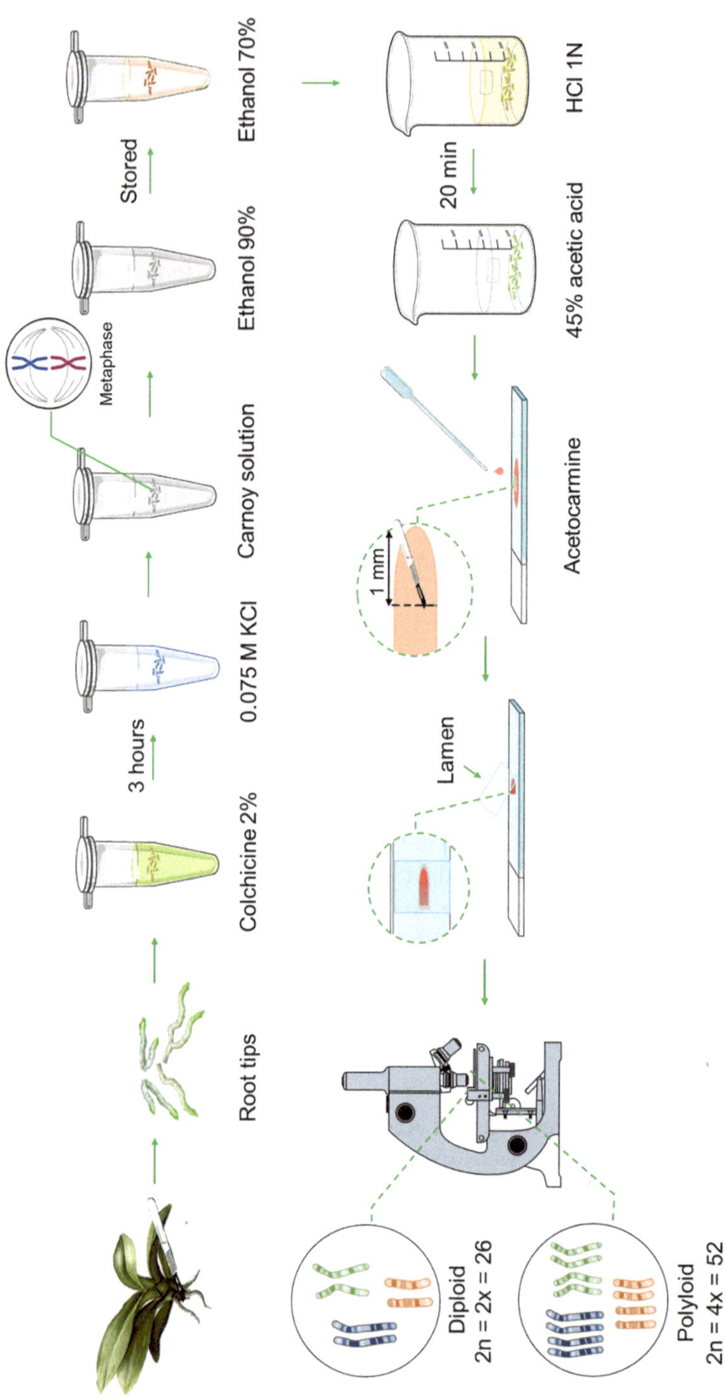

Fig. 16.4 Chromosome counting procedure

16.3.2 Results

16.3.2.1 Shoot Elongation of In Vitro P. villosum in the Darkness

The results in this study showed that darkness condition was exploited for in vitro *P. villosum* shoot elongation. The young-shoot placed in the darkness in 30, 60, 90, or 120 days showed a significant difference in shoot height and number of new leaf in comparison with control treatment (light condition) (Fig. 16.5). The first node initially elongated with young green leaves formed after 30 days of culture. After 60 days, shoot height increased with internodes at second position (from top) elongated. The results observed after 90 days of culture showed that shoot length rapidly increasing. The highest shoot height and number of newly formed leaves were recorded after 120 days of culture in the darkness though the leaves did not grow considerably and with the formation of new roots (Fig. 16.5). A number of studies reported that have various factors such as BA, IAA, phytochrome system, or a novel triazole derivative with coumarin moiety (YCZ-FL) impacting on shoot elongation in dark condition (Tomoki et al. 2014)

16.3.2.2 In Vitro Shoot Production Process

The shoot was elongated to provide the original material for shoot regeneration and polyploidy induction. Various cytokinins in the culture medium stimulated axillary bud formation of *P. villosum* whose shoot induction rate and number of shoots per

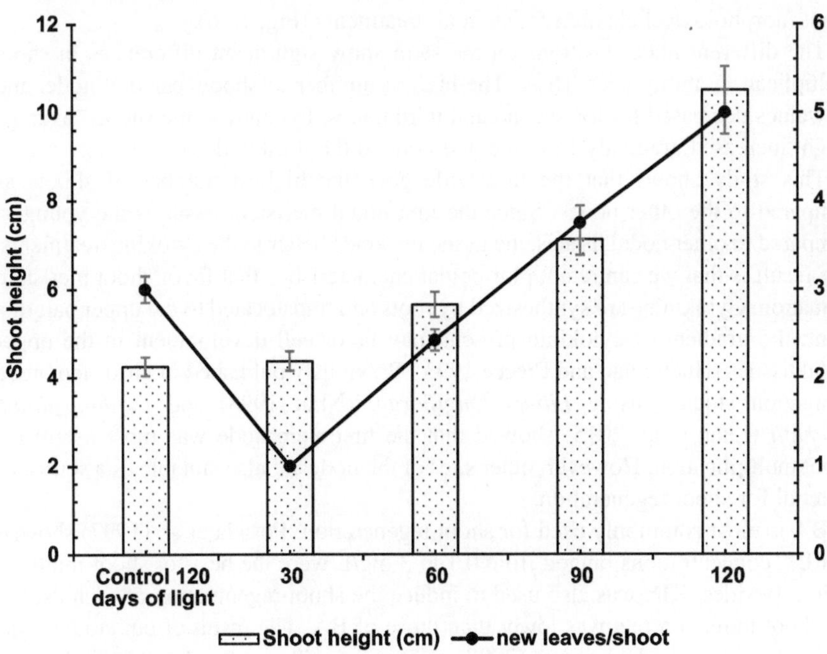

Fig. 16.5 Effects of darkness condition on shoot elongation

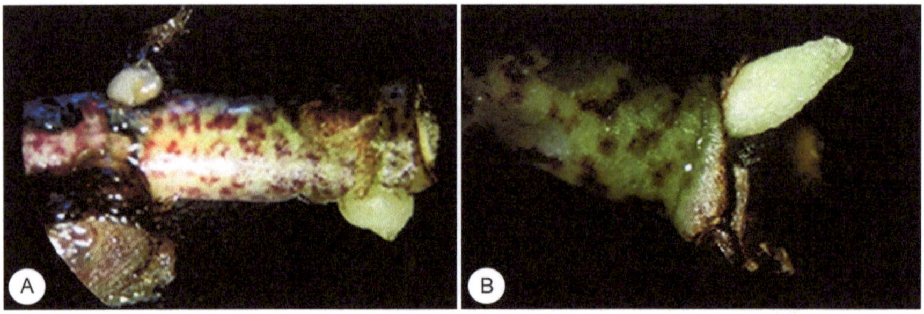

Fig. 16.6 Axillary bud formed on the stem node. (**a**) 1.0 mg/L BA; (**b**) 0.5 mg/L TDZ

node were significantly higher than the control treatment. Accordingly, the different concentrations of cytokinin resulted in significant differences in 2-month shoot regeneration. The number of shoots and shoot regeneration rate increased proportionally to the increase of BA (from 0 to 1.0 mg/L) and reached the optimum at 1.0 mg/L BA treatment. At a higher concentration of BA, shoot regeneration was decreased. The shoot induction rates in KIN treatment tended to be lower than that of BA treatment while the mean values of a number of shoots per node were higher in KIN. The regeneration shoot from stem node was optimal on medium supplemented with 0.5 mg/L TDZ, and shown the highest shoot induction rate, number of shoots, and shoot multiplication coefficient in all treatments. There were no abnormalities in shoot morphological characteristics in all treatments (Fig. 16.6).

The different node positions on the stem show significant differences in shoot multiplication ability (Fig. 16.7). The highest number of shoots per first node, and the values decreased for the second and third nodes. In contrast, the shoots' average height increased gradually from the first node to the third node (Fig. 16.7).

This study shows that the first node gave the highest number of shoots as compared to the other nodes. Since the first nodal meristem tissue is the youngest compared to other nodal tissues, the tissue responds better to the cytokine treatments. The result is that we can retain pluripotent characteristics that favor shoot meristem formation. Cytokinins are synthesized by roots and translocated to the upper part of a plant; the gradient of cytokinin present may favor cell development in the upper nodal tissues (Huetteman and Preece 1993). Previously published in vitro stem node elongation studies, using *Lilium longiflorum* (Nhut 1998) and *Paphiopedilum delenatii* (Luan et al. 2015) showed that the first stem node was most useful for shoot multiplication. However, other sites of the node are also suitable as a source of material for shoot regeneration.

BA is most commonly used for shoot regeneration. Posada et al. (1997) showed that BA concentrations ranged from 0.1 to 3 mg/L were the best for shoot multiplication. Besides, KIN was also used to induce the shoot regeneration of orchids, but the shoot induction rate was lower than those of BA. The result of our studies was similar to a study by Hong et al. (2008) on *Paphiopedilum* Alma Gavaert, indicating the concentration of 1.0 mg/L of KIN had the best effect on shoot multiplication.

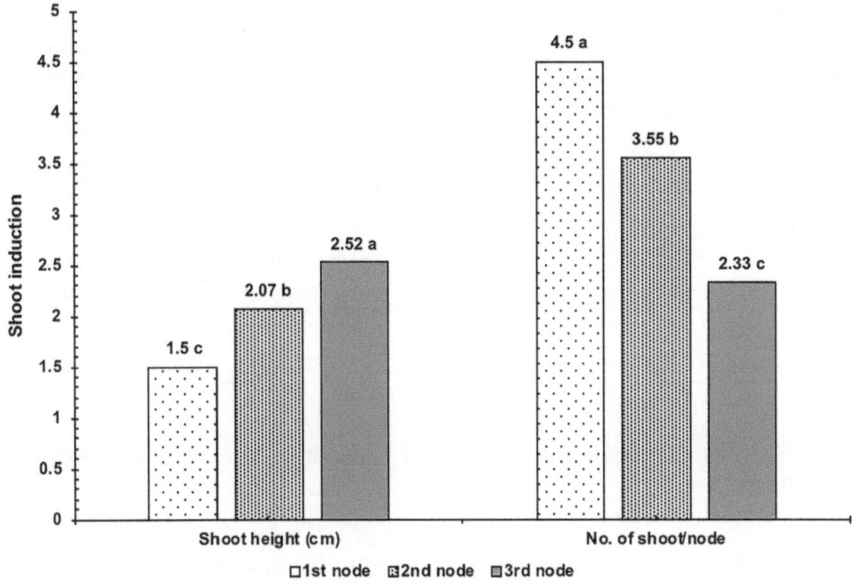

Fig. 16.7 Effects of node position on shoot induction

Ahmad and Faisal 2018 indicated that TDZ was a very effective synthetic cytokinin for shoot regeneration. It is a derivative of urea commonly used in the breeding of woody plants (Huetteman and Preece 1993). A high concentration of TDZ can cause some adverse effects (abnormal shoot morphology and in vitro rooting) (Nhut et al. 2005). Lu (1993) records that explant treated with TDZ over a long time can cause hyperhydricity. In this study, the reduction in the number of shoots was observed along with the increase in TDZ concentrations (better than 0.5 mg/L). The results were similar to a study on shoot micropropagation of *Paphiopedilum* sp. (Huang et al. 2001).

In previous reports on *P. delenatii*, the highest shoot multiplication coefficient was 3.9 using the wounding method combined with a culture in a liquid medium supplemented with 1 mg/L TDZ (Nhut et al. 2005). The regeneration rate reached 33.5% using ex vitro stem node elongation method to obtain materials for in vitro shoot regeneration (Luan et al. 2015). In this study, shoot regeneration rate and shoot multiplication coefficient were higher than previous methods by using in vitro stem nodes.

16.3.2.3 The Colchicine Treatment on Polyploid Production

The colchicine treatment at different concentrations and durations on young shoots significantly impacted shoots' survival rate and polyploid induction (Table 16.1). In general, the survival rate of shoots decreased as the concentration of colchicine and the length of treatment increased. However, there was almost no significant

Table 16.1 Effects of colchicine treatment on shoot polyploid plantlet after 2 months of culture

Colchicine treatment (day)	Concentration of colchicine (μM)	Number of treated shoots per repetition	Percentage of survival shoots (%)	Polyploid induction (%)
0	0	45	100 [a*]	0 [e]
	2	45	97.03 [ab]	0.74 [e]
3	10	45	83.70 [c]	2.97 [e]
	50	45	64.44 [e]	9.67 [c]
	2	45	90.37 [bc]	0.74 [e]
6	10	45	85.19 [c]	13.80 [b]
	50	45	65.19 [e]	19.88 [a]
	2	45	74.07 [d]	2.70 [e]
9	10	45	66.67 [de]	5.93 [d]
	50	45	37.04 [f]	10.67 [c]

[*]Different lower-case letters in the column represent significant differences at $p < 0.05$ by Duncan's multiple range test

difference between treated samples and the control on shoot height, leaf shape, leaf area, and leaf color after 2 months (data not shown).

The observations of stomata showed that the average length of colchicine treated plantlets' stomatal guard cells (50.21 μm) was greater than 1.25 × the value of the control treatment (40.17 μm). The polyploidy determination based on the stomatal parameter indicated that the polyploid induction rate were not the similar at different concentrations and treatment times of colchicine (Table 16.1). The number of polyploid plantlets increased proportionally with the increase in the concentration of colchicine. The polyploid rate was highest at 50 μM treatment in 6 days treatment, although the survival rate decreased in this treatment.

The result also showed that stomata were oval shaped in the leaves of untreated plants, while there were large-sized spherical stomata in leaves of polyploid plantlets. The optimal colchicine treatment for polyploid plantlet were 50 μM and 6 days, respectively (Table 16.1).

Induced polyploidy can be achieved by exposure in vitro shoots of *P. villosum* with colchicine. Young tissues containing many actively dividing cells are preferred (Franzke and Ross 1957). As this study, the in vitro young shoots (1.5 cm in height and 2–3 leaves) derived from the stem nodes were the optimal explants for treated colchicine. Meristems of explants need to be exposed to colchicine at appropriate concentrations and durations to interfere with the mitotic process. Colchicine treatment (concentration and duration) which affect the polyploidy induction and was inversely proportional to the percentage of survival shoots (Nakasone 1960; Sikdar 1994). This study showed that a low concentration of colchicine (2 μM) was not sufficient to cause polyploidy despite increased treatment time. Thus, colchicine concentrations and treatment durations need to be optimized for each species. Watrous and Wimber (1988) obtained a tetraploid explant among shoots of Vanda Miss Joaquim treated with 0.5% colchicine in 3 days. Besides, Sarathum et al.

(2010) indicated that 0.025–0.1% colchicine treatment in 3–21 days produced tetraploids in *Dendrobium scabrilingue* (Sarathum et al. 2010).

16.3.2.4 Ploidy Determination

The method for determining the ploidy level of plantlets is flow cytometry analysis, which is frequently used. The results shown that most polyploid plantlets were pure tetraploids, and the rest were mixoploids. The best way to determine and confirm ploidy in plants is to count the chromosomes as described earlier. The chromosome numbers of the diploid plantlet were $2n = 2x = 26$ (Fig. 16.8a) and those of the tetraploid plantlet were $2n = 4x = 52$ (Fig. 16.8b). Observation of stomatal guard cell length was a simple, effective, and economical method to determine polyploid mutants from the original diploid clones during tissue culture of Orchidaceae such as *Stanhopea* (Ferry et al. 2000), and selected polyploid individuals of *Cymbidium*, *Phalaenopsis*, *Dendrobium*, *Epidendrum*, *Odontioda*, etc. (Chen et al. 2009; Miguel and Leonhardt 2011). In our studies we determined the ploidy levels of *P. villosum* by using a combination of three different methods.

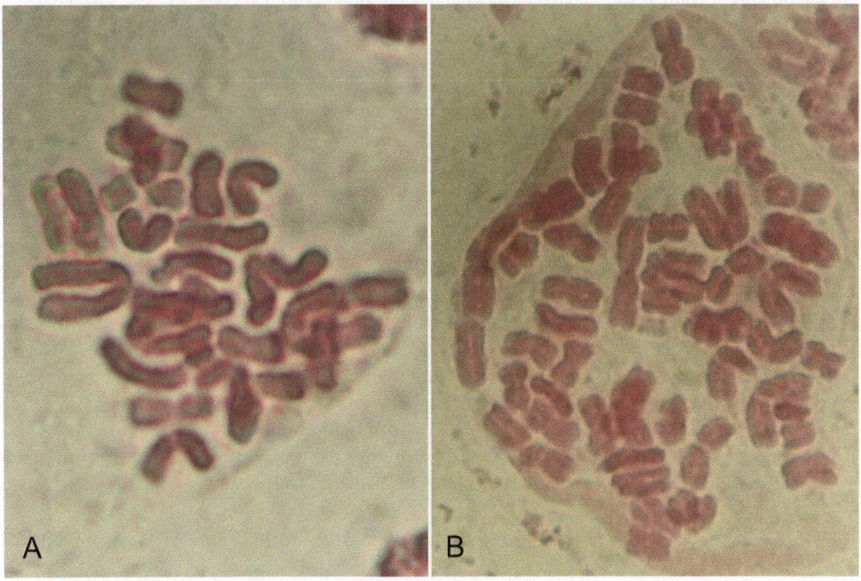

Fig. 16.8 The chromosome squashes of *Paphiopedilum villosum* root tip cells: (**a**) Diploid plantlet-control ($2n = 2x = 26$); (**b**) Tetraploid plantlet ($2n = 4x = 52$). Photomicrographs were observed under 1000× magnification of light microscope

16.4 Conclusion

In vitro adventitious shoots of *P. villosum* were readily produced by the shoot regeneration from stem nodes. The shoot induction rate and number of shoots in 0.5 mg/L TDZ treatment were highest among all treatments. The first node was the most responsive position for shoot regeneration by generating 4.5 shoots/node. Polyploid induction rate reached 19.88% when *P. villosum* shoots were treated in vitro with 50 μM colchicine for 6 days. Stomatal measurements, flow cytometric analysis, and chromosome counting indicated that the obtained polyploidy plantlets included tetraploids and mixoploids. In conclusion, the nodal explants recorded from the elongated shoot procedure and combination with the colchicine treatment, results demonstrated that an effective protocol in the production of *P. villosum* polyploid plantlets (Fig. 16.4).

References

Abu-Qaoud H, Shtaya MJY (2014) The effect of colchicine on adventitious shoot regeneration from cultured leaf explants of *Petunia hybrida*. Br J Biotechnol 4:531–540

Adaniya S, Shirai D (2001) In vitro induction of tetraploid ginger (*Zingiber officinale* Roscoe) and its pollen fertility and germinability. Sci Hortic 88:277–287

Ahmad N, Faisal M (2018) Thidiazuron: from urea derivative to plant growth regulator. Springer Nature Singapore Pte Ltd, Singapore, Singapore, p 489p

Azmi T, Sukma D, Aziz S, Syukur M (2016) Polyploidy induction of moth orchid (*Phalaenopsis amabilis* (L.) Blume) by colchicine treatment on pollinated flowers. J Agric Sci 11:62–73

Blakeslee AF, Avery AG (1937) Methods of inducing doubling of chromosomes in plants by treatment with colchicine. J Hered 28:393–411

Chen WH, Tang CY, Kao YL (2009) Ploidy doubling by in vitro culture of excised protocorms or protocorm-like bodies in *Phalaenopsis* species. Plant Cell Tissue Organ Cult 98:229–238

Dhawan OP, Lavania UC (1996) Enhancing the productivity of secondary metabolites via induced polyploidy: a review. Euphytica 87:81–89

Dpooležel J, Binarová P, Lcretti S (1989) Analysis of nuclear DNA content in plant cells by flow cytometry. Biol Plant 31:113–120

Duncan DB (1955) Multiple range and multiple F tests. Biometrics 11:1–42

Ferry RJ, Foroughbakhch R, Contreras S, Badil MH, Verde-Star M, Hauad L (2000) Leafprints and statistical analyses: findings and implications. Orchid Digest 51:21–26

Franzke C, Ross J (1957) A lineal series of mutants induced by colchicine treatment. J Hered 48:47–50

Ganga M, Chezhiyan N (2002) Influence of the antimitotic agent colchicine and oryzalin on in vitro regeneration and chromosome doubling of diploid bananas (*Musa* spp.). J Hort Sci Biotechnol 77:572–575

Hong PI, Chen JT, Chang WC (2008) Plant regeneration via protocorm-like body formation and shoot multiplication from seed-derived callus of a maudiae type slipper orchid. Acta Physiol Plant 30:755–759

Huang LC, Lin CJ, Kuo CI, Huang BL, Murashige T (2001) *Paphiopedilum* cloning in vitro. *Sci Hortic* 91: 111–121

Huetteman CA, Preece JE (1993) Thidiazuron: a potent cytokinin for woody plant tissue culture. Plant Cell Tissue Organ Cult 33:105–119

Huy NP, Luan VQ, Tung HT, Hien VT, Ngan HTM, Duy PN, Nhut DT (2019) In vitro polyploid induction of *Paphiopedilum villosum* using colchicine. Sci Hortic 252:283–290

Islam S (2010) The effect of colchicine pretreatment on isolated microspore culture of wheat (*Triticum aestivum* L.). Aust J Crop Sci 4:660–665

Jiang JL, Ye W, Li YQ, Zhou JJ, Lei FG, Wei DZ (2014) Growth and polysaccharides accumulation in autotetraploid *Dendrobium officinale*. J Plant Physiol 50:519–526

Knudson L (1946) A new nutrient solution for the germination of orchid seed. Am Orchid Soc Bulletin 15:214–217

Liu Z, Gao S (2007) Micropropagation and induction of autotetraploid plants of *Chrysanthemum cinerariifolium* (Trev.) Vis. In Vitro Cell Dev Biol – Plant 43:404–408

Long B, Niemiera AX, Cheng ZY, Long CL (2010) In vitro propagation of four threatened *Paphiopedilum* species (Orchidaceae). Plant Cell Tissue Organ Cult 101:151–162

Lu CY (1993) The use of thidiazuron in tissue culture. In Vitro Cell Dev Biol – Plant 29:92–96

Luan VQ, Huy NP, Nam NB, Huong TT, Hien VT, Hien NTT, Hai NT, Thinh DK, Nhut DT (2015) Ex vitro and in vitro *Paphiopedilum delenatii* Guillaumin stem elongation under light-emitting diodes and shoot regeneration via stem node culture. Acta Physiol Plant 37:136–146

Miething F, Hering S, Hanschke B, Dressler J (2006) Effect of fixation to the degradation of nuclear and nitochondrial DNA in different tissues. J Histochem Cytochem 54:371–374

Miguel TP, Leonhardt KW (2011) In vitro polyploid induction of orchids using oryzalin. Sci Hortic 130:314–319

Murashige T, Skoog F (1962) A revised medium for rapid growth and bio assays with tobacco tissue cultures. Physiol Plant 15:473–497

Nakasone HY (1960) Artificial induction of polyploidy in orchids by the use of colchicine. (PhD thesis). University of Hawaii, USA, p 29p

Nhut DT (1998) Micropropagation of lily (*Lilium longiflorum*) via in vitro stem node and pseudo-bulblet culture. Plant Cell Rep 17:913–916

Nhut DT, Trang PTT, Vu NH, Thuy DTT, Khiem DV, Binh NV, Van KTT (2005) A wounding method and liquid culture in *Paphiopedilum delenatii* propagation. Propag Ornam Plants 5:158–163

Niazian M, Nalousi AM (2020) Artificial polyploidy induction for improvement of ornamental and medicinal plants. Plant Cell Tissue Organ Cult 143:477–490

Posada M, Ballesteros N, Obando W, Angarita A (1997) Micropropagation of gerberas from floral buds. In: International symposium on cut flowers in the tropics, vol 482, pp 329–332

Predieri S (2001) Mutation induction and tissue culture in improving fruits. Plant Cell Tissue Organ Cult 64:185–210

Ramsey J, Schemske DW (1998) Pathways, mechanisms, and rates of polyploid formation in flowering plants. Annu Rev Ecol Evol Syst 29:467–501

Ranney TG (2006) Polyploidy: From evolution to new plant development. In: Combined proceedings, international plant propagators' society, pp 137–142

Russell G (2004) Stomatal guard cell measurements using leaf prints. CSA J 4:137–139

Sanguthai O, Sanguthia S, Kamemoto H (1973) Chromosome doubling of a *Dendrobium* hybrid with colchicine in meristem culture. Hawaii Orchid J 2:12–16

Sarathum S, Hegele M, Tantiviwat S, Nanakorn M (2010) Effect of concentration and duration of colchicine treatment on polyploidy induction in *Dendrobium scabrilingue* L. Eur J Hortic Sci 75:123–127

Sattler MC, Carvalho CR, Clarindo WR (2016) The polyploidy and its key role in plant breeding. Planta 243:281–296

Schenk RU, Hildebrandt AC (1972) Medium and techniques for induction and growth of mono-cotyledonous and dicotyledonous plant cell cultures. Can J Bot 50:199–204

Sikdar A (1994) Induced polyploidy in mulberry (*Morus* spp.): induction of tetraploids. Sericologia 34:105–116

Song C, Liu S, Xiao J, He W, Zhou Y, Qin Q, Zhang C, Liu Y (2012) Polyploid organisms. Sci China Life Sci 55:301–311

Sun Q, Sun H, Li L, Bell RL (2009) In vitro colchicine-induced polyploid plantlet production and regeneration from leaf explants of the diploid pear (*Pyrus communis* L.) cultivar, 'Fertility'. J Hortic Sci Biotechnol 84:548–552

Thao NTP, Ureshino K, Miyajima I, Ozaki Y, Okubo H (2003) Induction of tetraploids in ornamental *Alocasia* through colchicine and oryzalin treatments. Plant Cell Tissue Organ Cult 72:19–25

Tomoki H, Tadashi M, Kazuhiro Y, Yuko Y, Keimei O (2014) Synthesis of new brassinosteroid biosynthesis inhibitor with coumarin moiety as a fluorescent probe. Int J Chem Eng Appli 5(5):379–383

Udall JA, Wendel JF (2006) Polyploidy and crop improvement. Crop Sci 46:3–14

Wang F, Fu J, Dong L (2009) Inducing autotetraploid pea with colchicine and DMSO. J Nuc Agri Sci 23:203–208

Watrous S, Wimber D (1988) Artificial induction of polyploidy in *Paphiopedilum.* Lindleyana 3: 177–183

Wimber D, Van Cott A (1966) Artificially induced polyploidy in Cymbidiums. In: Proceedings 5th World Orchid Conference, Long Beach, CA, USA, pp 11–22

Yang XM, Cao ZY, An LZ, Wang YM, Fang XW (2006) In vitro tetraploid induction via colchicine treatment from diploid somatic embryos in grapevine (*Vitis vinifera* L.). Euphytica 152:217–224

Zeng S, Huang W, Wu K, Zhang J, Teixeira da Silva JA, Duan J (2016) In vitro propagation of *Paphiopedilum* orchids. Crit Rev Biotechnol 36:521–534

In Vitro Growth and Development of Plants Under Stimulated Microgravity Condition 17

Duong Tan Nhut, Hoang Dac Khai, Nguyen Xuan Tuan, Le The Bien, and Hoang Thanh Tung

Abstract

In vitro culture is a useful tool for physiological, biochemical, and genetic studies of plants. In the context of experiments under real microgravity conditions are limited in quantity and high cost, the evaluation of plant growth and development under simulated microgravity conditions (MG) using in vitro culture models has received more attention in recent years. In the framework of the Space Seed for Asian Future program 2010–2011, our laboratory participated as a member, the results obtained have shown that Rose balsam seeds derived from real MG pretreatment for 6 months had significant changes in germination, growth, and morphology under nursery conditions. However, changing the level Rose balsam plants was dependent on the genotype tested. Based on these results, we applied in vitro culture and simulated MG conditions using clinostats two dimensional (2D) to study the germination, growth, development, and secondary compound accumulation of some medicinal plants. The results showed that under clinostating (2 rpm), *Hibiscus sagittifolius* Kurz., *Phyllanthus amarus*, and *Catharanthus roseus* plants showed different growth responses. In vitro germination of *H. sagittifolius* Kurz. and *C. roseus* seeds was significantly increased under clinostating compared with gravity conditions. Furthermore, all three cases of the present study showed a significant increase in the number of shoots in shoot multiplication stage under clinostating. In particular, microgravity simulation conditions stimulated the accumulation of total coumarin compounds, total saponins in *H. sagittifolius* Kurz seedlings and total phyllanthin in *P. amarus* plantlets. In addition, the in vitro flowering time of *P. amarus* plantlets was significantly shortened compared with plantlets grown under normal gravity

D. T. Nhut (✉) · H. D. Khai · N. X. Tuan · L. T. Bien · H. T. Tung
Tay Nguyen Institute for Scientific Research, VAST, Dalat City, Vietnam

D. T. Nhut et al. (eds.), *Plant Tissue Culture: New Techniques and Application in Horticultural Species of Tropical Region*,
https://doi.org/10.1007/978-981-16-6498-4_17

conditions. These results have provided important information for further studies on the effects of real and simulated MG conditions on plant growth and development, thereby developing effective strategies to exploit the potential of plants as a biological life support system during long-term space flights.

Keywords

Clinorotation · Flowering · Germination · In vitro · Medicinal plants · Microgravity · Secondary compound

17.1 Introduction

17.1.1 Plant Growth and Development Are Associated with Gravity

An irreversible increase in the mass of an organism is called growth (Srivastava 2002). An increase in biomass means an increase in size (cell expansion) and the number of cells (cell division). Simply dividing cells is not enough to lead to the rebuilding of an organism. In addition, differentiation ensues to specialize cells that share a common ancestor (development). In other words, development is the differentiation of function, structure, and morphology between cells in a tissue or between organs in a complete body; for example, epidermis, sub-epidermis, cortex, and vasculature in the roots. The process of alternating between growth and development creates a complete plant body with strict organization and characteristic morphology is called morphogenesis. Therefore, the control of cell reproduction and differentiation is a key to morphogenesis studies.

During million years of evolution and development of life on the Earth, all living things are constantly under the influence of gravity. Thus, plants have used gravity as the most stable and reliable source of signal for their survival (Hoson and Wakabayashi 2015; Vazquez et al. 2019). Based on this signal, plants have different gravity sensing strategies to control their growth direction and body structure in favor of growth and development, which is called gravitropism (Morita 2010; Nakamura et al. 2019). For example, the root organs of plants develop in the direction of gravity to facilitate efficient water absorption and nutrient uptake (positive gravitropism), or shoots grow in the opposite direction of gravity to capture the light energy needed for growth (negative gravitropism), etc. Plants sense gravity signals through cells with specialized structures and functions called gravity sensing cells (statocytes).

Cytological studies have shown that the columella cells located in the root cap and the endodermal cells located in the stem are the two receptors for gravity signals (Nakamura et al. 2019). In addition to the usual cell organelles, the structure of columella and endodermal cells is further enhanced by the presence of starchy plastids called amyloplasts (statoliths) (Fig. 17.1). Compared with the amyloplasts of columella cells, amyloplasts located in stem endodermal cells have a well-developed thylakoid membrane system and photosynthetic pigments in addition to

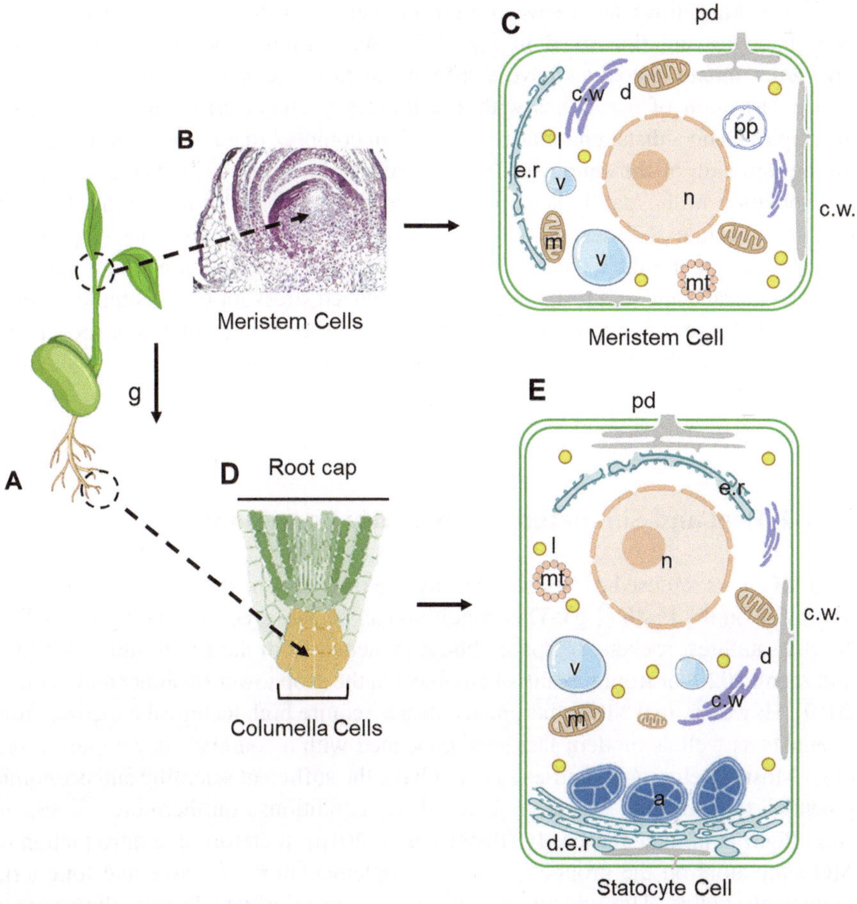

Fig. 17.1 Schematic structure of a shoot meristem cell and a columella cell in the root cap. (**a**) Illustration of seedlings germinated under gravity conditions; (**b**) the apical meristem cells showed negative gravitropism; (**c**) the meristem cells have a symmetrical distribution of organelles, in which the nucleus is centrally located, the remaining organelles are randomly distributed; (**d**) columella cells located in the root cap showed positive gravitropism; (**e**) the statocytes are polarized organelles in the cell, where the nucleus (n) represents the proximal cell pole and the distal endoplasmic reticulum complex (d.e.r. complex) represents the distal cell pole; the amyloplasts (a) are sedimented onto d.e.r. complex induced primary signaling that regulates both normal and gravitropic root growth; (c.w.—cell wall; d—dictyosomes; e.r.—endoplasmic reticulum complex; 1—lipid droplets; m—mitochondria; mt—microtubules; pd—plasmodesma; pp—proplastid; v—vacuole)

the starch granules (Morita 2010). Despite these differences, the presence of amyloplasts in statocytes helps plants sense gravity; that is, naturally, amyloplasts are sedimented under the influence of gravity thereby providing the important physical signal for gravity sensing in plants.

The striking difference between meristem cells (non-functional gravity sensing) and statocytes are illustrated in Fig. 17.1. Accordingly, the polarization of the organelles makes statocytes have a special function—sensing gravity (Fig. 17.1e). The polarization of organelles within statocytes is characterized by sedimentable amyloplasts and a distal endoplasmic reticulum complex (d.e.r. complex) located on the bottom side of the statocytes (Sievers and Volkmann 1977). The dense starch accumulation in the statoliths causes them to be sedimented down in the direction of gravity, while sedimentation is prevented by d.e.r. complex. The stopping of amyloplasts on the d.e.r. complex affected the structural and functional states of d.e.r. membrane and as a result amyloplasts induced stress for d.e.r. complex. These stress signals are converted into biochemical signals to redirect auxin transport in the plant body; thereby directing their development in response to gravity (Nakamura et al. 2019).

17.1.2 Real and Simulated Microgravity Conditions

Real MG is a term used to describe gravity at extremely small values (μg) compared to gravity on the Earth (1 g). This condition can be achieved in orbit flights such as NASA satellites, rockets, or space shuttle launched from the Earth into space (Xie and zheng 2020) or from free fall of an object in the drop tower (Böhmer and Schleiff 2019). However, real MG experiments clearly require high technical expertise from scientists as well as modern facilities associated with a country's developing economy. Most developing countries may not have the sufficient scientific and economic potential to carry out research under these conditions. Furthermore, access to spaceflight is limited and costly (Brungs et al. 2016); therefore, the introduction of MG simulators on the ground allows the implementation of active and long-term biological studies. The widely known devices are clinostat 2D (two-dimensional clinostat), clinostat 3D (three-dimensional clinostat), Rotating Wall Vessel (RWV), Random Positioning Machine (RPM), Magnetic Levitation, etc. In particular, clinostat 2D is the most commonly used device for cell culture experiments or small organisms (Kamal et al. 2015) or experiments related to molecular biology (Hemmersbach et al. 2006) and the experiments investigating in vitro growth of plants in our present work.

Investigating the effects of gravity or MG on plants allows understanding the interactions between plants and gravity; thus, providing us with important answers revolves around the physiological role of gravity in plant growth, development, and reproduction (Krikorian et al. 1992; Böhmer and Schleiff 2019; Malarvizhi et al. 2020). Furthermore, plant studies under MG are essential to space exploration programs because plants can provide food, recycling water, oxygen, and wastes (El-Nakhel et al. 2019). In particular, the plant–crew interaction has been known as an important drug for improving the psychology of astronauts during stressful flights (Malarvizhi et al. 2020). Therefore, successful design and construction of biological life support systems (BLSS) are required in long-term flights (Matía et al. 2010), in which plants are central of the system. Going further, our ambition is toward space agriculture for human migration to the Moon or Mars (Kitaya 2019).

17.1.3 Microgravity Simulation Using Clinostat 2D

The clinostat 2D is a device that rotates around the horizontal axis, samples placed in the center of the rotation in which the rotation takes place continuously to produce a uniform acceleration (Kordyum 1997) (Fig. 17.2). Obviously, the rotation of the clinostat takes place in the gravitational field, so it does not change the magnitude and scalar of gravity. On the other hand, like other MG simulators, clinostat only prevents plants from sensing gravity signals; therefore, plants exhibit growth response similar to those under real MG conditions.

It should be noted when using clinostat 2D for MG simulation purpose:

1. The distance from the center of the rotation to the position of the sample under study (rotation radius) is very important. It is well known that the radius of rotation is directly proportional to the centrifugal acceleration. In the case of large radius, the value of the centrifugal acceleration can reach near the gravitational threshold (1 g); thus, reducing simulation efficiency on the sample. For

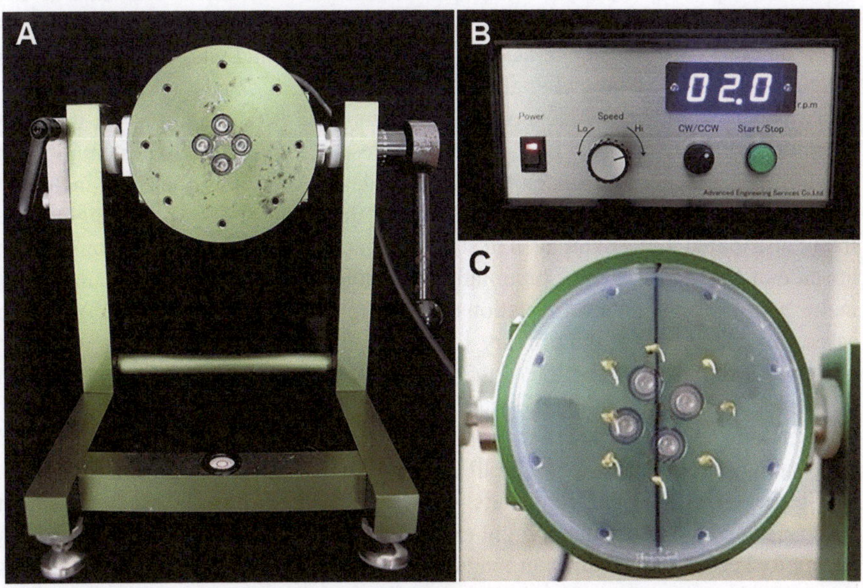

Fig. 17.2 Structure of a clinostat 2D (Advanced Engineering Services Co., Ltd. Japan) consists of 3 main parts. (**a**) An AC motor is fitted with a disc (diameter 10 cm) and the supporting frame (height 20 cm) with 4 adjustable feet; (**b**) an amplifier controlled the rotation speed uses 1 phase AC with rated voltage from 100 to 240 V (AC 100–240 V); (**c**) germination of rice seeds under clinostating 3 days of culture (2 rpm)

example, plants may grow away from the center of rotation during the experiment and this leads to an accelerated change that affects the sample (Kiss et al. 2019).
2. The rotation speed setting is also a factor that has a direct effect on the MG simulation efficiency because fast rotation leads to the formation of multidirectional gravity (centrifugal force); on the other hand, if the clinostat slowly rotates, gravity affects the sample; therefore, the rotation speed of the simulator must be faster than the biological process studied by the specimen, but not so fast that other forces are applied (Van Loon 2007).
3. Consideration for sample weight distribution while rotating is also very important; for example, over time an excessive increase in cell biomass results in non-random mechanical stimulation of the specimen itself (mechanical stress) and on the load-bearing capacity of the simulator (Kiss et al. 2019).

In general, it can be seen that simulation quality depends on rotation speed, effective diameter and sample processing time (Dedolph and Dipert 1971). Furthermore, this must be clearly defined for the specific subjects being tested (Herranz et al. 2013).

It has been shown that regardless of starch accumulation in the amyloplasts and not, the sedimentation of these plastids still occurs based on the mass of the amyloplasts (Weise and Kiss 1999; Nakamura et al. 2019). Therefore, any change in gravity affects the amyloplasts sedimentation and the sensing of gravity signals required for plant growth and development (Kamal et al. 2015; Hassanpour and Ghanbarzadeh 2021). Typically, the potential effects of microgravity conditions (MG) on plants have been evaluated at both the cellular and whole organism levels (Matía et al. 2010; Malavizhi et al. 2020).

To better explain the effects of real and simulated MG conditions on the sedimentation of amyloplasts, Fig. 17.3 was presented with the sedimentation of amyloplasts in statocytes under real and simulated MG conditions and including gravity conditions. Under real MG conditions, the amyloplasts were thought to be freely distributed in the cytoplasm in statocytes, so it did not stress d.e.r. complex. In the case of MG simulations (clinorotation), the amyloplasts moved in circular orbits in the cytoplasm and no sedimentation occurred to d.e.r. complex (Fig. 17.3). In both cases, due to the lack of the primary signal of sedimentation, plants were unable to navigate to control the direction of their growth or structure as under gravity, which is referred to as automorphogenesis (Hoson et al. 2001).

There have been many reviews comparing results between space experiment and clinostating (Brown et al. 1996; Kraft et al. 2000; Herranz et al. 2013; Huang et al. 2018; Kiss et al. 2019). The results showed similarities only to a certain extent because of specific elements of flight experiments such as vibration and acceleration during rocket launch, changes in magnetic fields and radiation in space. All of this has caused non-uniform results between space and clinostating experiments (Vandenbrink and Kiss 2016). Furthermore, experimental results depend on physiological state and complexity of experimental subjects (Kordyum 1997). Therefore, Kiss et al. (2019) proposed that identical experimental conditions should be designed to optimize the results comparison between space and clinostating experiments, for

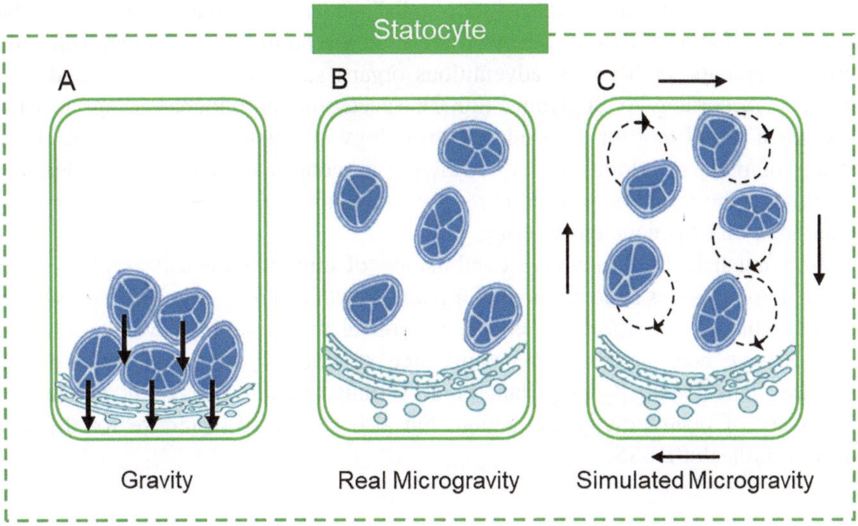

Fig. 17.3 Schematic of the sedimentation of amyloplasts under gravity, real MG and simulated MG conditions. (**a**) The sedimentation of amyloplasts by gravity; (**b**) free dispersion of amyloplasts under real MG conditions; (**c**) orbital movement of amyloplasts under simulated MG conditions

examples light intensity and quality, humidity, temperature, vibration, or even growth substrate. Despite these differences, the results obtained from clinostat are still important because it is necessary to predict and supplement the results obtained in real MG conditions (Kamal et al. 2015; Kiss et al. 2019).

17.1.4 Application of In Vitro Culture Technology to Study Plant Growth Under Clinorotation

Plant growth and development are complex biological processes that require a long time to study. While space flights are limited in time and experimental space. Therefore, the establishment of research models poses a challenge for early scientists in this field (Krikorian et al. 1992). Recently, the development of models that combine MG simulators and aseptic culture technology is known to be a simple and efficient system for biological studies.

The advantages of in vitro culture in this case include: (i) Allow the researcher to observe biological events that take place in the primitive period. (ii) Allows separation of plant organs and assess the response of each to MG. In other words, the in vitro system provides the conditions of sustaining growth to the plant organ without interaction (inhibition or stimulation) and dependence between tissues and other organs. Whereas this is unavoidable when studying the system in vivo (dominated by the whole plant organism). (iii) Allows to obtain morphologically diverse results through reprogramming from differentiated cells. Initially, dedifferentiation occurs to form "stem cells" under the influence of controllable exogenous factors. This is followed by reprogramming of totipotent cells that produce somatic embryos or adventitious organs such as shoots, roots, and even flowers. According to Sugiyama (2000), successful morphogenesis in vitro has become a useful tool for the study of physiology, biochemistry and genetics in all stages of plant development. (iv) Allows easy control of sample mass and size because of the limits on sample mass per simulation rotation. (v) Allows the easy manipulation of genetic engineering.

In a nutshell, the innovative combination of clinorotation and aseptic culture technology provides an ideal model for plant growth and development studies under MG conditions. Therefore, based on this model the results of several promising studies have been reported and contributed significantly to the database of this science. This knowledge of plant biology under MG not only encourages the discovery of plant–gravity interaction, but also serves as a premise for the next design studies of BLSS.

17.1.5 Application of Clinorotating in Plant Growth and Development Studies

It is well known that changing gravity (MG or hypergravity) is a type of abiotic stress that can positively/negatively affect plants. Altered gravity vectors (gravistimulation) are perceived by the plant through statocytes, then signaling induced a series of intracellular physiological and biochemical transformations, and ultimately in response to changes in growth, development, and morphology of plants (Malarvizhi et al. 2020). The lack of gravity signaling has caused an imbalance between redox reactions in cellular metabolism, that is, producing reactive oxygen species (ROS) such as hydrogen peroxide (H_2O_2), singlet oxygen (O^{2-}), and hydroxyl radicals (OH^-). According to Zhao et al. (2007), an overproduction of ROS has significantly affected cell viability, division, reproduction, differentiation, and programming of cell death. Interestingly, in vitro calli of *Matricaria chamomilla* plants grown under MG simulated conditions by a 2D clinostat at 10 rpm showed an increase in ROS breakdown enzymes such as peroxidase, catalase (CAT), and superoxide dismutase (SOD). In addition, the slow rotation speed of clinostat was shown to have a positive effect on the growth of calli of *M. chamomilla* which is expressed through indicators such as biomass, cell division, and energy related substances (sugars and starches) (Hassanpour and Ghanbarzadeh 2021). This result implied that MG caused ROS; on the other hand, MG stimulated biosynthesis of beneficial compounds to prevent ROS and this also benefited plant growth.

Previously, Soleimani et al. (2019a, b) obtained similar results when tested on Tobacco calli; in addition, the results showed that MG stimulates the biosynthesis of radical scavenging compounds such as phenolic and peroxide neutralizing amino acids (His, Pro, Ser, and Asp) to protect cells from membrane lipid peroxidation. Furthermore, clinorotating has been shown to alter the metabolic direction of Tobacco cells along an energy-saving pathway to service the Krebs cycle; therefore, produce more ATP instead of redirecting in energy-consuming roads in the presence of gravity.

Plant growth and development depend entirely on the correct functioning and coordination of many cellular processes including proliferation. Cell proliferation is a biological process triggered by signals transmitted between plant organs (Matía et al. 2010). One of the key signals is gravity (Xie and Zheng 2020). However, experiments have shown that simulated MG stimulates cell proliferation; in other words, gravity could negatively control the increase in plant biomass. There have been many reports contributing to clarify this issue such as: increased proliferation of the root meristem Arabidopsis under clinorotating (Matía et al. 2010) compared with the gravitational control; significant increase in biomass of Tobacco calli as mentioned above by Soleimani et al. (2019a, b) or calli of *M. chamomilla* reported by Hassanpour and Ghanbarzadeh (2021); etc.

To date, there have been many studies surrounding the question of cell proliferation under simulated MG (Boucheron-Dubuisson et al. 2016; Soleimani et al. 2019a, b). Soleimani et al. (2019a) demonstrated that on the ground, plant cells must consume energy to maintain homeostasis in position against gravity; on the contrary, under MG, that energy can be saved for other processes such as biosynthesis of growth beneficial metabolites. In addition, under clinorotation cells were confirmed to enter the mitochondrial process before reaching the critical size; however, abnormally rapid cell division may also cause mutations (Boucheron-Dubuisson et al. 2016). Another mechanism is explained based on the structure of the cell wall. It is well known that the plant cell wall is responsible for controlling cell division and proliferation and cell wall biosynthesis is controlled by gravity signals (Hoson and Wakabayashi 2015). Therefore, the lack of gravity caused unpredictable effects on the cell wall. Soleimani et al. (2019b) reported that it is true that clinorotating caused cell wall loosening and this promoted proliferation.

In 2016, Professor Duong Tan Nhut had the opportunity to attend the Asian Winter School held in Japan (Fig. 17.4).

17.2 Germination, Growth, and Phenotypic Variation of 5 Genotypes of Rose Balsam (*Impatiens balsamina*) Derived From Real Microgravity

Selecting mutant varieties by space technology was a new direction in many countries with developed space technology such as Russia, the United States, China, and Japan. From the Space Seed for Asian Future program 2010–2011, Vietnam has directly participated as a research member with 3 species: Rose balsam (*Impatiens balsamina*), *Antirrhinum majus*, and *Salvia splendens* Ker-Gawl. Our laboratory belonging to Tay Nguyen Institute for Scientific Research has evaluated the germination and growth of real MG-derived Rose balsam seeds. The obtained results have been briefly presented in this section, thereby opening an important premise for research on plant breeding using space technology—a new direction of the breeding industry in developing countries.

Fig. 17.4 (**a**) Japan Aerospace Exploration Agency (JAXA, Tsukuba Space Center); (**b**) clinostat 3D; (**c**) Professor Duong Tan Nhut at Asian Winter School 2006 (Space environment and Space Environment Utilization Researches, JAXA); (**d**) processing the shuttle for flight

17.2.1 Germination of Real Microgravity-Derived Rose Balsam Seeds

The seeds of Rose balsam (*Impatiens balsamina* spp.) including 5 flower colors: purple, pink, pink, orange, and white were processed according to the steps of the Space Seed for Asian Future program 2010–2011. Then the seeds were stored and divided into 2 plastic bags: 1 control bag (placed on the ground) and 1 bag was taken to the international space station for 6 months by H-2B rocket (Japan Aerospace Exploration Agency). After returning to Earth, these seeds were sown in plots with dimensions of 10 × 10 cm. Fifty plots (each treatment) were surveyed. Each plot was sown 10 seeds on the soil mixed with coal and coir (2:1:1). Then, evaluated the germination of seeds after 9, 11, 13, 15, 17, 19, 21, 23, and 25 days. Seeds are considered germinated when the sprout appeared (1 cm in long) above the ground. Germination rate (%) = 100 × Total number of seeds germinated/Total seeds sown.

The results showed that the white Rose balsam seeds of the MG treatment germinated after 9 days of sowing, while the control was 11 days (Fig. 17.5). On

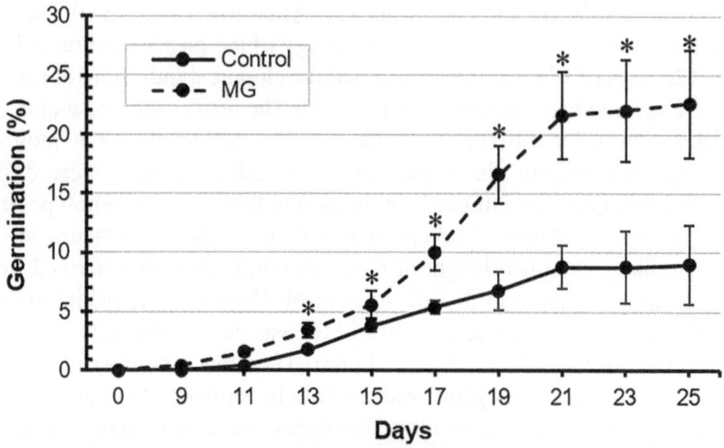

Fig. 17.5 Germination of white rose balsam seeds derived from 6 months real MG exposure and control after 25 days of sowing under normal gravity conditions. Control; seedlings derived from MG; comparison of germination rates of seeds derived from MG and control. Asterisks represent statistically significant differences with Duncan's test ($\alpha = 0.05$)

the other hand, purple, pink, rouge, and orange flower genotypes showed no difference in germination between the two treatments (data not shown). For the white flower genotype, germination was slowed down after 21 days of sowing and no further seed germination was observed thereafter. After 25 days of sowing, the germination rate of the MG seeds (22.9%) was 2.5-fold higher than that of the control (9%) (Fig. 17.5). In summary, the germination of the MG-derived seeds was significantly higher than that of the ground control. However, the results varied among the genotypes examined. These results are similar to the study of Ren et al. (2010), *Medicago sativa* seeds that were placed in space for 15 days gave a 6.2% higher germination rate than the ground control. According to research by Yan et al. (2008), *Dactylis glomerata* and *Lolium multiflorum* seeds showed increased germination rates after being brought back from space. This can be explained by the effect of radiation and space MG environment on plant gene expression (Nelson 2003; Schimmerling 2003). Cosmic ion radiation has also been identified to cause changes in grain structure such as the formation of porous molecular structures that were permeable to water and oxygen that may have led to early seed germination (Ren et al. 2010). The results indicated that cosmic influences including MG increased the germination rate of some plant seeds and Rose balsam seeds.

17.2.2 Growth and Development of Real Microgravity-Derived Rose Balsam Plants

After 45 days of sowing, the seedlings were transferred to 18 × 25 cm plastic bags containing mixed soil (mixing ratio was similar to germination experiment). The plants were spaced 30 cm apart in the nursery. After flowering, growth and development indicators as well as abnormal morphology of the plants were recorded.

After 120 days of sowing, the growth and development indicators of rose balsam plants were recorded and presented in Fig. 17.6 The results showed that there were statistically significant differences in the growth and development indicators of MG-derived Rose balsam plants and control. In addition, there were differences between the genotypes investigated. In the purple Rose balsam genotype, the plant height of the MG treatment (54.7 cm) was 1.7-fold higher than that of the control (32 cm); similarly, the plant height of rouge and pink genotypes derived from MG was significantly higher than that of the control. However, the plant height of the white and orange genotypes derived from MG was not significantly different from that of the control was observed (Fig. 17.6a). The branch number of MG-derived rouge balsam (10.7 branches/plant) was 2.4-fold lower than that of the control (25.7 branches/plant) while the remaining genotypes were not show the difference (Fig. 17.6c). Total chlorophyll content of MG-derived white balsam (54.5 nmol/cm^2) was significantly higher than that of the control (42.1 nmol/cm^2). This difference was also observed in rouge Rose balsam plants of MG origin (52.3 nmol/cm^2) while in the control group it was 42.7 nmol/cm^2. In the purple, orange and pink Rose balsam genotypes there was no significant difference in total chlorophyll content between the two treatments (Fig. 17.6b). The results suggested that MG pretreatment significantly altered the growth and development of Rose balsam plants and the extent of the change was dependent on the genotypes tested.

17.2.3 Flowering of Real Microgravity-Derived Rose Balsam Plants

Flowering time is an important factor for many ornamental plants. The flowering time of Rose balsam plants of the MG treatment and the control was shown in Fig. 17.6d. After 100 days of sowing, the first flowering was recorded in the purple Rose balsam of the MG treatment while the flowering time of the plants in the control treatment was 16 days; followed by flowering in rouge, pink, orange, and white Rose genotypes, respectively. Flowering of all MG-derived balsam genotypes was statistically significantly earlier than the control by 16–23 days. The difference in growth process between MG and control treatments could be explained by the serious influence of MG medium and cosmic radiation on the regulation of gene expression in plants (Nelson 2003; Schimmerling 2003). High-energy rays were responsible for the increase in height in Flax (Bari 1971) or Rice (Katoch et al. 1992). The influence of spatial conditions on plant polymorphism has been mentioned in a number of previous studies through the evaluation of growth and development parameters. As reported by Luo et al. (2006), 201 Rice plants grown from spatially treated seeds showed polymorphism with the difference in phenotype and yield compared with the ground control was 30.2%; besides, some growth indices were reduced in MG treatment was also recorded. Wei et al. (2006) investigated the phenotypic variation in the F1 generation in Rice of MG origin; the results showed the difference in plant height with two groups, which was the

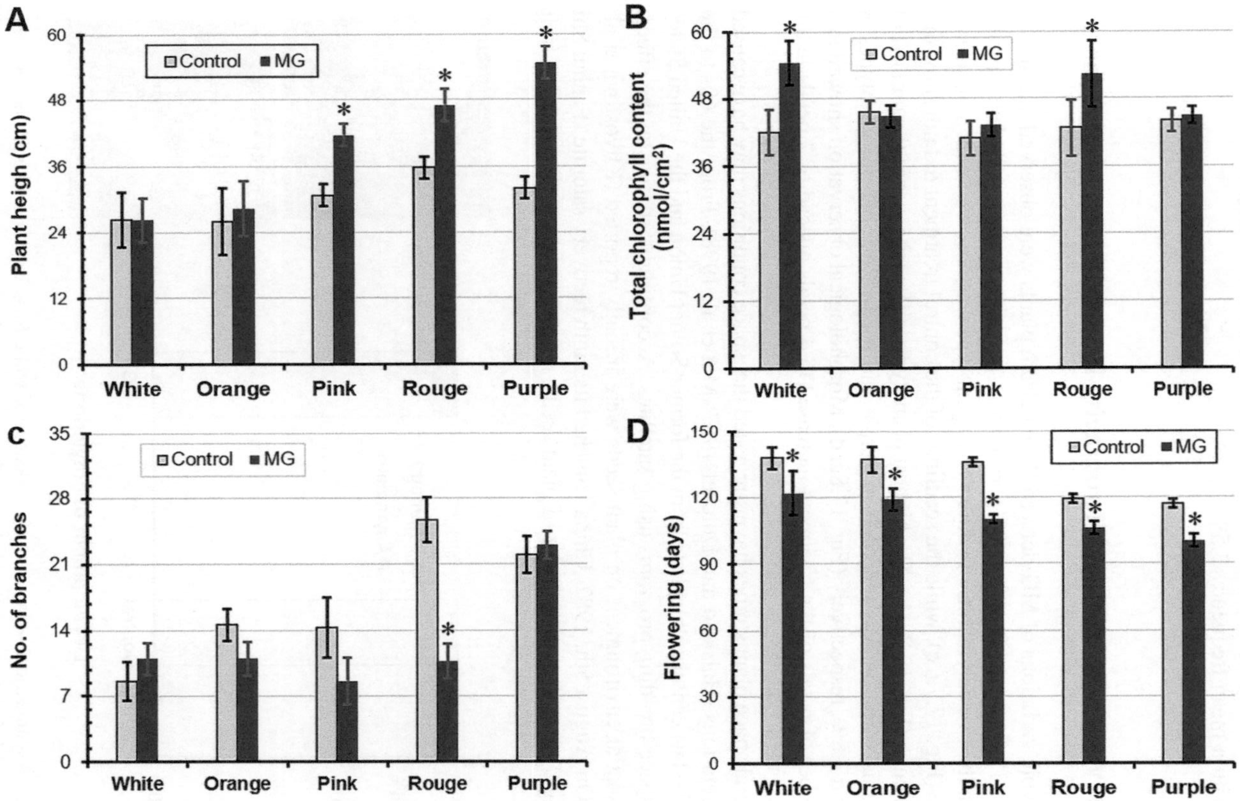

Fig. 17.6 Growth and development of white, orange, pink, rouge, and purple Rose balsam plants derived from 6 months real MG exposure. (**a**) Plant height of Rose balsam plants derived from MG and control after 90 days in the greenhouse; (**b**) total chlorophyll content in leaves of Rose balsam plants derived from MG and control after 90 days in the greenhouse; (**c**) number of branches of Rose balsam plants derived from MG and control after 90 days in the greenhouse; (**d**) the flowering time of the Rose balsams derived from MG and control

taller group (accounting for 34.9%) and the shorter group (accounting for 39.1%) compared to the control. In a study on the genetic change of Rice plants under the influence of MG, Ou et al. (2010) reported that after 18 days in spaceflight conditions, the analysis showed that genomic alterations ranged from 0.7% to 6.8% with a mean frequency 3.5%.

17.2.4 Variations of Real Microgravity-Derived Rose Balsam Plants

Phenotypic variations of MG-derived Rose balsam plants were observed and shown in Figs. 17.7 and 17.8. After the germination period, asymmetrical development between the two cotyledons of white flower genotype was observed with a frequency of 0.8% (Fig. 17.7a, c), while the seedlings of the control treatment did not show this phenomenon (Fig. 17.7a, b). The highest phenotypic difference was in the number of petals; this variation was recorded in purple and rouge Rose balsam genotypes at 4.5% and 3.8%, respectively (Fig. 17.8c, d). Morphological observation showed that the number of petals of the treatment increased to 4 petals instead of 2 petals in the control (Fig. 17.8a, b).

Several space plant studies have suggested that space conditions may have caused plant mutations (Halstead and Dutcher 1987; Mei et al. 1998). From the 60s to the 80s of the last century, scientists from the former Soviet Union and the United States began space breeding programs using satellites. Accordingly, studies on the effects of the spatial environment on plant seeds were initially reported (Khvostova et al. 1963). On August 5th, 1987, China launched its ninth recovery satellite (Fanhui Shi Weixing 9), which carried 36 kg of plant seeds, mushrooms, shrimp eggs, and fruit

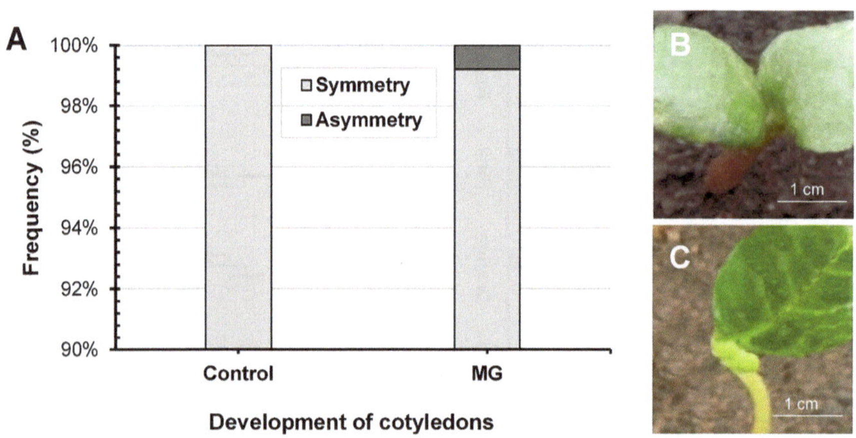

Fig. 17.7 Asymmetric development of cotyledons of white Rose balsam seedlings derived from 6 months real MG exposure after 15 days in the greenhouse. (**a**) Frequency variation; (**b**) normal development of cotyledons of control white Rose balsam seedlings; (**c**) asymmetric development of cotyledons of white Rose balsam seedlings derived from MG

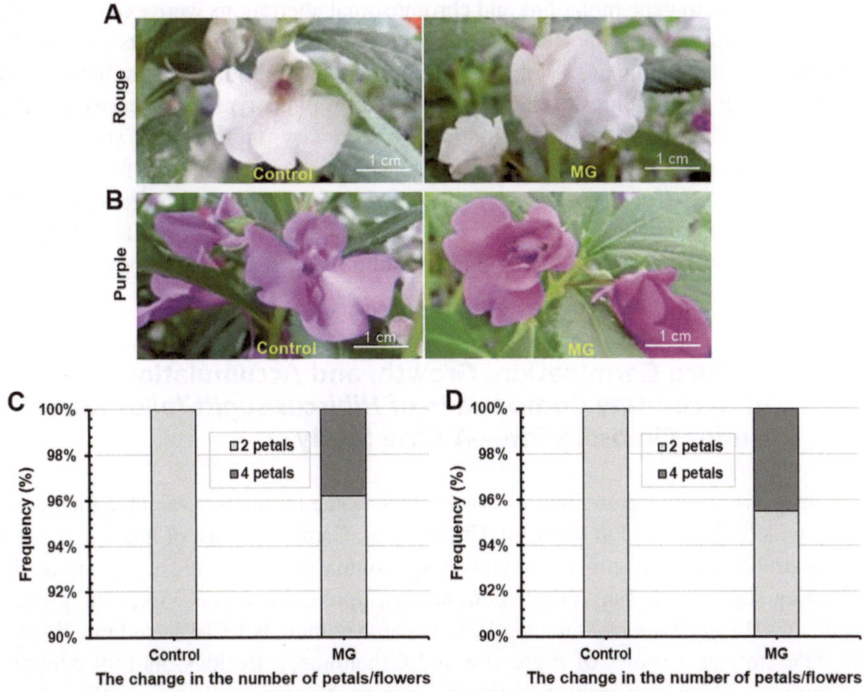

Fig. 17.8 The change in the number of petals/flowers of rouge and purple Rose balsam plants derived from 6 months real MG exposure. (**a**) Variation in rouge rose balsam genotype; (**b**) variation in purple rose balsam genotype; (**c**) frequency of variation in rouge rose balsam genotype; (**d**) frequency of variation in purple rose balsam genotype

flies into space and it returned to Earth on August 10th, 1987. Following that on September 9th, 1987; on November 5th, 1990 and August 8th, 1998, China carried out a series of similar seed experiments. After returning to Earth, some interesting phenotypic variations and genetic differences at the cellular and molecular levels were observed; since then, scientists have begun to propose methods of plant breeding related to the space environment (Wen et al. 2004). According to Gu and Shen (1989), in Wheat, many chromosomal aberrations were observed under the influence of extraterrestrial conditions; in addition, this anomaly frequency has increased proportionally with the contact time between the seed and the space environment; therefore, this study came to the conclusion that MG and radiation were two spatial of the organismal variation at the molecular level. According to Toyota (2007a, b), altered gravity led to increased intracellular calcium concentrations, decreased saccharides, and fatty acids that caused DNA damage and its replication. The resulting disruption of Ca^{2+}, inhibited DNA repair leading to the presence of nucleotide and chromosomal abnormalities. Kostina et al. (1984) suggested that abnormal manifestations such as increased germination rate, inhibition of root development were due to chromosomal aberrations; however, some expression due to gene mutations and chromosomal aberrations were expressed only in later generations. In this study, the MG-derived Rose balsam plants exhibited a mutant phenotype compared with the control plants. MG and cosmic radiation may have acted on the genomes of the seeds resulting in phenotypic differences and this could be considered a form of mutation. In summary, these results showed that cosmological conditions had a certain influence on germination, growth and development in the 5 Rose balsam cultivars investigated. Depending on the genotypes, the level of growth and phenotypic changes were different. These results will be an important premise for research on plant breeding using space technology.

17.3 In Vitro Germination, Growth, and Accumulation of Secondary Compounds of *Hibiscus sagittifolius* Kurz. Under Clinorotation—A Case Study

Hibiscus sagittifolius Kurz. belongs to the Malvaceae family is a valuable medicinal plant widely distributed in Vietnam, China, Laos, Cambodia, east of India, and north of Australia. Its roots contain phytosterols, coumarins, fatty acids, organic acids, reducing sugars and many uronic compounds; lipid content is 3.96%; total protein content is 0.23%; protein content is 1.26%; starch content is 15.14%, and mucilage is 18.92% (mucus consists of D-glucose and L-rhamnose). Besides, its root contains 11 amino acids, which include histidine, arginine, threonine, alanine, proline, tyrosine, valine, phenylalanine, and leucine. In addition, 13 other mineral elements beneficial to the human body have been confirmed (Loi 2004). Therefore, *H. sagittifolius* Kurz. has been used in traditional medicines to support the body's weakness, cure diseases such as poor appetite, insomnia, back pain, abdominal pain, dizziness, cough fever, constipation (Loi 2004).

In this section, the effect of MG-simulated conditions on the germination, growth, development, and accumulation of secondary compounds of *H. sagittifolius* Kurz. was presented. The results showed that the application of MG-simulated conditions produced mechanical stimulation on the in vitro growth and secondary compound accumulation of this plant. In addition, it has initially provided some background information on the physiological and biochemical changes of medicinal plants under clinorotation.

17.3.1 Germination of H. sagittifolius Kurz. Under Clinorotation

This experiment was performed to evaluate the effect of MG-simulated conditions on in vitro germination of *H. sagittifolius* Kurz. Accordingly, the experiment was designed following the same steps as shown in Fig. 17.9. The germination rate and root length of the seedlings were obtained after 7, 14, and 20 days of culture. For histological evaluation, the seedlings after 20 days of culture were obtained and histologically performed as described by Peterson et al. (2008).

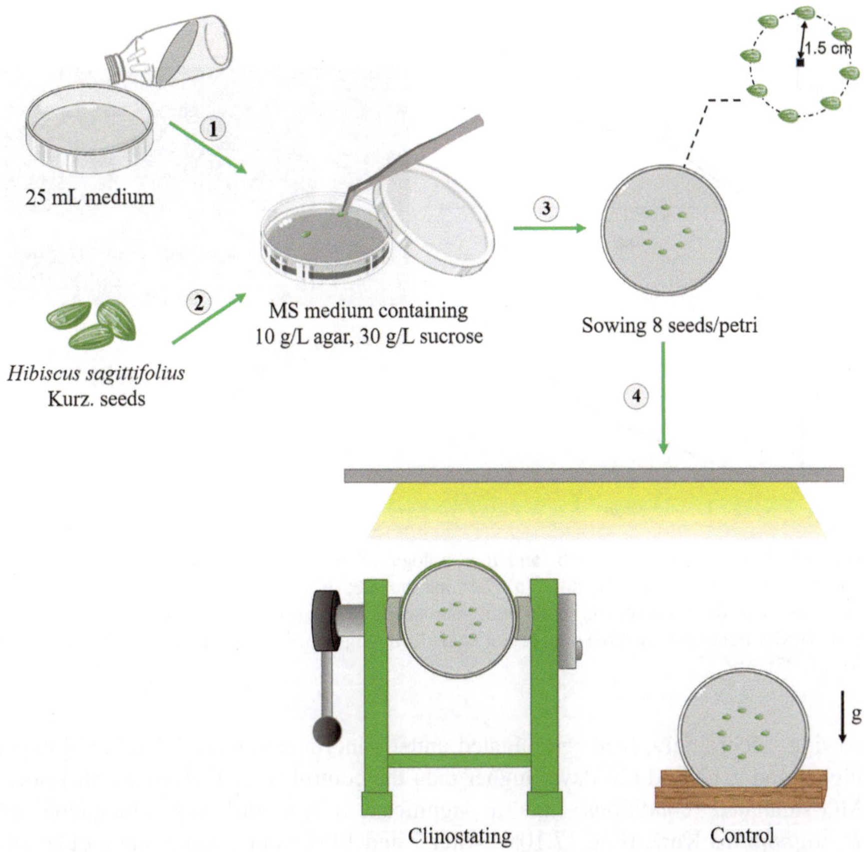

Fig. 17.9 Scheme of experimental design for in vitro germination of *Hibiscus sagittifolius* Kurz. seeds under clinostating. (**1**) MS (Murashige and Skoog 1962) culture medium containing 8 g/L agar, 30 g/L sucrose was prepared in plastic petri dishes (8.6 cm in diameter); (**2**) *Hibiscus sagittifolius* Kurz. seeds uniform sizes were sterilized with 0.1% $HgCl_2$ for 5 min and rinsed with sterile distilled water 3 times thereafter; (**3**) next, the seeds were sown in a circular pattern with 8 seeds/petri, that is, the seeds were 1.5 cm from the center of the clinostat axis of rotation; (**4**) then explants were grown under slow rotation of clinostat 2D with rotational speed set to 2 rpm, that is, maximum centrifugal acceleration of 1.9×10^{-4} g (Nakajima et al. 2019), for the control, the samples were grown under normal gravity. Both treatments were grown at 25 ± 2 °C, 55–60% relative humidity, illuminated by fluorescent lamp with photoperiod of 16 h/day and light intensity of 40–45 μmol m^{-2} s^{-1}

After 20 days of sowing, the germination and growth of the seedlings under MG and the control were significantly different, which was demonstrated by the indicators of germination rate, root length and root morphology (Fig. 17.10). The results showed that the seeds began to break the dormancy after 3 days of sowing in both treatments. After 7 days of sowing, the germination rate of seeds in the MG treatment (41.67%) was significantly higher than that in the control (33.33%). The difference in germination rate continued to be maintained after 14 and 20 days of

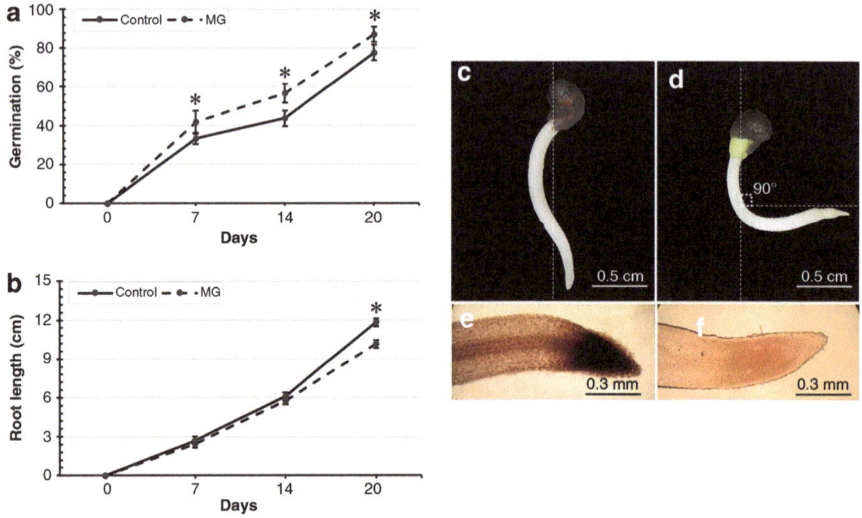

Fig. 17.10 Germination, growth, and morphology of *H. sagittifolius* Kurz seedlings. Under clinorotation after 20 days of culture. (**a**) Germination rate; (**b**) root length; (**c**) root tip morphology of control and (**d**) MG treatment under light microscope (×10); (**e**) seedling morphology of control and (**f**) MG treatment. Asterisks represent statistically significant differences with Duncan's test ($\alpha = 0.05$)

sowing; accordingly, seeds germinated under clinorotation were 1.3-fold (14 days) higher and 1.12-fold (20 days) higher than the control (Fig. 17.10a). Furthermore, MG-simulated conditions had a significant effect on root elongation of *H. sagittifolius* Kurz. (Fig. 17.10b). After 7 and 14 days of sowing, the root length of the MG treatment was not significantly lower than that of the control. While after 20 days of culture there was a significant difference in root length between the two treatments that was 11.83 ± 0.06 cm in the control and 10.16 ± 0.15 cm in the MG treatment.

Root histological morphology of two treatments after 7 days of sowing was obtained as shown in Fig. 17.10e, f. The results showed that the root meristem of the control treatment was dark pink after staining with carmine (Fig. 17.10e), while the roots of the MG treatment showed a lighter color (Fig. 17.10f). The results suggested that there was a change in the number of cells in the root meristem region under clinorotation. In addition, root morphology showed differences between two treatments; in the control seedlings, the roots developed straight in the direction of gravity (Fig. 17.10c); in the MG seedlings, the roots were elongated no directional with a curved shape (Fig. 17.10d).

The change in biomass of seedlings under MG simulation conditions was obtained (Fig. 17.11). After 7 days of sowing, fresh weight (FW) and dry weight (DW) of seedlings were not statistically significant between two treatments. After 14 days of culture, the FW of MG seedlings (42.97 ± 1.10 mg) was significantly lower than that of the control (53.47 ± 0.87 mg) (Fig. 17.11a–c); however, the DW

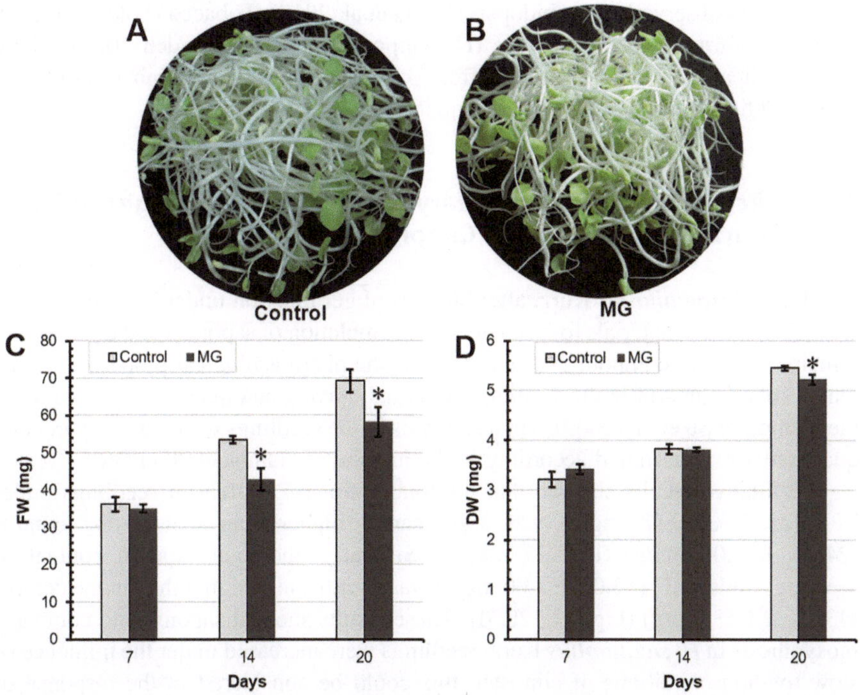

Fig. 17.11 Biomass accumulation of *H. sagittifolius* Kurz. seedlings under clinorotation after 20 days of culture. (**a**) Growth of seedlings of control and (**b**) MG treatment, (**c**) fresh weight of seedlings (FW) and (**d**) dry weight of seedlings (DW). Asterisks represent statistically significant differences with Duncan's test ($\alpha = 0.05$)

of two treatments did not differ (Fig. 17.11d). After 20 days of sowing, the FW and DW of the MG treatment (58.28 ± 0.36 mg and 5.23 ± 0.08 mg, respectively) were significantly lower than that of the control (69.24 ± 0.77 mg and 5.46 ± 0.05 mg, respectively). In summary, the results showed that the slow rotation conditions of clinostat stimulated the germination rate of *H. sagittifolius* Kurz. seeds; however, root length, FW and DW of seedlings were significantly reduced under gravity deprivation.

Current results in *H. sagittifolius* Kurz. showed similarities with studies on *Lepidium sativum* (Hensel and Sievers 1980) and on Soybeans (Howard and William 2005) when the seeds were sown under simulated MG conditions. Biomass of *H. sagittifolius* Kurz. was lower under MG may be explained by the limited bioavailability of oxygen in the rhizosphere under gravity deprivation (Porterfield 2002); which has affected the absorption of nutrients in roots and caused a deficiency of minerals necessary for plant growth. On the other hand, the underlying cause of the reduction in the bioavailability of oxygen to the plants may have been related to the increased temperature production of the plants under MG stress (Kitaya et al. 2003). Contrary to the present case in *H. sagittifolius* Kurz., a significant increase in biomass was observed in Arabidopsis (Matía et al. 2010), Tobacco (Soleimani et al. 2019a, b), *Matricaria chamomilla* (Hassanpour and Ghanbarzadeh 2021). These results indicated that the extent of effects of MG on plants and their responses to gravity-deficit stress was species dependent.

17.3.2 Accumulation of Secondary Compounds of H. sagittifolius Kurz Seedlings. Under Clinorotation

Seedlings *H. sagittifolius* Kurz. after 20 days of germination under MG and control were used as raw materials to evaluate the accumulation of saponins and coumarin—2 main secondary compounds that determine the pharmacological properties of this plant. Total coumarin in the seedlings was qualitatively and quantified according to the method of Su et al. (2009). Total saponins in the seedlings were qualitatively and quantitatively determined according to the method of Namba et al. (1974).

Quantitative results showed that the total coumarin content in seedlings under MG (25.67 ± 0.58 mg/g) was significantly higher than that in the control (24.00 ± 1.00 mg/g) (Fig. 17.12a, c). Similarly, the total saponin content in seedlings under MG (53.00 ± 1.00 mg/g) was significantly higher than in the control (43.33 ± 1.53 mg/g) (Fig. 17.12b, d). These results showed saponin and coumarin biosynthesis in *H. sagittifolius* Kurz. seedlings were increased under the influence of slow rotation conditions of clinostat; this could be considered as the response of plants to MG stress.

17.3.3 In Vitro Shoot Multiplication of H. sagittifolius Kurz. Under Clinorotation

This experiment was performed to evaluate the effect of MG-simulated conditions on in vitro shoot multiplication of *H. sagittifolius* Kurz. Accordingly, the experiment was designed following the same steps as Fig. 17.13. Growth indicators such as length of shoot (cm), number of nodes, number of shoots, FW (mg), DW (mg), chlorophyll a and b content (μg/g), and total chlorophyll ab content (μg/g) in leaves were recorded after 30 days of culture. In which, the chlorophyll a and b contents were evaluated by spectrophotometric analysis of leaf extracts in acetone solution using a UV-2900 spectrophotometer (Hitachi, Japan). The absorbance was measured at 662 and 645 nm (Lichtentaler and Wellburn 1985).

The results showed that shoots cultured under MG simulation gave the length of shoots (3.07 ± 0.21 cm), number of nodes (6.33 ± 0.58), number of shoots (3.33 ± 0.58), FW (401.33 ± 53.27 mg), DW (37.00 ± 5.20 mg) were significantly higher than the control treatment (Fig. 17.14a, b, d). In addition, chlorophyll a, b, and total chlorophyll were significantly increased under MG (Fig. 17.14c). These results revealed that slow rotation of clinostat had a positive effect on *H. sagittifolius* Kurz. shoot multiplication. Morphological observation of shoots propagated under MG did not show any abnormal morphology; in addition, shoots with large and dark green

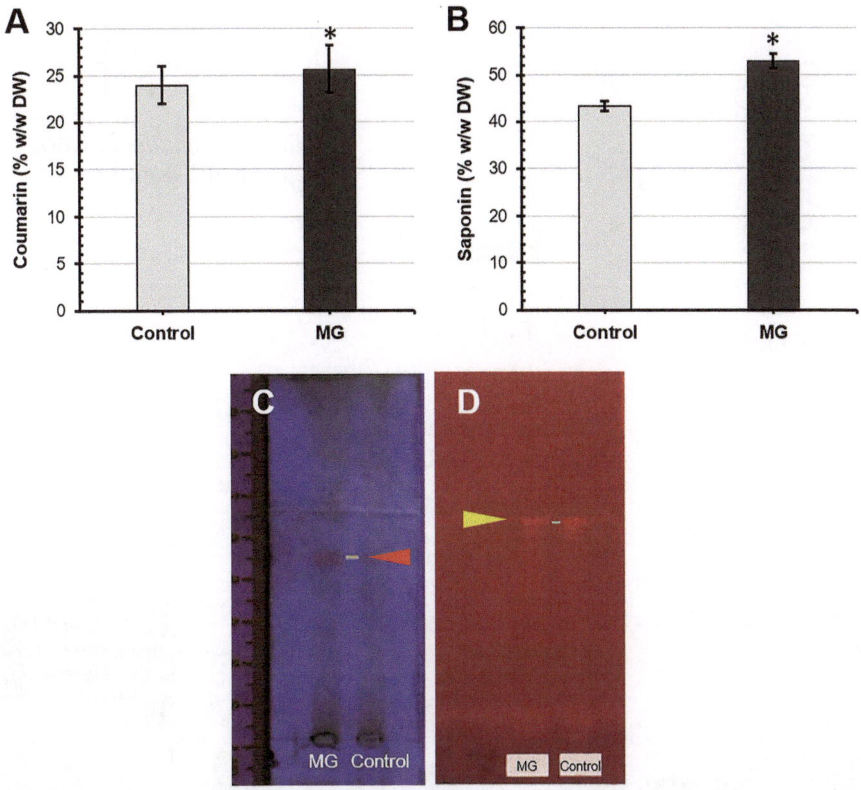

Fig. 17.12 Qualitative and quantitative determination of total saponins and coumarins accumulated in the seedlings of *H. sagittifolius* Kurz. under clinorotation after 20 days of culture. (**a**) Quantification of coumarin; (**b**) quantification of saponins; (**c**) qualitative thin layer chromatography of coumarin and (**d**) saponin. Asterisks represent statistically significant differences with Duncan's test ($\alpha = 0.05$)

leaves under MG were observed. An increase in plant height under MG was observed in *Veronica arvensis* (Aarrouf et al. 1999; Driss-École et al. 1994).

17.3.4 In Vitro Rooting of H. sagittifolius Kurz. Under Clinorotation

This experiment was performed to evaluate the effect of MG-simulated conditions on the in vitro rooting of *H. sagittifolius* Kurz. Accordingly, the shoot tips (1 cm in length with 1 pair of leaves) of *H. sagittifolius* Kurz. 2-month-old in vitro were cultured into test tubes (12 × 120 mm) containing 5 mL of MS medium containing with 1.0 mg/L NAA. The rotation speed of the clinostat was set to 2 rpm; for the control treatment the explants were grown in stationary condition. In vitro culture conditions were established similar to the shoot multiplication experiment. Growth indicators such as length of shoots (cm), root length (cm), number of leaves, number

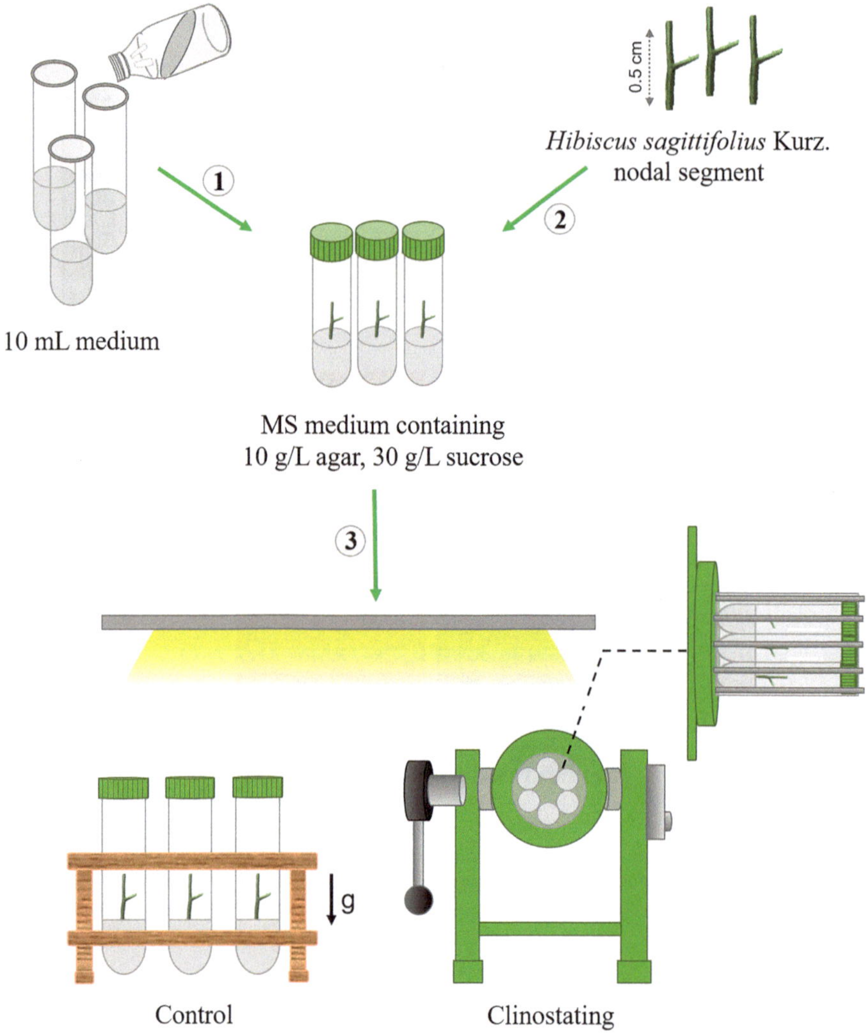

Fig. 17.13 Scheme of experimental design for evaluation of shoot multiplication of *H. sagittifolius* Kurz. under clinostating. (**1**) MS culture medium containing 8 g/L agar, 30 g/L sucrose, and 0.5 mg/L BA was prepared in test tubes (12 × 120 mm); (**2**) after that the nodal segment (0.5 cm in length) of *H. sagittifolius* Kurz. in vitro 2-month-olds were inoculated into test tubes; (**3**) then explants were grown under slow rotation of clinostat 2D with rotational speed set to 2 rpm, that is, maximum centrifugal acceleration of 1.9×10^{-4} g (Nakajima et al. 2019), for the control, the samples were grown under normal gravity. Both treatments were grown at 25 ± 2 °C, 55–60% relative humidity, illuminated by fluorescent lamp with photoperiod of 16 h/day and light intensity of 40–45 μmol m^{-2} s^{-1}

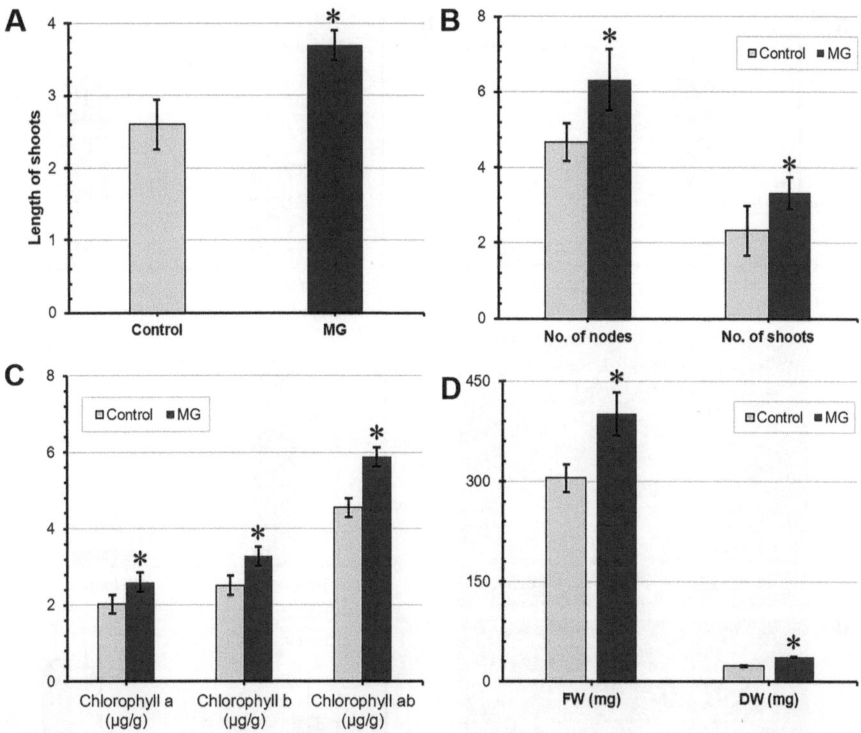

Fig. 17.14 Shoot multiplication of *H. sagittifolius* Kurz. under clinorotation after 30 days of culture. (**a**) Length of shoots; (**b**) number of nodes and number of shoots; (**c**) the content of chlorophyll a, b, and ab in the leaves; (**d**) fresh and dry weight of shoots. Asterisks represent statistically significant differences with Duncan's test ($\alpha = 0.05$)

of roots, FW (mg), DW (mg), chlorophyll a and b content (μg/g), and total chlorophyll ab content (μg/g) in leaves were recorded after 30 days of culture.

As shown in Fig. 17.15, the results showed that, under MG condition, plant height (12.17 ± 1.04 cm), number of leaves (5.67 ± 0.58), root length (1.77 ± 0.25 cm), FW of plantlet (419.00 ± 15.87 mg), DW of plantlet (36.00 ± 2.00 mg) was significantly higher than the control treatment (Fig. 17.15a–c, e). However, MG did not affect the number of roots (9.00 ± 2.00), chlorophyll a content (82 ± 6.66 μg/g), chlorophyll b content (8.37 ± 1.76 μg/g), and total chlorophyll ab content (30.19 ± 8.41 μg/g) in leaves compared with the control (Fig. 17.15b, f). Morphological observations showed that the roots formed under slow rotation of clinostat were longer than those of the control (Fig. 17.15c, d); in addition, the roots have grown in multidirection and even antigravity directions (Fig. 17.15d). In summary, the in vitro growth of the plantlet *H. sagittifolius* Kurz. was significantly increased under MG. These results were similar to those obtained on *Brassica napus* (Aarrouf et al. 1999), *Veronica arvensis* (Driss-École et al. 1994). As reported by Leather et al. (1972) increased ethylene signaling under MG stress has caused a

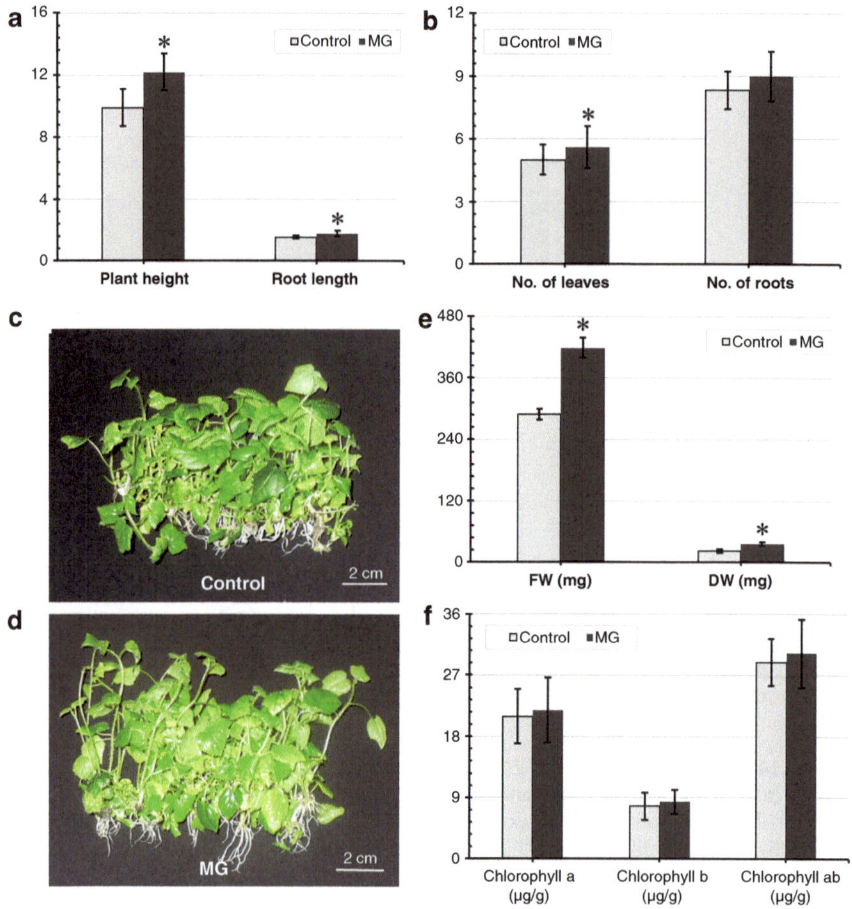

Fig. 17.15 In vitro rooting of *H. sagittifolius* Kurz shoots under control and clinorotation after 30 days of culture. (**a**) Plant height and root length; (**b**) number of leaves and number of roots; (**c**) plantlet under control condition; (**d**) plantlet under simulated microgravity condition (MG); (**e**) fresh and dry weight of plantlets; (**f**) the content of chlorophyll a, b, and ab in the leaves. Asterisks represent statistically significant differences with Duncan's test ($\alpha = 0.05$)

series of physiological, biochemical, and developmental changes in plants. For example, the stem elongation of *B. napus* was abnormal under simulated MG conditions compared with control (Aarrouf et al. 1999). In addition, MG accelerated cell growth and differentiation and senescence (Kordyum 1994). In this case, the results revealed that gravity deprivation did not negatively affect the in vitro growth of *H. sagittifolius* Kurz. plantlet; in contrast, MG acted as a mechanical stimulant to increase biomass, root elongation and stem elongation—which needs further study at the molecular level to elucidate.

17.4 In Vitro Growth, Flowering, and Accumulation of Secondary Compounds of *Phyllanthus amarus* Under Clinorotation

Phyllanthus amarus is a precious medicinal plant belonging to the *Phyllanthaceae* family. It is a tropical herb that has been found in the Americas, Africa, India, China, India, Malaysia, the Philippines, Indonesia, Myanmar, Thailand, and Vietnam. Extracts from *P. amarus* protected hepatocytes against toxins such as paracetamol, carbon tetrachloride, ethanol, aflatoxin B1, and galactosamine (Wongnawa et al. 2005; Harish and Shivanandappa 2006; Bhattacharjee and Sil 2007). In addition, *P. amarus* had analgesic effects, increased uric acid excretion, inhibited some viruses (hepatitis B virus, human immunodeficiency syndrome virus-HIV) (Qian-Cutrone et al. 1996; Sane et al. 1997).

In this section, the effects of MG simulation conditions using clinostat 2D (2 rpm) on the growth, flowering, and accumulation of secondary compounds of *P. amarus* cultured in vitro were reported. The results showed a significant effect of MG conditions on in vitro shoot multiplication, biomass, and shoot elongation of *P. amarus*. In particular, the slow rotation of clinostat stimulated the in vitro flowering of *P. amarus* earlier than that of the control. In addition, an increase in Phyllanthin synthesis under the influence of gravity deprivation was obtained. Current results on the effect of MG on *P. amarus* and *H. sagittifolius* Kurz. (the case study has been shown above) have initially provided important information on the growth, development and accumulation of secondary compounds of medicinal plants—on-site drug supplies for astronauts during long-term space exploration missions.

17.4.1 In Vitro Shoot Multiplication of P. amarus Under Clinorotation

In vitro nodal segments of 1-month-old *P. amarus* cultured on MS medium were used as initial material for this experiment. Nodal segments were cultured on MS medium containing 0.8 mg/L BA and fixed on clinostat (2 rpm); for the control treatment the explants were grown in stationary condition. The explants in both treatments were grown at 25 ± 2 °C, 55–60% relative humidity, illuminated with fluorescent lamp light with a photoperiod of 16 h/day and light intensity 40–45 μmol m^{-2} s^{-1}. Growth indicators such as length of shoot (cm), number of shoots, number of leaves, FW (mg), DW (mg) were recorded after 30 days of culture.

As shown in Fig. 17.16, the shoot multiplication of *P. amarus* under MG was significantly different from that of the control. Accordingly, the number of shoots under the influence of simulated MG conditions (11.00 shoots) was significantly higher than the control (7.00 shoots) (Fig. 17.16b). In contrast, plant height (0.33 cm), FW (135.67 mg), DW (13.33 mg) of the MG treatment were significantly lower than that of the control (0.57 cm; 181.33 mg; 17.33 mg; respectively) (Fig. 17.16c, d), while the number of leaves was not statistically significant between

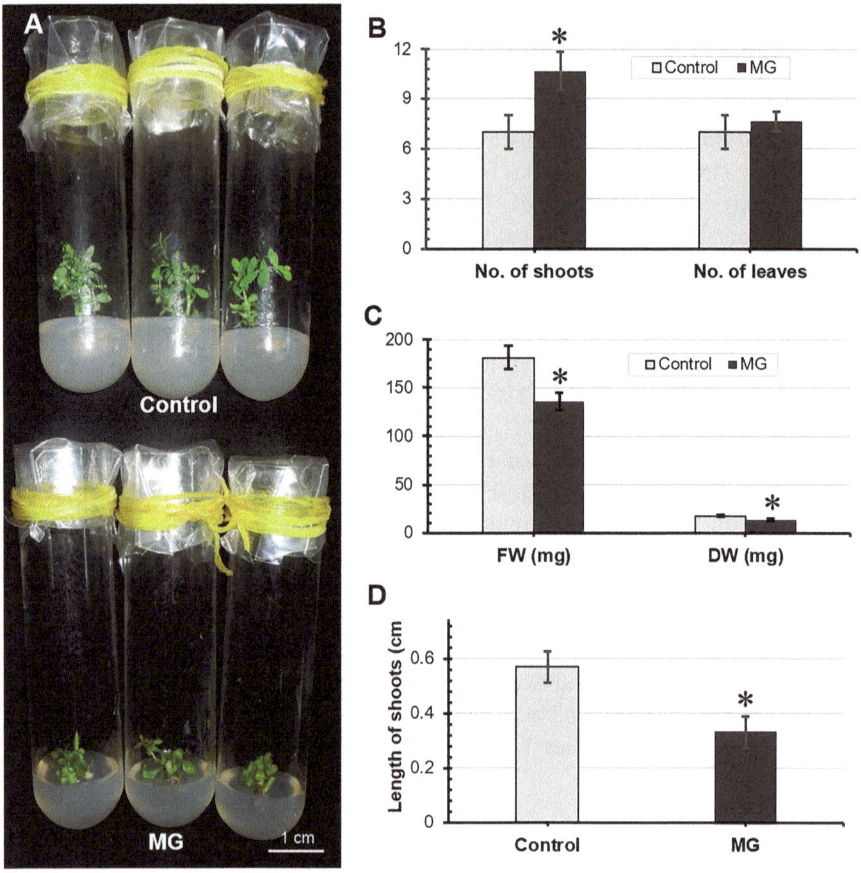

Fig. 17.16 *P. amarus* shoot multiplication under clinorotation after 30 days of culture. (**a**) Growth of control shoots and MG treatment shoots; (**b**) number of shoots and number of leaves; (**c**) fresh and dry weight of shoots; (**d**) length of shoots. Asterisks represent statistically significant differences with Duncan's test ($\alpha = 0.05$)

two treatments (Fig. 17.16b). Morphological observations showed that the shoots of *P. amarus* under MG were more stunted than those of the control (Fig. 17.16a). The results suggested that MG-simulated conditions had a positive effect on the number of shoots of *P. amarus*; this result showed similarity with the case of *H. sagittifolius* Kurz. (the case study was shown above). It is well known that exogenous cytokinins are an important source of hormones for shoot proliferation; while cytokinin uptake by plants was increased under clinorotation (Fabio et al. 2002) thus the shoot proliferation of both *H. sagittifolius* Kurz. and *P. amarus* significantly increased under MG simulation conditions. However, shoot height and biomass of *P. amarus* were significantly reduced under slow rotation of clinostat which was in contrast to the results in *H. sagittifolius* Kurz (case study) and in *Mentha piperita* L. (Fabio et al. 2002). This difference was probably due to the limitation of bioavailability of oxygen under MG conditions which reduced plant growth (Porterfield 2002). In summary, the results revealed that the response of each plant species to MG stress was different.

17.4.2 In Vitro Rooting and Flowering of *P. amarus* Under Clinorotation

To evaluate the in vitro rooting and growth of *P. amarus* shoots under simulated MG conditions, in vitro shoot tips (0.5 cm) of 1-month-old *P. amarus* derived from MS medium were cultured on MS medium containing 1.5 mg/L IBA and fixed on clinostat (2 rpm), for the control treatment the explants were grown in stationary condition. In vitro culture conditions were established similar to the shoot multiplication experiment. Growth indicators such as length of shoot (cm), root length (cm), internode length (cm), number of leaves, number of nodes, number of roots, FW (mg), DW (mg) were recorded after 30 days of culture.

The results showed that the plantlets of the MG treatment achieved plant height (4.40 cm), internode length (0.63 cm), number of roots (18.33 roots), FW (181.33 mg), and DW (17.67 mg) was significantly lower than the control (6.33 cm; 1.03 cm; 23.00 roots; 286.00 mg; 26.33 mg; respectively) (Fig. 17.17). However, the root length of the MG treatment (1.87 cm) was 1.4-fold (1.3 cm) higher than that of the control (Fig. 17.17a). Meanwhile, the number of nodes (8.33 nodes) and the number of leaves (8.67 leaves) did not have a statistically significant difference between two treatments (Fig. 17.16d). Interestingly, the plantlets on clinostat showed significantly earlier flowering than the control 8 days (17.7 days and 24.33 days, respectively) (Fig. 17.18a–c).

Morphological observations showed that roots under MG conditions grew in the opposite direction of gravity; in addition, primary roots under clinorotation showed underdevelopment instead of secondary roots formation and their over elongation (Fig. 17.16b, c). These results were similar to the results of Aarrouf et al. (1999), the secondary roots of *B. napus* increased significantly in number and length after 15 days of clinostat treatment. According to Leather et al. (1972), the slow rotation of clinostat promoted the plants to twofold higher ethylene production after 120 min of treatment. Similarly, MG stress has been confirmed to cause ethylene-mediated accelerated plant senescence (Kordyum et al. 2019). In the present case, the early flowering of *P. amarus* plantlets under clinorotation could be related to the altered accumulation of ethylene gas in the culture tubes. In contrast to the present results on shoot length of *P. amarus*, shoot length was significantly increased under clinorotation in *H. sagittifolius* Kurz. (case study); Arabidopsis (Brown et al. 1976); *Coleoptiles avena* (Brown et al. 1995). The results of the reduction in biomass of *P. amarus* plantlet under MG were consistent with the results in *Hemerocallis flava* L. (Levine and Krikorian 1992); however, it was contrary to the results in *H. sagittifolius* Kurz.

In summary, these results demonstrate the profound effects of MG conditions not only on the growth but also on the reproduction of *P. amarus* cultured in vitro. In

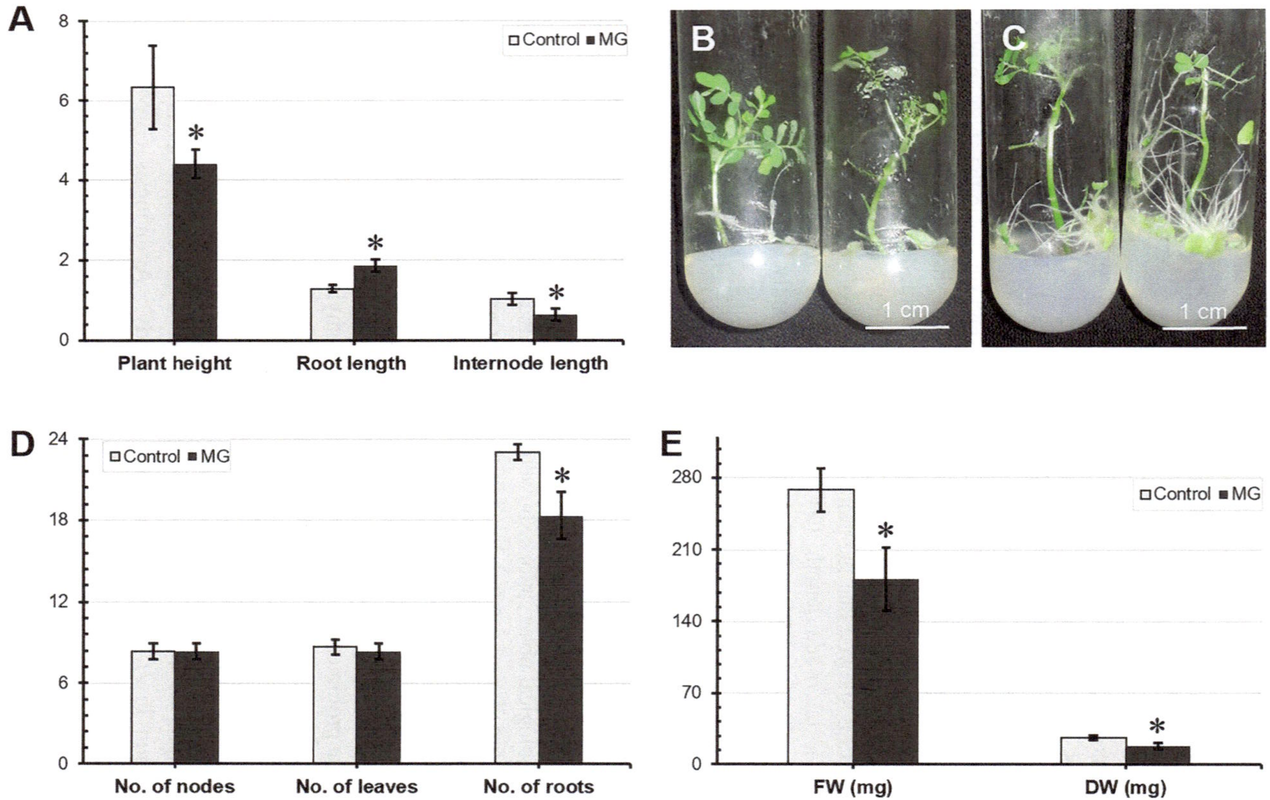

Fig. 17.17 In vitro rooting of *P. amarus* shoots. Under clinorotation after 30 days of culture. (**a**) Length of shoots, length of roots and internode length; (**b, c**) growth of control plantlets and MG treatment plantlets; (**d**) number of nodes, number of leaves, and number of roots; (**e**) fresh and dry weight of plantlets. Asterisks represent statistically significant differences with Duncan's test ($\alpha = 0.05$)

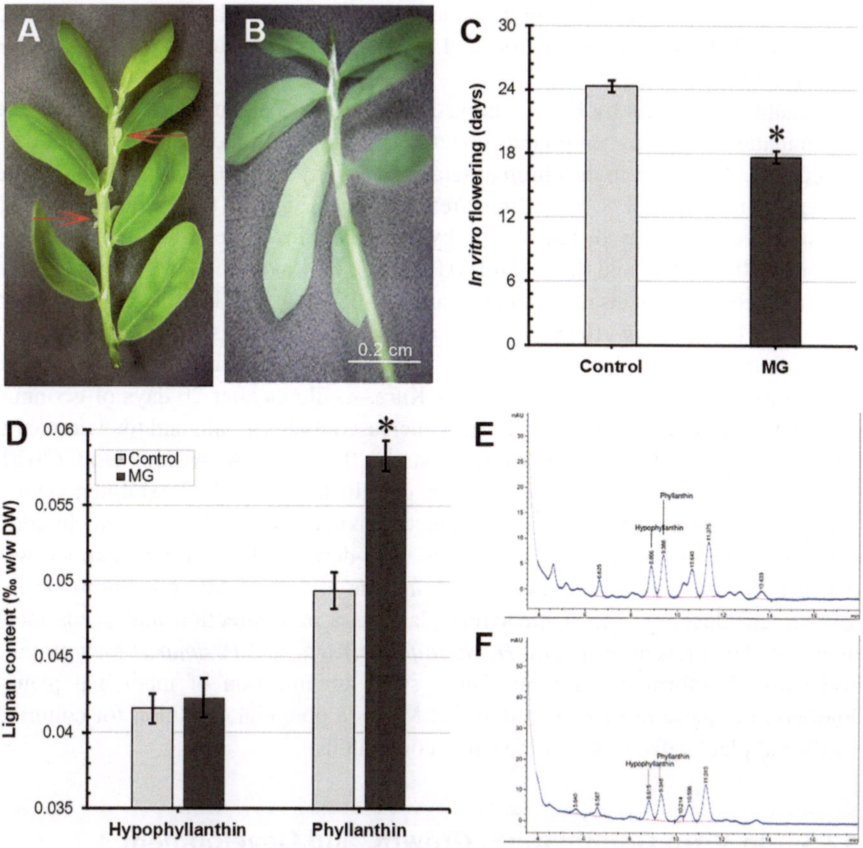

Fig. 17.18 In vitro flowering and secondary accumulation of *P. amarus* plantlets under clinorotation. (**a**) Flowering of plantlets under clinorotation after 18 days of culture (arrow); (**b**) no flowering of plantlets under gravity (control) after 18 days of culture; (**c**) flowering time; (**d**) total content of hypophyllanthin and phyllanthin in plantlets after 30 days of culture; (**e**) peak hypophyllanthin, phyllanthin on HPLC chromatogram for total lignan analysis of control plantlets; and (**f**) plantlets under clinostating after 30 days of culture. Asterisks represent statistically significant differences with Duncan's test ($\alpha = 0.05$)

addition, these alterations are the directions for further studies at the molecular level for the purpose of understanding the potential impact of MG conditions on higher plants.

17.4.3 Accumulation of Secondary Compounds of P. amarus Under Clinorotation

Two bioactive lignans of *P. amarus* (phyllanthin, hypophyllanthin) were evaluated by high performance liquid chromatography (HPLC) (Sharma et al. 1993).

Accordingly, 1-month-old *P. amarus* plantlets cultured on MS medium containing 1.5 mg/L IBA under clinorotation and control were collected and evaluated for lignans content.

Quantitative results by HPLC showed that there was a significant change in the accumulation of lignan compounds under MG simulation conditions compared with the control. As shown in the chromatograms, hypophyllanthin and phyllanthin were present at intervals of 8 and 9 min, respectively (Fig. 17.18e, f). There was no statistically significant difference in hypophyllanthin content between the MG treatment (0.0432‰) and the control (0.0416‰). However, the level of phyllanthin accumulated in the plants under clinorotation (0.0583‰) was 1.2-fold higher than in the control treatment (0.0494%) (Fig. 17.18d). The results suggested that MG-simulated conditions stimulated increased phyllanthin biosynthesis in *P. amarus*. Similarly, the *H. sagittifolius* Kurz. seedlings after 20 days of germination under clinorotation had significantly increased total saponin and total coumarin content compared with the control (case study). Recently, Al-Awaida et al. (2020) reported that *Triticum aestivum* seedlings germinated under MG exhibited significantly increased antioxidant potential and antioxidant metabolite content; In addition, the antidiabetic bioactivity of the MG-derived *T. aestivum* extract was enhanced in rat tests. In summary, studying the effects of MG conditions on the growth and development of medicinal plants is a new direction that needs more attention. The present results on *H. sagittifolius* Kurz. and *P. amarus* may provide some useful information for the future space colonization of medicinal plants. Furthermore, these results suggested that MG is a potential condition for culturing medicinal plant cells to obtain bioactive compounds.

17.5 In Vitro Germination, Growth, and Development of *Catharanthus roseus* under Clinorotation

Catharanthus roseus is a multipurpose medicinal species of the genus Catharanthus. In the wild, *C. roseus* is critically endangered by the destruction of its habitat due to human farming activities. However, it has been widely cultivated and conserved in many tropical and subtropical regions around the world (Huxley 1992). Recent studies have shown that in *C. roseus* there are many pharmaceutical substances such as vinblastine, vincristine, vindoline, and catharanthine; these drugs have strong mitotic inhibition (Wendell et al. 1993); it is therefore used in tumor suppression in blood cancers, cervical cancers, lung carcinomas, lymphomas, and endometrial tumors (Magnotta et al. 2006). Besides, *C. roseus* is also used to treat malaria, hypertension, diabetes, anti-inflammatory, and dysentery in traditional Chinese medicine, Bangladesh, and Vietnam (Mallikq et al. 2013). In particular, recent studies have shown that secondary compounds present in *C. roseus* leaf extracts are effective against dangerous diseases caused by bacteria *Pseudomonas aeruginosa*, *Salmonella typhimurium*, or *Staphylococcus aureus* (Prajakta and Jai 2010) and pathogenic fungi such as *Fusarium moniliforme*, *Candida albicans*, *Aspergillus fumigatus*, or *Aspergillus niger* (Kumari and Gupta 2013).

In this section, the effects of MG simulation conditions on the germination, growth and development of *C. roseus* were presented. The results showed that the application of conditions simulating MG stimulated the germination and in vitro growth of this plant. These results have contributed some background information needed for the development of a strategy to effectively exploit the medicinal potential of *C. roseus* during long-term space flights.

17.5.1 Germination of *C. roseus* Under Clinorotation

The seeds of *C. roseus* were sterilized in $HgCl_2$ (0.1%) for 4 min. Then, the seeds were sown in a plastic petri dish (8.6 cm in diameter) containing 25 mL of MS medium, 10 g/L agar, and 30 g/L sucrose. The inoculation density was 8 seeds/dish and the seeds were sown in a circular pattern, that is, the seeds were 1.5 cm from the center of the clinostat axis of rotation. For the control treatment the seeds were grown in stationary condition. For the MG treatment the seeds were grown under slow rotation of clinostat 2D and the rotational speed was set to 2 rpm. The explants in both treatments were grown at 25 ± 2 °C, 55–60% relative humidity, illuminated with fluorescent lamp light with a photoperiod of 16 h/day and light intensity 40–45 μmol m^{-2} s^{-1}.

After 30 days of sowing, germination and growth of *C. roseus* seedlings under simulated MG and control conditions showed significant differences (Fig. 17.19). The results showed that the seeds in both treatments started to germinate after 7 days of culture. However, the germination rate under MG after 7, 14, and 20 days of

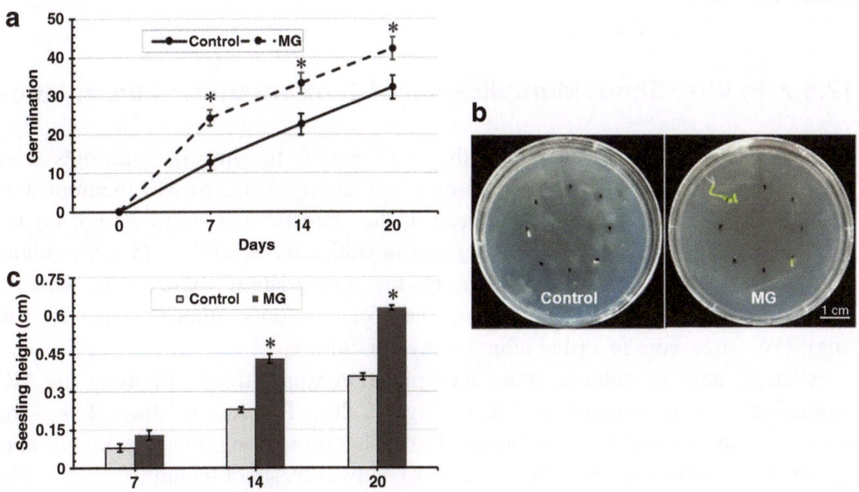

Fig. 17.19 Germination, growth and morphology of *C. roseus* seedlings under clinorotation after 20 days of culture. (**a**) Germination rate; (**b**) seedling height; (**c**) seedling growth of control and MG treatment. Asterisks represent statistically significant differences with Duncan's test ($\alpha = 0.05$)

culture (24.33%, 33.33%, and 42.33% respectively) was significantly higher than the control (12.67%, 22.67%, and 32.33% respectively). After 20 days of culture, no further seed germination was observed in both treatments (Fig. 17.19a, b). In addition, the slow rotation of clinostat stimulated an increase in plant height of *C. roseus* seedlings (Fig. 17.19c). After 14 days of culture, there was no statistically significant difference in plant height between two treatments. However, after 20 days of culture, the plant height under MG was (0.43 cm) twofold higher than the control (0.23 cm) (Fig. 17.19b, c); moreover, after 30 days of culture, differences in plant height were still observed under clinorotation (0.63 cm) and control (0.36 cm) (Fig. 17.19c).

The results showed that MG simulation conditions (2 rpm) stimulated germination and increased height of *C. roseus* seedlings. Similarly, the germination rate of *H. sagittifolius* Kurz. seeds were significantly increased under clinorotation (case study); in contrast, the root length of *H. sagittifolius* Kurz. seedlings under MG were significantly lower than that of the control 20 days after sowing. Emmanuel et al. (1996) reported that slow rotation of clinostat (1 rpm) stimulated increased germination of Soybean seeds and significantly increased root length and FW of seedlings after 7 days of sowing. According to Nakajima et al. (2019) Clinorotation (2 rpm) stimulated shoot and root elongation of Mung bean seedlings 4 days after sowing. It is well known that seed germination is a complex biological process involving many biochemical reactions with the participation of the starch-degrading enzyme amylase. Under the influence of MG simulation conditions, the α-amylase enzyme activity was confirmed to be higher than under normal gravity conditions (Nakajima et al. 2019). Therefore, it is possible that MG-simulated conditions stimulated disruption of the seed dormancy through a pathway that regulates the activity of related enzymes.

17.5.2 In Vitro Shoot Multiplication of *C. roseus* Under Clinorotation

The nodal segment (0.3 cm in length) of *C. roseus* in vitro 1-month-olds were inoculated into test tubes (12 × 120 mm) containing 5 mL of MS medium with 0.8 mg/L BA added. The rotation speed of the clinostat was set to 2 rpm; for the control treatment the explants were grown in stationary condition. In vitro culture conditions were established similar to the germination experiment. Growth indicators such as length of shoot (cm), number of shoots, number of leaves, FW (mg), DW (mg) were recorded after 30 days of culture.

After 30 days of culture, shoot multiplication was different between the MG simulation and the control treatment (Fig. 17.20). The results showed that the number of shoots under clinorotation (11.67 shoots) was significantly higher than that of the control treatment (Fig. 17.20a, b). However, shoot length (0.23 cm), FW (590.33 mg), and DW (55.76) were significantly lower than that of the control (0.43 cm; 766.33 mg and 76.67 mg, respectively) (Fig. 17.20c, d). Meanwhile, there was no statistically significant difference in the number of leaves between two treatments (2.76 and 3.76 leaves, respectively) (Fig. 17.20b).

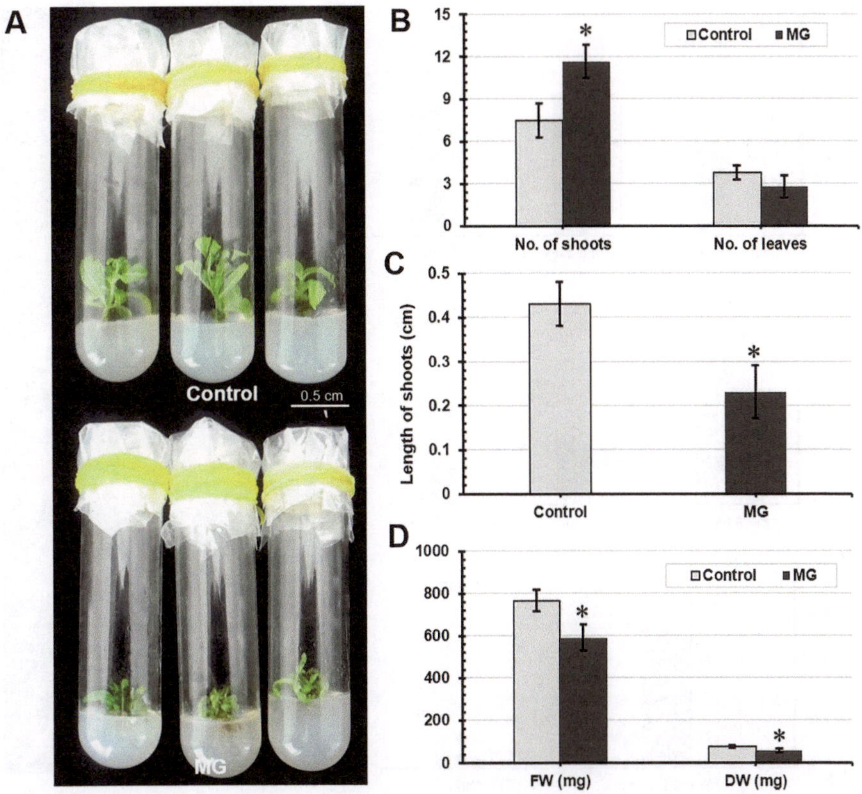

Fig. 17.20 Shoot multiplication of *C. roseus* under clinorotation after 30 days of culture. (**a**) Growth of control shoots and MG treatment shoots; (**b**) number of shoots and number of leaves; (**c**) length of shoots; (**d**) fresh and dry weight of shoots. Asterisks represent statistically significant differences with Duncan's test ($\alpha = 0.05$)

These results showed that conditions simulating MG had a positive effect on the number of adventitious shoots of *C. roseus*; similar results have been reported in *H. sagittifolius* Kurz. (Case study) and *P. amarus*. However, in the case of *H. sagittifolius* Kurz., the MG condition stimulated more shoot elongation than the gravity condition; while on *C. roseus* and *P. amarus* shoot length was lower than under clinorotation. The results suggested that different plant species responded differently to the same conditions.

17.5.3 In Vitro Rooting of *C. roseus* Under Clinorotation

In vitro shoot tips (0.5 cm) of 1-month-old *C. roseus* derived from MS medium were cultured on MS medium containing 1.5 mg/L NAA and fixed on clinostat (2 rpm), for the control treatment the explants were grown in stationary condition. In vitro

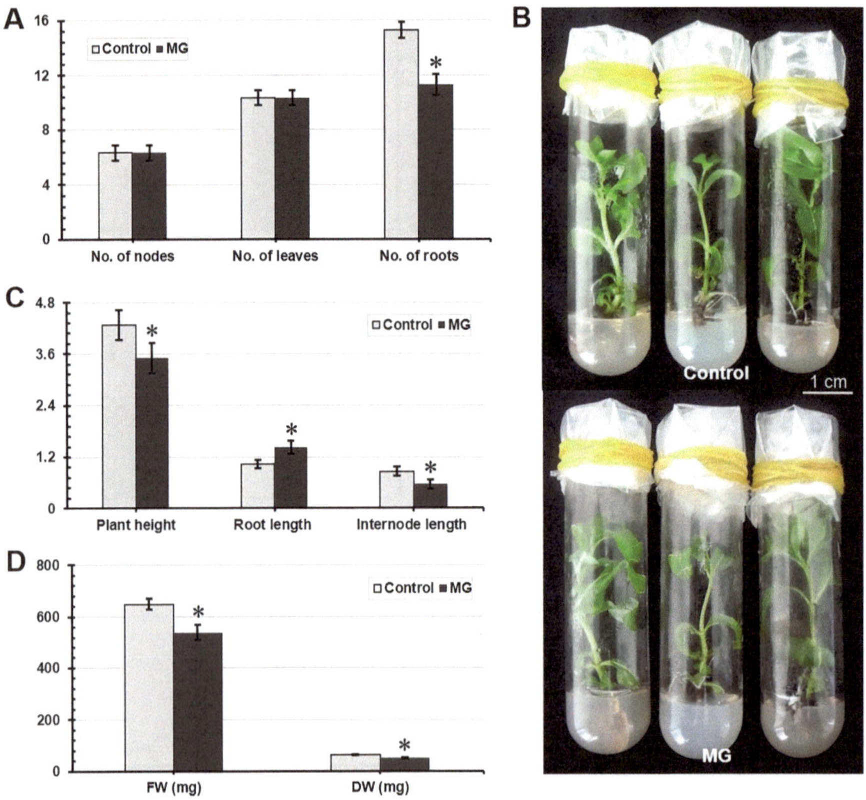

Fig. 17.21 In vitro rooting of *C. roseus* shoots. Under clinorotation after 30 days of culture. (**a**) Number of nodes, number of leaves and number of roots; (**b**) growth of control plantlets and MG treatment plantlets; (**c**) plant height, root length and internode length; (**d**) fresh and dry weight of plantlets. Asterisks represent statistically significant differences with Duncan's test ($\alpha = 0.05$)

culture conditions were established similar to the shoot multiplication experiment. Growth indicators such as length of shoot (cm), root length (cm), internode length (cm), number of leaves, number of nodes, number of roots, FW (mg) and DW (mg) were recorded after 30 days of culture.

The results showed that there was a difference in the growth and development indicators of plants under simulated MG conditions compared with the control (Fig. 17.21). Accordingly, the plantlets under clinorotation reached plant height (3.50 cm), internode length (0.56 cm), number of roots (11.30 roots), FW (539.00 mg), and DW (49. 67 mg) were lower than the control (4.27 cm; 0.86 cm; 15.30 roots; 648.00 mg and 62.00 mg, respectively) (Figs. 17.21a, c, d). However, the root length of the MG treatment (1.43 cm) was higher than that of the control (1.03 cm). Meanwhile, the number of nodes and the number of leaves were not statistically different between the MG treatment and the control (Fig. 17.21a). Similar to the results in *H. sagittifolius* Kurz. (case study) and *P. amarus*, adventitious root length was significantly increased under clinorotation; in addition, the roots have grown in different directions and include antigravity (Fig. 17.21b).

In the present case, a similar response under clinorotation of *C. roseus* and *P. amarus* was shown by the decrease in internode length, number of roots, plant height and biomass under simulated MG stress. In contrast, plant height and biomass of *H. sagittifolius* Kurz. plantlets were significantly increased under clinorotation, while the number of roots that were unaffected by MG conditions was also observed. In summary, the results showed different responses of 3 medicinal plants *H. sagittifolius* Kurz., *P. amarus* and *C. roseus* under the influence of MG simulation conditions. In addition, these results show the efficiency of studying many plant species under the same MG simulation conditions, thereby building a usable database for further research and BLSS construction purposes.

17.6 Conclusion

In summary, results from the Space Seed for Asian Future program 2010–2011 in Rose balsam have shown that Space breeding by spaceflight is a potential way for breeding work in the field of Biotechnology. In addition, in vitro growth and development studies of plants under simulated MG conditions showed that different plant species reacted differently under clinostating. However, consensus results indicated that MG-simulated conditions stimulated increased seed germination and increased accumulation of secondary compounds in medicinal plants. These results are important basis for studies to design and construction of BLSS for space flights, in which medicinal plants are abundant source of medicine for astronauts. Moreover, it has also opened up the potential applications of clinostating in biomass production and obtaining valuable medicinal substances to serve the pharmaceutical industry. Although current technical limitations did not allow us to conduct further experiments on the mechanism of plant growth and development under clinostating, current information may have provided important ideas and a reference for future research.

References

Aarrouf J, Schoëvaërt D, Maldiney R, Perbal G (1999) Changes in hormonal balance and meristematic activity in primary root tips on the slowly rotating clinostat and their effect on the development of the rapseed root system. Physiol Plant 105:708–718

Al-Awaida WJ, Sharab AS, Al-Ameer HJ, Ayoub NY (2020) Effect of simulated microgravity on the antidiabetic properties of wheatgrass (*Triticum aestivum*) in streptozotocin-induced diabetic rats. NPJ Microgravity 6(1):1–10

Bari G (1971) Effects of chronic and acute irradiation on morphological characters and seed yield in flax. Radiat Bot 11:293–302

Bhattacharjee R, Sil PC (2007) Protein isolates from the herb, *Phyllanthusniruri* L. (Euphorbiaceae), plays hepatoprotective role against carbon tetrachloride induced liver damage via its antioxidant properties. Food Chem Toxicol 45:817–826

Böhmer M, Schleiff E (2019) Microgravity research in plants: A range of platforms and options allow research on plants in zero or low gravity that can yield important insights into plant physiology. EMBO Rep 20(7):e48541

Boucheron-Dubuisson E, Manzano AI, Le Disquet I, Matía I, Sáez-Vasquez J, Van Loon JJ, Medina FJ (2016) Functional alterations of root meristematic cells of *Arabidopsis thaliana* induced by a simulated microgravity environment. J Plant Physiol 207:30–41

Brown AW, Chapman DK, Dahl AO (1976) Morphology of Arabidopsis grown under chromic centrifugation and on the clinostat. Plant Physiol 57:358–364

Brown AH, Chapman DK, Johnsson A, Heathcote D (1995) Gravitropic responses of the Avena coleoptile in space and on clinostats. I Gravitropic response thresholds. Physiol Plant 95(1): 27–33

Brown AH, Johnsson A, Chapman DK, Heathcote D (1996) Gravitropic responses of the *Avena coleoptile* in space and on clinostats. IV. The clinostat as a substitute for space experiments. Physiol Plant 98:210–214

Brungs S, Egli M, Wuest SL, Christianen PC, Van Loon JJ, Anh TJN, Hemmersbach R (2016) Facilities for simulation of microgravity in the ESA ground-based facility programme. Microgravity Sci Tec 28(3):191–203

Dedolph RR, Dipert MH (1971) The physical basis of gravity stimulus nullification by clinostat rotation. Plant Physiol 47(6):756–764

Driss-École D, Cottignies A, Jeune B, Corbineau F, Rbal G (1994) Increased mass production of *Veronica arvensis* grown on a slowly rotating clinostat. Environ Exp Bot 34:303–310

El-Nakhel C, Giordano M, Pannico A, Carillo P, Fusco GM, De Pascale S, Rouphael Y (2019) Cultivar-specific performance and qualitative descriptors for butterhead Salanova lettuce produced in closed soilless cultivation as a candidate salad crop for human life support in space. Life 9(3):61

Emmanuel H, Barbara VP, James AG, Christopher SB (1996) Clionrotation affects morphology and ethylene production in Soybean. Plant Cell Physiol 37(7):929–934

Fabio P, Anna MS, Franco T (2002) Effect of clinorotation on in vitro cultured explants of *Mentha piperita* L. Sci Hort 92(3–4):305–315

Gu RQ, Shen HM (1989) Effects of space flight on the growth and some cytological characteristics of wheat seedlings. Acta Phytophysiol Sin 15(4):403–407

Halstead TW, Dutcher FR (1987) Plants in space. Annu Rev Physiol 38:317–345

Harish R, Shivanandappa T (2006) Antioxidant activity and hepatoprotective potential of Phyllanthus niruri. Food Chem 95:180–185

Hassanpour H, Ghanbarzadeh M (2021) Induction of cell division and antioxidative enzyme activity of *Matricaria chamomilla* L. cell line under clino-rotation. Plant Cell Tiss Org 146: 215–224

Hemmersbach R, von der Wiesche M, Seibt D (2006) Ground-based experimental platforms in gravitational biology and human physiology. Signal Transd 6(6):381–387

Hensel W, Sievers A (1980) Effects of prolonged omnilateral gravistimulation on the ultrastructure of statocytes and on the graviresponse of roots. Planta 150:338–346

Herranz R, Anken R, Boonstra J, Braun M, Christianen PCM, de Geest M, Hauslage J, Hilbig R, Hill RJA, Lebert M, Medina FJ, Vagt N, Ullrich O, van Loon JJWA, Hemmersbach R (2013) Ground-based facilities for simulation of microgravity, including terminology and organism-specific recommendations for their use. Astrobiology 13(1):1–17

Hoson T, Wakabayashi K (2015) Role of the plant cell wall in gravity resistance. Phytochemistry 112(1):84–90

Hoson T, Saiki M, Kamisaka S, Yamashita M (2001) Automorphogenesis and gravitropism of plant seedlings grown under microgravity conditions. Adv Space Res 27(5):933–940

Howard GL, William CP (2005) Growth patterns for etiolated soybeans germinated under spaceflight conditions. Adv Space Res 36(7):1237–1243

Huang B, Li DG, Huang Y, Liu CT (2018) Effects of spaceflight and simulated microgravity on microbial growth and secondary metabolism. Military Med Res 5:18

Huxley A (1992) The new royal horticultural society dictionary of gardening. Macmillan Press, London, UK, p 3200p

Kamal KY, Hemmersbach R, Medina FJ, Herranz R (2015) Proper selection of 1 g controls in simulated microgravity research as illustrated with clinorotated plant cell suspension cultures. Life Sci Space Res 5:47–52

Katoch PC, Massar JE, Plaha P (1992) Effect of gamma irradiation on variation in segregating generations of F2 seeds of rice. Ind J Genet 52:213–218

Khvostova VV, Mozhaev VS, Aigaes NS, Valeva SA (1963) A further study of the effect of spaceflight conditions on chromosomes in the radicals of pea and wheat seeds. Cosm Res 1(1): 153–157

Kiss JZ, Wolverton C, Wyatt SE, Hasenstein KH, van Loon JJ (2019) Comparison of microgravity analogs to spaceflight in studies of plant growth and development. Front Plant Sci 10:1577

Kitaya Y (2019) Plant factory and space development, "space farm". In: Anpo M, Fukuda H, Wada T (eds) Plant factory using artificial light adapting to environmental disruption and clues to agricultural innovation. Elsevier, Osaka, Japan, pp 363–379

Kitaya Y, Kawai M, Tsuruyama J, Takahashi H, Tani A, Goto E, Saito T, Kiyota M (2003) The effect of gravity on surface temperatures of plant leaves. Plant Cell Environ 26(4):497–503

Kordyum EL (1994) Effects of altered gravity on plant cell processes: Results of recent space and clinostatic experiments. Adv Space Res 14:77–85

Kordyum EL (1997) Biology of plant cells in microgravity and under clinostating. Int Rev Cytol 171:1–78

Kordyum E, Chapman D, Brykov V (2019) Plant cell development and aging may accelerate in microgravity. Acta Astronaut 157:157–161

Kostina JL, Anikeeva I, Vaulina E (1984) The influence of space flight factors on viability and mutability of plants. Adv Space Res 4(10):65–70

Kraft TFB, van Loon JJWA, Kiss JZ (2000) Plastid position in Arabidopsis columella cells is similar in microgravity and on a random-positioning machine. Planta 211:415–422

Krikorian AD, Levine HG, Kann RP, O'Connor SA (1992) Effects of spaceflight on growth and cell division in higher plants. Adv Space Biol Med 2:181–209

Kumari K, Gupta S (2013) Antifungal properties of leaf extract of *Catharanthus roseus* l (g.) Don. Am J Phytomedicine Clin Ther 1(9):698–705

Leather GR, Forrence LE, Abeles FB (1972) Increased ethylene production during clinostat experiments may cause leaf epinasty. Plant Physiol 49:183–186

Levine HG, Krikorian AD (1992) Shoot growth in aseptically cultivated daylily and haplopappus plantlets after 5-day spaceflight. Physiol Plant 86:349–359

Lichtentaler HK, Wellburn AR (1985) Determination of total carotenoids, chlorophyll a và b of leaf in different solvents. Biolchem Soc Trans 11:591–592

Loi DT (2004) Vietnamese medicinal plants and herbs. Medicine Publishing House, Vietnam, p 813p

Luo Y, Wang X, Mei M, Zhuang C, Zhou F, Wei Z, Xie H, Yao J, Zhao J (2006) Genomic pholymorphism in consecutive generation rice plants from seeds on board spaceship and their relationship with space HZE particles. Acta Biol Sin 22(2):131–137

Magnotta M, Murata J, Chen J, De Luca V (2006) Identification of a low vindoline accumulating cultivar of *Catharanthus roseus* (L.) G. Don by alkaloid and enzymatic profiling. Phytochemistry 67(16):1758–1764

Malarvizhi S, Hosamani R, Swamy B, Sreejalekshmi K (2020) Mechano-biological plant responses to microgravity. Life Sci Space Res 28:74–86

Mallikq J, Chowdhuryq HB, Al Faruq A, Das S (2013) Pharmacological profile of *Catharanthus roseus* (apocynaceae): a detailed review. Asian J Pharm Res Dev 1(2):1–6

Matía I, González-Camacho F, Herranz R, Kiss JZ, Gasset G, van Loon JJ, Roberto Marco R, Medina FJ (2010) Plant cell proliferation and growth are altered by microgravity conditions in spaceflight. J Plant Physiol 167(3):184–193

Mei M, Qin Y, Sun Y (1998) Morphological and molecular changes of maize plants after seeds been flown on recoverable satellite. Adv Space Res 22:1691–1697
Morita MT (2010) Directional gravity sensing in gravitropism. Annu Rev Plant Biol 61:705–720
Murashige T, Skoog F (1962) A revised medium for rapid growth and bioassays with tobacco tissue culture. Plant Physiol 15:473–479
Nakajima S, Ogawa Y, Suzuki T, Kondo N (2019) Enhanced antioxidant activity in mung bean seedlings grown under slow clinorotation. Microgravity Sci Tec 31(4):395–401
Nakamura M, Nishimura T, Morita MT (2019) Gravity sensing and signal conversion in plant gravitropism. J Exp Bot 70(14):3495–3506
Namba T, Yoshizaki M, Tomimori T, Kobashi K, Mitsui K (1974) Fundamental studies on the evaluation of the crude drugs. 3. Chemical and biochemical evaluation of ginseng and related crude drugs. Yakugaku Zasshi 94(2):252–260
Nelson GA (2003) Fundamental space radiobiology. Gravity Space Biol B 16:29–35
Ou X, Long L, Wu Y, Yu Y, Lin X, Qi X, Liu B (2010) Spaceflight-induced genetic and epigenetic changes in the rice (*Oryza sativa* L.) genome are independent of each other. Genome 53(7): 524–532
Peterson RL, Peterson CA, Melville LH (2008) Teaching plant anatomy through creative laboratory exercises. NRC Research Press, Ottawa, Canada, p 154p
Porterfield DM (2002) The biophysical limitations in physiological transport and exchange in plants grown in microgravity. J Plant Grow Reg 21(2):177–190
Prajakta JP, Jai SG (2010) Antimicrobial activity of Catharanthus roseus – A detailed study. Br J Pharm Toxicol 1(1):40–44
Qian-Cutrone J, Huang S, Trimble J, Li H, Lin PF, Alam M, Klohr SE, Kadow KF (1996) Niruside, a new HIV REV/RRE binding inhibitor from *Phyllanthus niruri*. J Nat Prod 59:196–199
Ren WB, Zhang Y, Deng B, Guo H, Cheng L, Liu Y (2010) Effect of space flight factors on alfalfa seeds. Afr J Biotechnol 9(43):7273–7279
Sane RT, Chawla JL, Kuber VV (1997) Studies on *Phyllanthus amarus*. Indian Drugs 34(11): 654–655
Schimmerling W (2003) Overview of NASA's space radiation research program. Gravity Space Biol B 16:5–10
Sharma A, Singh RT, Handa SS (1993) Estimation of phyllanthin and hypophyllanthin by high performance liquid chromatography in *Phyllanthus amarus*. Phytochem Anal 4(5):226–229
Sievers A, Volkmann D (1977) Ultrastructure of gravity-perceiving cells in plant roots. Proc R Soc Lond 199(1137):525–536
Soleimani M, Ghanati F, Hajebrahimi Z (2019a) The role of phenolic compounds in growth improvement of cultured tobacco cells after exposure to 2-D clinorotation. Iranian J Plant Physiol 9(4):2921–2929
Soleimani M, Ghanati F, Hajebrahimi Z, Hajnorouzi A, Abdolmaleki P, Zarinkamar F (2019b) Energy saving and improvement of metabolism of cultured tobacco cells upon exposure to 2-D clinorotation. J Plant Physiol 234-235:36–43
Srivastava LM (2002) Plant growth and development: hormones and environment. Elsevier, CA, USA, p 757p
Su J, Zhang C, Zhang W, Shen YH, Li HL, Liu RH, Zhang X, Hu XJ, Zhang WD (2009) Qualitative and quantitative determination of the major coumarins in Zushima by high performance liquid chromatography with diode array detector and mass spectrometry. J Chromatogr A 1216(11): 2111–2117
Sugiyama M (2000) Genetic analysis of plant morphogenesis in vitro. Inte Rev Cytol 196:67–84
Toyota M, Furuichi T, Tatsumi H, Sokabe M (2007a) Cytoplamic calcium increases in response to changes in the gravity vector in hypocotyls and petioles of Arabidopsis seedlings. Plant Physiol 146:505–514
Toyota M, Furuichi T, Tatsumi H, Sokabe M (2007b) Hypergravity stimulation induces changes in intracellular calcium concentration in Arabidopsis seedlings. Adv Space Res 39(7):1190–1197

Van Loon JJWA (2007) Some history and use of the random positioning machine, RPM, in gravity related research. Adv Space Res 39(7):1161–1165

Vandenbrink JP, Kiss JZ (2016) Space, the final frontier: a critical review of recent experiments performed in microgravity. Plant Sci 243:115–119

Vazquez MM, Casalongué CA, París R (2019) Nitrate reductase mediates nitric oxide-dependent gravitropic response in *Arabidopsis thaliana* roots. Plant Sign Behav 14(4):e1578631

Wei LJ, Xu JL, Wang JM, Yang Q, Luo RL, Zhang MX, Bao GL, Sun XQ (2006) A comparative study on mutagenic effects of space flihgt and irradiation of r-rays on rice. Agr Sci China 5(11): 812–819

Weise SE, Kiss JZ (1999) Gravitropism of inflorescence stems in starch-deficient mutants of Arabidopsis. Int J Plant Sci 160:521–527

Wen X, Zhang L, Dai W, Li C (2004) Study of space mutation breeding in china. Acta Agric Nucleatae Sinica 18(4):241–246

Wendell KL, Wilson L, Jordan MA (1993) Mitotic block in HeLa cells by vinblastine: ultrastructural changes in kinetochore-microtubule attachment and in centrosomes. J Cell Sci 104(2): 261–274

Wongnawa M, Thaina P, Bumrungwong N, Nitiruangjarat A, Muso A, Prasartthong V (2005) Effect of *Phyllanthus amarus* Schum and Thonn. and its protective mechanism on paracetomol hepatotoxicity in rats. Acta Hort 680:195–201

Xie J, Zheng H (2020) Arabidopsis flowering induced by photoperiod under 3-D clinostat rotational simulated microgravity. Acta Astronaut 166:567–572

Yan H (2008) Study on the space mutagenic effect of two Herbage seeds. J Anhui Agr Sci 2:945–949

Zhao MG, Zhao X, Wu YX, Zhang LX (2007) Enhanced sensitivity to oxidative stress in Arabidopsis nitric oxide synthesis mutant. J Plant Physiol 164:737–745

Wireless Light-Emitting Diode System for Micropropagating Chrysanthemum and Strawberry

18

Duong Tan Nhut, Nguyen Ba Nam, and Hoang Thanh Tung

Abstract

The primary objective of this chapter was to compare the effectiveness of various light-emitting diode systems for plant tissue culture applications. The usefulness of the wireless power transmission—LED uni-Pack (WPT-LP) is emphasized. This system utilizes the physical space efficiently when compared with the conventional illumination system with fluorescent lamps. More plantlets can be cultured per unit space with the WPT-LP system. The growth of Chrysanthemum and strawberry plantlets cultured in vitro under different lighting systems, that is, wireless power transmission (WPT-LP) system, LED-Uni-Pack (LP) system, LED tubes, and fluorescent lights (FL) were compared. After examining various growth parameters, the results showed that the growth of Chrysanthemum plantlets under the WPT-LP system was better than those under LED tubes and FL. The development of strawberry plantlets gave similar results. Future improvements in the design and automation procedures of the WPT-LP system will ensure further energy savings and plant quality.

Keywords

Chrysanthemum · Fluorescent lamps · Light-emitting diode · LED-Uni-pack · Photosynthetic photon flux density · Strawberry · Wireless power transmission—LED uni-Pack

D. T. Nhut (✉) · H. T. Tung
Tay Nguyen Institute for Scientific Research, VAST, Dalat City, Vietnam

N. B. Nam
University of Dalat, Dalat City, Vietnam

D. T. Nhut et al. (eds.), *Plant Tissue Culture: New Techniques and Application in Horticultural Species of Tropical Region*,
https://doi.org/10.1007/978-981-16-6498-4_18

18.1 Introduction

The process of regulation of plant growth and development is dependent on light. Plants convert light energy into chemical energy through photosynthesis and use light as an information source. Many responses by plants are influenced and regulated by the photosynthetic photon flux density (PPFD), photoperiod, and light quality (Taiz and Zeiger 2002). Therefore, the growth of plant is possible in an in vitro environment using artificial lights, providing the necessary conditions for growth.

Fluorescent lamp (FL) has been the primary light source commonly used for plant micropropagation. Recently, light-emitting diodes (LEDs) are becoming an alternative light source for in vitro culture. Unlike as FLs, LEDs have improved features, consist of smaller mass and volume, longer lifetime, and tailored spectrum (Bornwaber and Tantau 2012). Tanaka et al. (1998) indicated that the growth of *Cymbidium* plantlet under red plus blue LEDs was improved. Besides, both in vitro culture and subsequent growth of strawberry plantlet under 70% red LEDs combined with 30% blue LEDs also were increased. Moreover, many studies obtained that LEDs were optimal for both in vitro and ex vitro plantlet growth than those under FL (Olle and Virsile 2013; Gupta and Jatothu 2013). However, a disadvantage of traditional and many commercial LED lighting systems is the uneven light intensity distribution on the culture shelves (Bula et al. 1991; Bornwaber and Tantau 2012). Furthermore, some especially systems were designed to improve plant quality in which LEDs were used as the light source such as UNIPACK, BIOLED, LED PACK, and COMPACK (Okamoto et al. 1996). Among them, the UNIPACK system has improved usable space and increased plant growth. However, one of the main drawbacks of the UNIPACK system is that many connecting wires are needed to supply direct current for the LED boards, which complicates the setting-up of this system. A Massachusetts Institute of Technology team generated a wireless power transmission design—transmitting energy without wires (Kurs et al. 2007). In that study, a new LED system (70% red LEDs plus 30% blue LEDs) in combination with wireless power transmission was proposed.

In this study, we selected in vitro grown Chrysanthemum and strawberry shoots to evaluate the effectiveness of various lighting systems on plant micropropagation. Both Chrysanthemum and strawberry are widely grown worldwide and are economically important plants. More importantly, in this study, these plants were chosen because they have different growth characteristics. Chrysanthemum has erected stems with a well-defined alternate leaf arrangement, whereas, strawberry has a short-thickened stem, called a "crown," where new leaves and branches emerge. One of our objectives is to evaluate the suitability of different lighting systems on the growth of plants with varying developmental characteristics.

18.2 Light Sources and Culture Conditions

18.2.1 LED uni-Pack

The LED uni-Pack (LP) system consists of a plastic Magenta-type vessel with a rectangular acrylic plastic lid (8 cm in length, 2 cm in height, and 8 cm in width) (Fig. 18.1). An LED panel with LEDs is attached to the inside surface of the acrylic plastic lid. The LED arrangement of this board consists of four rows of 20 red LEDs and three rows of 9 blue LEDs with the ratio of red LED and blue LED is 7:3. The LED circuit is on the cover on the plastic box. A 12 V direct current (DC) is provided from the power supply (Source AXT 450—Golden Field Company), which converts 220 V alternating current (AC) to 12 V DC. The power supply is located at the back of the uni-Pack. The system PPFD can be adjusted using voltage control.

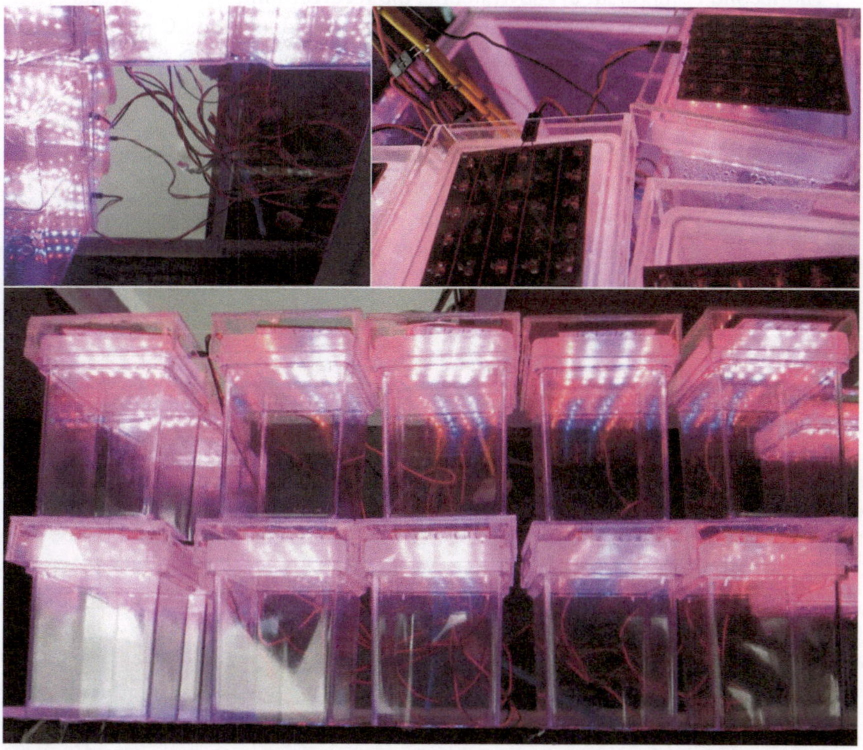

Fig. 18.1 The LED uni-Pack (LP) system

18.2.2 Wireless Power Transmission—LED uni-Pack

In order to reduce the clustering of electric wires, wireless power transmission is utilized. The key to the unit is the transmitter device, which includes a transformer, an oscillation circuit (OC), a rectifier unit, an amplifier, and an LC circuit (an electric circuit consisting of an inductor—L, and a capacitor—C) (Fig. 18.2). These devices are connected to a power source that converts the input power into oscillating electromagnetic fields. The transformer circuit converts the input power (220 V AC to 60 V AC and 24 V AC). The rectifier circuit converts the alternating current to direct current consists of two sources: *(1) a 15 V power supplied for the oscillation circuit* and *(2) a high-voltage 80 V power supplied for the LC circuit*. The OC creates oscillation for the LC circuit. The amplifier circuit delivers peak power to the load as efficiently as possible. The LC resonant circuit is used for recording resonance and signal transmit the signal. Signals are transmitted as electromagnetic waves (frequency range from 100 to 130 kHz).

The second component of the system is the receiver unit, which converts electrical waves into electrical energy. The receiver circuit includes the LC resonant circuit that resonates with the transmitted signal and loads by the LED board. The LED panel of the WPT-LP system is similar to the LP system; the lights are arranged in four rows (20 red LEDs) and three rows (9 blue LEDs). Also, the light intensity of system can be adjusted by changing the frequency of the transmitter device. Figures 18.2 and 18.3 illustrate the key components of the WPT-LP system.

The lights are placed on the culture vessels. The PPFD be adjusted into 45 μmol m^{-2} s^{-1} (checked at the bottom of the culture vessels) with a 16 h photoperiod. A light meter (LI-COR250A, Lincoln, USA) with a quantum sensor (LI-190, LI-COR Inc., USA) was used to check the PPFD.

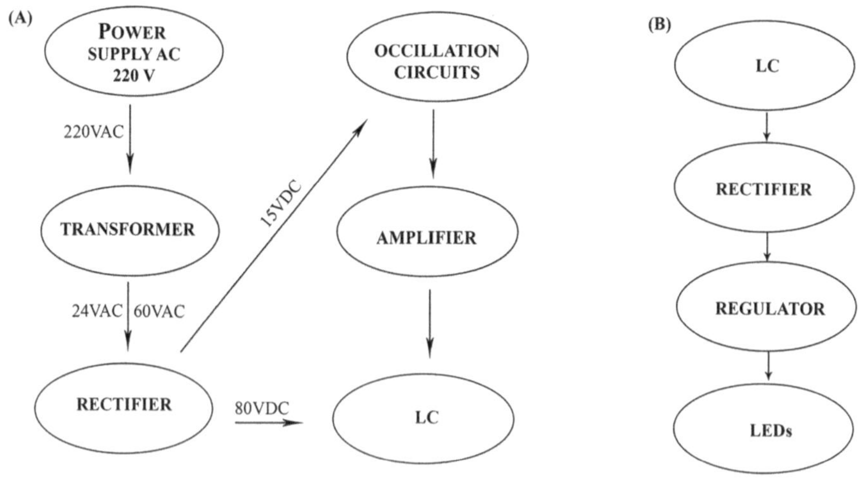

Fig. 18.2 The setup diagram of the circuit in the WPT-LP system. (**a**) Output circuit (Receiver); (**b**) input circuit (Transmitter)

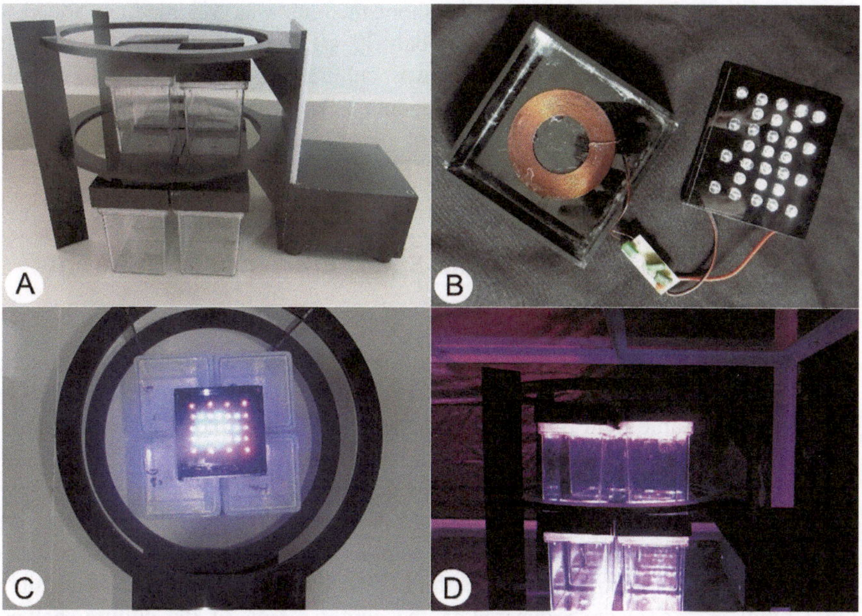

Fig. 18.3 The WPT-LP system. (**a**) A receiver device; (**b**) a transmitter device; (**c, d**) the operation of the WPT-LP system

18.3 Plant Materials

The following tissue culture protocols allow the generation of explants to illustrate the efficacies of the lighting systems. Axillary buds of 3-month-old *Chrysanthemum morifolium* cv. "Jimba" plants were washed under running tap water (30 min) and soaked in 70% ethanol (30 min), then sterilized with 0.1% w/v $HgCl_2$ (7 min). After that, the buds were rinsed three times with sterilized distilled water and cultured in 100 mL glass vessels containing 20 mL MS medium (Murashige and Skoog 1962) added to 0.5 mg/L 6-benzyladenine (BA), 30 g/L sucrose, and 8 g/L agar for adventitious shoots.

Young leaves of 1-year-old *Fragaria ananassa* plants were sterilized with 70% (v/v) ethanol (30 s) and 0.1% $HgCl_2$ (5 min), rinsing four times with sterile distilled water. These leaves were cut into round discs (0.8 cm diameter) and cultured in 250 mL glass vessels containing 40 mL MS medium added to 0.1 mg/L indole-3-butyric acid (IBA), 1 mg/L thidiazuron (TDZ), 30 g/L sucrose, and 8 g/L agar for adventitious shoot. The 8-week-old shoot clusters were transferred into the MS medium supplemented with 0.2 mg/L BA, 30 g/L sucrose, and 8 g/L agar for elongated shoot.

The Chrysanthemum shoots (1.5 cm in height) and strawberry shoots (2 cm in height) were cultured in Magenta GA-7 vessels (Sigma-Aldrich Co., US) containing

70 mL medium. The strawberry rooted shoot medium was MS medium added to 0.02 mg/L NAA, 1 g/L activated charcoal, 30 g/L sucrose, and 8 g/L agar. In contrast, Chrysanthemum rooted shoot was cultured in a MS medium (without hormone) for in vitro rooting. These Magenta vessels were placed under the LP and WPT-LP systems with a combination of 70% red and 30% blue LEDs. FLs were used as controls.

18.4 Distribution of PPFD and Space Efficiency of Different Lighting Systems

The PPFD is defined as the amount of photosynthetically active photons (400–700 nm) falling onto a surface per unit area per unit time. The value indicates the amount of usable light reaches the plants. It is a critical factor that has a direct effect on plantlet growth and photosynthetic activities (Kozai et al. 1997; Kim et al. 2004). Uniform distribution and a known value of PPFD play an important role in maintaining the plantlet quality (Chen 2005). When FLs or LED tubes are used as a light source, the PPFD varies at different positions of the culture shelves (Bornwaber and Tantau 2012). Maximum PPFD of FLs (45 μmol m^{-2} s^{-1}) and LED tubes (46.7 μmol m^{-2} s^{-1}) are registered in the center region of the culture-shelves. Moreover, PPFD decrease rapidly near the edge of the shelves. As a result, it is challenging to set up a uniform PPFD for the culture vessels. In this study, the PPFD of LP and WPT-LP systems were readily maintained at 45 μmol m^{-2} s^{-1} at the bottom of the culture vessels. Another advantage of the WPT-LP and LP systems is that their photosynthetically active photons are focused on the culture vessels.

Hayashi et al. (1992) introduced a sideward lighting system using FLs to improve space efficiency in the culture room. However, this system needs more space to install the light bulbs. Additional space is required between culture vessels and the light source due to the heat generated from the FLs (Tanaka et al. 1998; Chen 2005). The tubes are usually placed 30 to 50 cm (above the vessels) to avoid temperature increases inside the containers in the conventional downward lighting system (Kitaya et al. 1995).

Over the years, LED has become a more efficient light source for plant growth. LEDs generated very little heat; thus, the distance between the culture vessels and light sources can be shortened. LED tubes and panels on the market can be used to replace FLs as lighting in the in vitro culture room without increasing shelf space efficiency. Okamoto et al. (1996) introduced a UNIPACK tissue culture lighting system to increase the space and energy efficiency of a culture room. Each UNIPACK system consists of a culture vessel (11 × 11 × 14 cm) and a light source "LEDCAP"—36 red and 9 blue LEDs. In this study, the LED panel of the LP system was designed according to the system of Okamoto et al. (1996), containing 9 blue LEDs and 20 red LEDs. Similar to UNIPACK system, the WPT-LP and LP systems can be arranged with multi-layered to increase the space efficiency of shelves. For LP system, wires were needed to supply direct current for the LED boards. As a result, the system structure becomes quite complicated. Thus, a wireless power

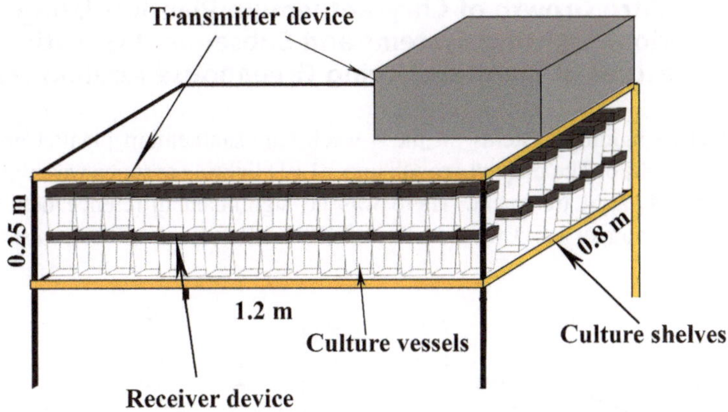

Fig. 18.4 A culture shelf with 192 culture vessels in the WPT-LP system

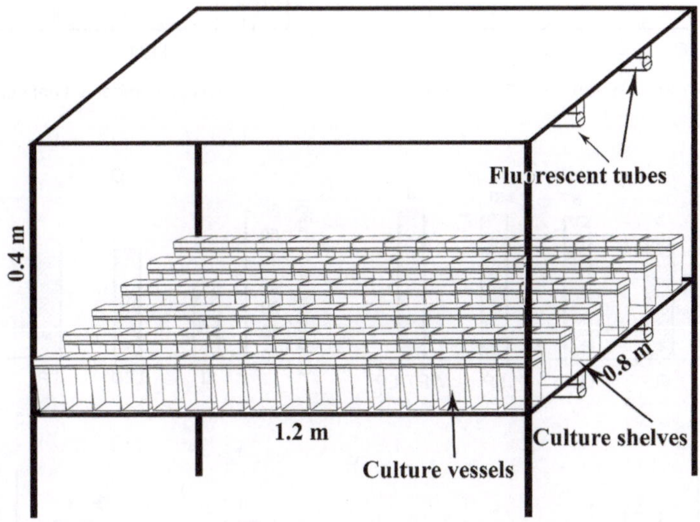

Fig. 18.5 A culture shelf with 96 vessels in the fluorescent lighting system

transfer is used in the WPT-LP system to solve this problem. The number of layers of the culture vessels depends on the number of transmitter coils. For instance, the volume of a culture shelf is 0.24 m^3 (1.2 × 0.8 × 0.25 m). A culture shelf with one transmitter coil in the WPT-LP system can house two layers of culture vessels. Since each layer consists of 96 vessels (77 × 77 × 97 mm each); thus, a shelf can accommodate 192 vessels in the WPT-LP system (Fig. 18.4).

In a conventional lighting system, a culture shelf has a dimension (1.2 × 0.8 × 0.4 m). The distance of the culture vessels to FLs is 20 cm; thus, a culture shelf can be accommodated 96 vessels (Fig. 18.5). For this reason, the space efficiency of the WPT-LP system increases by twofold in comparison with the conventional lighting system.

18.5 In Vitro Growth of Chrysanthemum Plantlets Using Various Lighting Systems and Subsequent Growth Features of Plantlets During Greenhouse Acclimatization

The effect of lighting systems on the 4-week Chrysanthemum plantlet growth is shown in Fig. 18.6. The plantlet growth under LED lighting systems was higher than that of the FL light treatment. There was no significant difference between the plantlet height under LED tubes, LP, and WPT-LP systems. The lowest plantlet

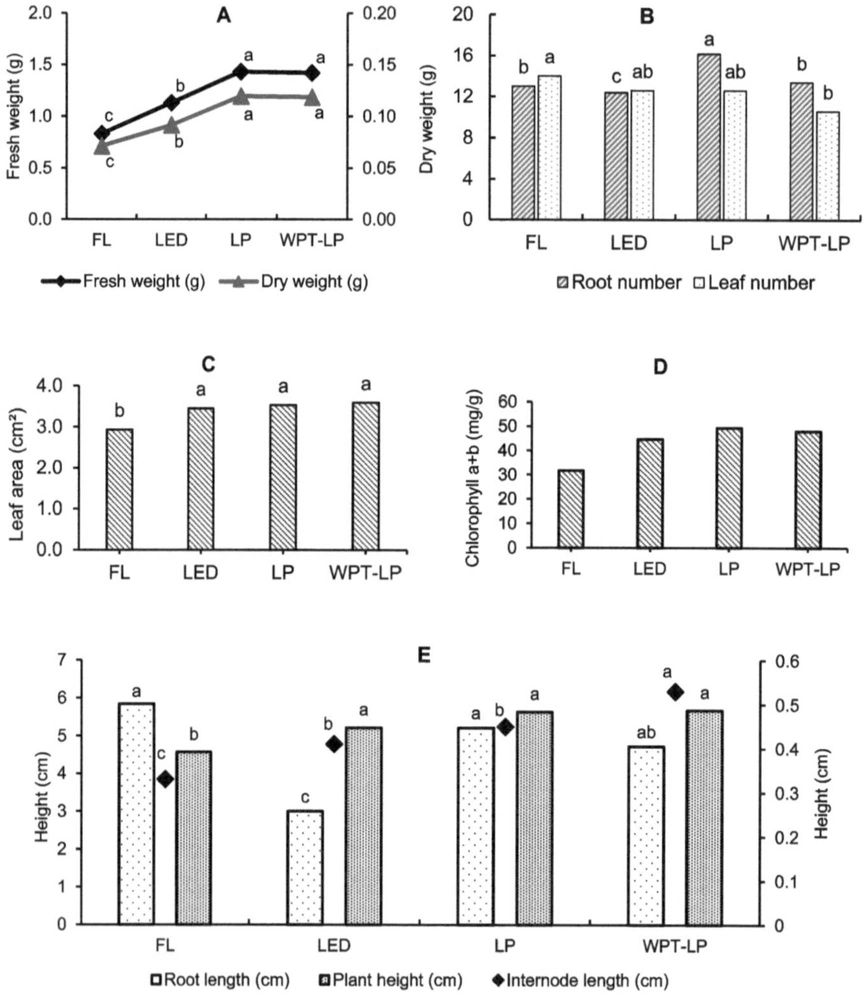

Fig. 18.6 Effects lighting systems on in vitro growth of Chrysanthemum. (**a**) Plantlet fresh and dry weight; (**b**); number of roots and number of leaves; (**c**) leaf area; (**d**) the chlorophyll (a + b) contents of leaves; (**e**) root length, plant height and internode length

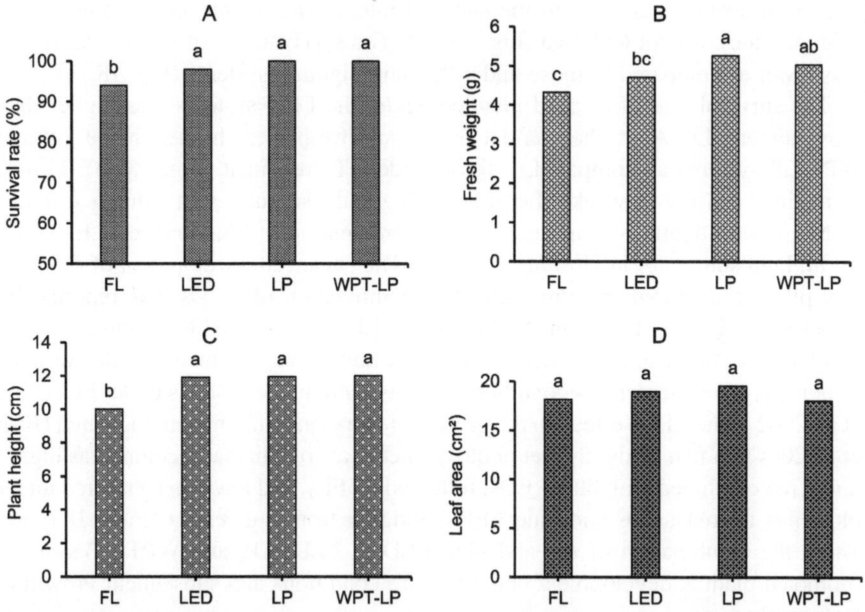

Fig. 18.7 Effects of lighting systems on acclimatization and growth of Chrysanthemum in a greenhouse after 6 weeks. (**a**) Survival rate; (**b**) fresh weight; (**c**) plantlet height; (**d**) leaf area

height was observed under FLs. For the plantlet weight, it was higher under LP and WPT-LP systems than under the LEDs and the FLs (Fig. 18.6a).

The number of leaves per plantlet and leaf area under LED conditions was higher than that under FLs (Fig. 18.6b, c). Moreover, the internode was elongated in plantlets under the WPT-LP system, intermediate values under LED tubes, and the LP system and shortest in plantlets under FLs (Fig. 18.7e). However, the number of leaves per plantlet under FLs was higher than those under the LEDs (Fig. 18.6b).

The rooted plantlets also affected by the lighting systems. The highest number of roots per plantlet was recorded under the LP system; meanwhile, the root length of plantlet under the FLs and LP systems was higher. In contrast, rooted plantlet under LED tubes was the poorest among the treatments. The chlorophyll (a + b) contents of leaves under the LEDs were higher than those of plantlets under FLs (Fig. 18.6d). However, there were no significant differences among chlorophyll content under WPT-LP and LP systems. This result showed that the WPT-LP and LP systems with ratio of 30% blue and 70% red LEDs were suitable for Chrysanthemum plantlet growth.

The lighting systems during cultured in vitro also affected the acclimatization and growth of Chrysanthemum in the greenhouse. Figure 18.7 shows the results of plantlet growth 6 weeks after transferring to the soil. These plantlets were cultured under LED tubes, LP, and WPT-LP systems, and FLs for 4 weeks before transferring to the soil. Plants acclimatized and grew well in all the treatments. Moreover, there were significant differences in the survival rate, fresh weight, and plantlet height (Fig. 18.7a-c), except leaf area (Fig. 18.7d). The survival rate of plantlet under FL was lower as compared to those under the other lighting systems (Fig. 18.7a).

The survival rate of derived plantlets under the LED systems was higher than those under FLs. Also, the plantlet height and fresh were higher in the LP and WPT-LP systems as compared to those under FL treatment (Fig. 18.6b). Flower formation occurred 16 weeks after transferring to the soil under all lighting systems.

Many investigations indicated that the combination of blue and red LEDs as a useful light source for in vitro plant growth. The enhanced development of *Cymbidium* plantlets cultured in vitro using this combination of LEDs was reported by Tanaka et al. (2009). Red combined with blue LEDs were suitable for the growth of bulblets of *Lilium* and the number of roots per bulblet and the fresh and dry weights of bulblets grown under this condition were higher than the bulblets under FLs (Lian et al. 2002). The LEDs effect on the growth of Chrysanthemum were indicated (Kim et al. 2004). In that study, the net photosynthetic rate of Chrysanthemum was higher under red combined with blue LEDs, followed by FLs, and low when cultured under blue plus far-red LEDs and blue LEDs. Judging from our results (Figs. 18.6 and 18.7), the combination of red and blue LED in LED, LP, and WPT-LP systems promoted plantlet development under in vitro conditions and subsequent growth in the greenhouse.

18.6 In Vitro Growth of Strawberry Plantlets Using Various Lighting Systems and Subsequent Growth Features of Plantlets During Greenhouse Acclimatization

Figure 18.8 summarizes the growth of in vitro strawberry plantlets under lighting systems after 4 weeks of culture. From the results, strawberry plantlets under LED systems were higher fresh and dry weights and plant height than those grown under FLs (Figs. 18.8a, b and 18.9a, c).

The number of leaves of explants among different treatments was no significant difference, but the leaf area was greatly affected by various lighting systems (Fig. 18.9b, d). The leaf area of plantlets was higher under the LP system, whereas those under FLs were the lowest (Fig. 18.9d). Also, the chlorophyll content exhibited significant differences among the different systems (Fig. 18.9e). The chlorophyll content of plantlets cultured under WTP-LP was higher, even though other growth parameters were lower than those under the LP system.

The in vitro strawberry plantlets under FL, LED tubes, LP, and WPT-LP systems after 4 weeks of acclimatization in the greenhouse are presented in Fig. 18.10. The acclimatization of strawberry plantlets had been affected by lighting systems. Strawberry plantlets cultured in the LED systems had a higher survival rate over the control treatment (Fig. 18.10a). Also, the plantlet height, fresh weight, and leaf area of strawberry plantlets in the LED systems were higher than plantlets from the

Fig. 18.8 Effects of various lighting systems on in vitro strawberry plantlet growth. (**a**) Plants in Magenta vessel cultured under various lighting systems (FL, LED tube, LP, and WPT-LP system, from left to right); (**b**) plants removed from culture vessels shown in A, for a closer comparison

control treatment (Fig. 18.10b–d). After approximately 10 weeks of vegetative growth, strawberry plantlets derived from LP, WPT-LP, FL, and LED tube systems started to flower, and fruit began to form after 1–2 weeks of planting.

18.7 Conclusion

Previous studies indicated that LEDs improved in vitro and ex vitro plant growth (Nhut et al. 2003). Studies of various growth parameters and photosynthetic characteristics have also shown that plants grown under LED systems are better when compared to fluorescent lighting. One can design LEDs with suitable wavelengths to maximize the growth and development of plants, maximizing plants yield.

The effects of LEDs in plant growth and development are apparent, and the influence of electromagnetic fields to plant growth appears to be limited. In our studies, an electromagnetic field was used at a low frequency, that is, 120 kHz, and seems not to cause any changes in explant physiology. However, the influence of electromagnetic fields on plant physiology is still controversial. Moreover, the field of weak electromagnetic inhibited plant growth and decreased cell division (Belyavskaya 2004). Ramezani Vishki et al. (2012) showed that electromagnetic

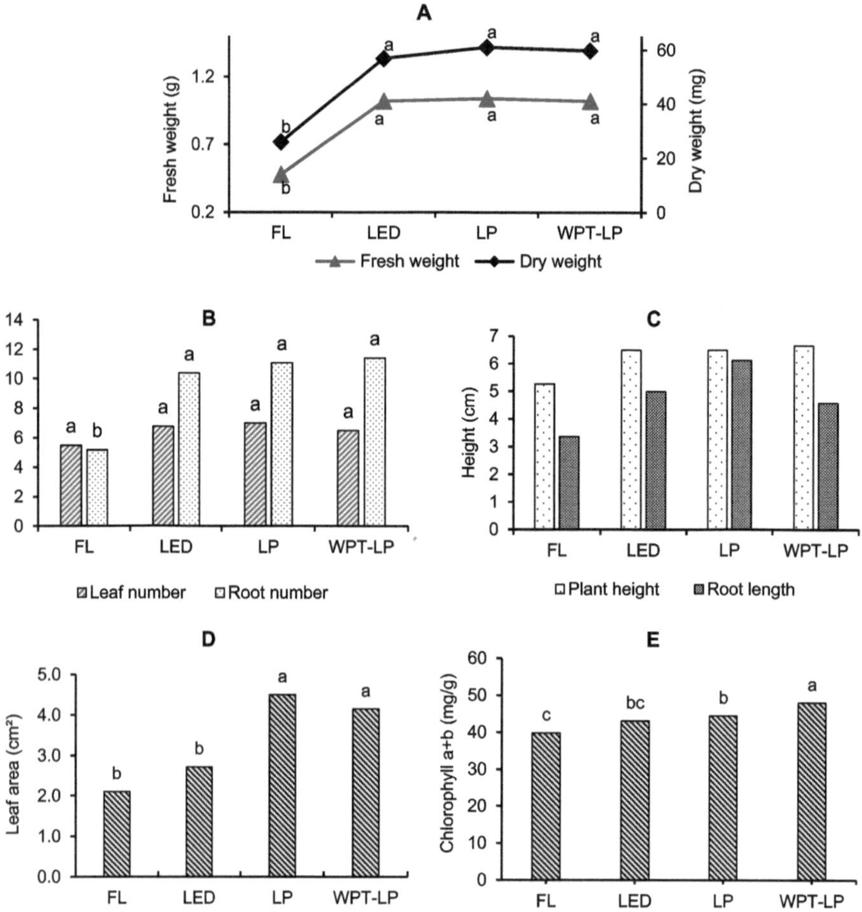

Fig. 18.9 Effects of lighting systems on the growth of strawberry cultured in vitro. (**a**) Fresh and dry weight; (**b**) number of leaves and number of roots; (**c**) plant height and root length; (**d**) leaf area; (**e**) chlorophyll content

fields increased seedling and elongated root in *Satureja bachtiarica*, but restrained leaf area, shoot length, fresh and dry weights. Dao-Liang et al. (2009) indicated that the electromagnetic field could be increased the numbers of sprout in *Prunus maritima*. With the varied response, further studies are needed to determine the effects of electromagnetic field on plant growth, especially with the in vitro growing process. Wireless power has been widely used in mobile devices; it has been applied to plant cultures in the first time. Further studies on interactions between the LEDs, electromagnetic field, and the plantlet physiology would be relevant. Our current design is highly suited to study the effects of electromagnetic field on plant growth and development.

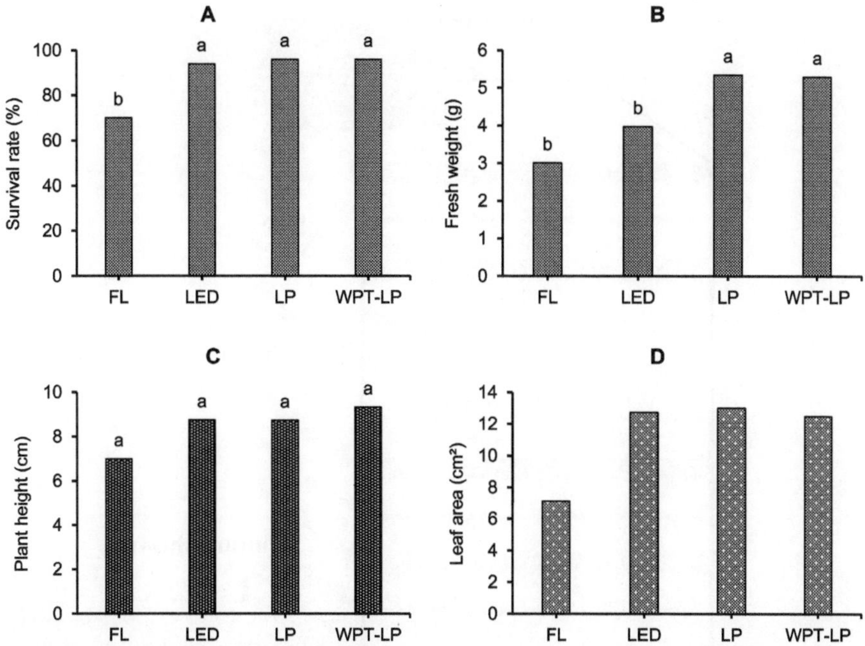

Fig. 18.10 Effects of lighting systems on the acclimatization and the growth of strawberry plantlets after 4 weeks in the greenhouse. (**a**) Survival rate; (**b**) plantlet height; (**c**) fresh weight; (**d**) leaf area

The results of our studies show that the WPT-LP system is equally or more effective than the LP system and other lighting conditions. Unlike conventional lighting systems, the WPT-LP system can accommodate the culture of more plantlets than other lighting systems having the same cultured area. This technology also offers the opportunity to study the effects of combining light and electromagnetic fields on plant growth.

With further investigations and improvements, we expect that the WPT-LP system will be suitable for use in most tissue culture rooms and plant production facilities for the growth of a wide variety of plants. By targeting improved system operation and automation control, it is possible to scale-up the system and reduce the costs of plant propagation. In the future, smartphones can control the operation of the wireless network for tissue culture rooms to save energy, ensuring plant quality. LED lighting control protocols are of particular interest to the research team. Figure 18.11 shows an improved design of the WPT-LP system.

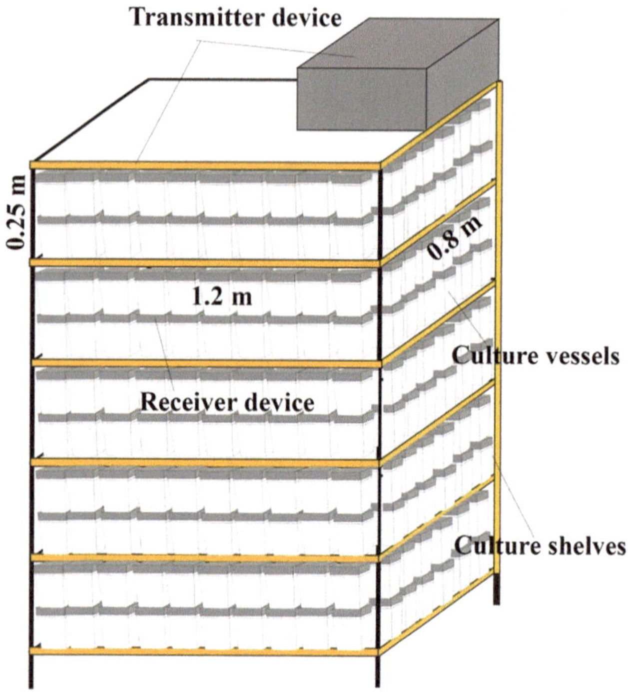

Fig. 18.11 The expected expansion of the WPT-LP system for tissue culture laboratories in the future

References

Belyavskaya NA (2004) Biological effects due to weak magnetic field on plants. Adv Spa Res 34(7):1566–1574

Bornwaber T, Tantau HJ (2012) Evaluation of LED lighting systems in in vitro cultures. Acta Hort 956:555–562

Bula RJ, Morrow RC, Tibbitts TW, Barta DJ, Ignatius RW, Martin TS (1991) Light-emitting diodes as a radiation source for plants. HortSci 26(2):203–205

Chen C (2005) Fluorescent lighting distribution for plant micropropagation. Biosyst Eng 90(3): 295–306

Dao-Liang Y, Yu-qi G, Xue-ming Z, Shu-wen W, Qin P (2009) Effect of electromagnetic fields exposure on rapid micro-propagation of Beach plum (*Prunus maritima*). Ecol Eng 35(4): 597–601

Gupta DS, Jatothu B (2013) Fundamentals and application of light-emitting diodes (LEDs) in in vitro plant growth and morphogenesis. Plant Biotech Rep 7:211–220

Hayashi M, Fujita N, Kitaya Y, Kozai T (1992) Effect of sideward lighting on the growth of potato plantlets in vitro. Acta Hort 19:163–166

Kim SJ, Hahn EJ, Heo JW, Paek KY (2004) Effects of LEDs on net photosynthetic rate, growth and leaf stomata of Chrysanthemum plantlets in vitro. Sci Hort 101(1-2):143–151

Kitaya Y, Fukuda O, Kozai T, Kirdmanee C (1995) Effects of light intensity and lighting direction on the photoautotrophic growth and morphology of potato plantlets in vitro. Sci Hort 62(1-2): 15–24

Kozai T, Kubota C, Jeong BR (1997) Environmental control for large-scale production of plants through in vitro techniques. Plant Cell Tissue Organ Cult 51(1):49–56

Kurs A, Karalis A, Moffatt R, Joannopoulos JD, Fisher P, Soljačić M (2007) Wireless power transfer via strongly coupled magnetic resonances. Science 317(5834):83–86

Lian ML, Murthy HN, Paek KY (2002) Effects of light emitting diodes (LEDs) on the in vitro induction and growth of bulblets of lilium oriental hybrid 'Pesaro'. Sci Horti 94(3-4):365–370

Murashige T, Skoog F (1962) A revised medium for rapid growth and bio assays with tobacco tissue culture. Physiol Plant 15:473–497

Nhut DT, Takamura T, Watanabe H, Okamoto K, Tanaka M (2003) Responses of strawberry plantlets cultured in vitro under superbright red and blue light-emitting diodes (LEDs). Plant Cell Tissue Organ Cult 73(1):43–52

Okamoto K, Yanagi T, Takita S, Tanaka M, Higuchi T, Ushida Y, Watanabe H (1996) Development of plant growth apparatus using blue and red LED as artificial light source. Acta Horti 440: 111–116

Olle M, Viršile A (2013) The effects of light-emitting diode lighting on greenhouse plant growth and quality. Agr Food Sci 22(2):223–234

Ramezani VF, Majd A, Nejadsattari T, Arbabian S (2012) Study of effects of extremely low frequency electromagnetic radiation on biochemical changes in *Satureja bachtiarica* L. Inter J Sci Tech 1(7):77–82

Taiz L, Zeiger E (2002) Plant Physiology. Sinauer Associates, New York, USA, p 690p

Tanaka M, Takamura T, Watanabe H, Endo M, Yanagi T, Okamoto K (1998) In vitro growth of *Cymbidium* plantlets cultured under superbright red and blue light-emitting diodes (LEDs). J Hort Sci Biotech 73(1):39–44

Tanaka M, Norikane A, Watanabe T (2009) Cold cathode fluorescent lamps (CCFL): revolutionary light source for plant micropropagation. Biotech Biotech Equip 23(4):1497–1503